Neuroscience

Springer
New York
Berlin
Heidelberg
Barcelona
Hong Kong
London
Milan
Paris
Singapore
Tokyo

Alwyn Scott

Neuroscience
A Mathematical Primer

With 58 Figures

 Springer

Alwyn Scott
Department of Mathematics
University of Arizona
Tucson, AZ 85721
USA

Library of Congress Cataloging-in-Publication Data
Scott, Alwyn, 1931–
    Neuroscience : a mathematical primer / Alwyn C. Scott.
      p. cm.
    Includes bibliographical references and index.
    ISBN 0-387-95403-1 (hardcover: alk. paper) ISBN 0-387-95402-3 (softcover : alk. paper)
      1. Neurosciences—Mathematical models.   I. Title.
    QP357.5 .S36 2002
    573.8′1—dc21   2001058975

Printed on acid-free paper.

Production managed by Louise Farkas; manufacturing supervised by Jeffrey Taub.
Photocomposed copy prepared from the author's LaTeX files.
Printed and bound by Edwards Brothers, Inc., Ann Arbor, MI.
Printed in the United States of America.

9 8 7 6 5 4 3 2 1

ISBN 0-387-95403-1   SPIN 10860575   (hardcover)
ISBN 0-387-95402-3   SPIN 10860567   (softcover)

Springer-Verlag   New York Berlin Heidelberg
A member of BertelsmannSpringer Science+Business Media GmbH

A star we perceive. The energy scheme deals with it, describes the passing of radiation thence into the eye, the little light-image of it formed at the bottom of the eye, the ensuing photochemical action of the retina, the trains of action potentials travelling along the nerve to the brain, the further electrical disturbance in the brain, the action-potentials streaming thence to the muscles of eye-balls and of the pupil, the contraction of them sharpening under the light-image and placing the seeing part of the retina under it. The 'seeing'? That is where the energy-scheme forsakes us. It tells us nothing of any 'seeing'. Much, but not that.

—Charles Scott Sherrington in *Man on His Nature*

How far even then mathematics will suffice to describe, and physics to explain, the fabric of the body, no man can foresee. It may be that all the laws of energy, and all the properties of matter, and all the chemistry of all the colloids are as powerless to explain the body as they are to comprehend the soul. For my part, I think it is not so. Of how it is that the soul informs the body, physical science teaches me nothing; and that living matter influences and is influenced by mind is a mystery without a clue. Consciousness is not explained to my comprehension by all the nerve-paths and neurones of the physiologist; nor do I ask of physics how goodness shines in one man's face, and evil betrays itself in another. But of the construction and growth and working of the body, as of all else that is of the earth earthy, physical science is, in my humble opinion, our only teacher and guide.

—D'Arcy Wentworth Thompson in *On Growth and Form*

Preface

Arguably the most intricate dynamic object in the universe, the human brain is an unsounded source of wonder for the scientific community. The primary aim of this book is to provide both students and established investigators in the growing area of neuroscience with an appreciation of the roles that mathematics may play in helping to understand this enigmatic organ. Along with discussions of results obtained by the neuroscience community, emphasis is placed on suggesting fruitful research problems for those planning to embark on mathematical studies in neuroscience.

To make the overall perspectives understandable to philosophers and psychologists, essential features of the discussions are presented in ordinary English, with more detailed mathematical comments in appendices and footnotes. Although it attempts to maintain both clarity and biological relevance, this is not a text on the anatomy of nerve systems; thus readers should bring some knowledge of neurophysiology through other courses, associated studies, or laboratory research.

It is a guiding theme throughout the book that the brain is organized into several quite different levels of dynamic activity. As will be seen, these levels are hierarchically structured, beginning with the molecular dynamics of intrinsic membrane proteins and proceeding upward, through the switching properties of active membrane patches and synapses, the emergence of impulses on active fibers, overall properties of individual neurons, and the growth of functional assemblies of interacting neurons, to the global dynamics of a brain. At each level of description, reality turns different facets of her mystery to us, and diverse phenomena make their contributions to the brain's collective behavior.

My intention in presenting these ideas is to avoid overemphasizing the importance of mathematics in neuroscience; thus, the *limitations* of mathematical analysis—which should be of interest to both theoretical neuroscientists and philosophers of the mind—are not ignored. Although analysis for its own sake has a place in the broad spectrum of academic activities, the neuroscientist need not jump through every hoop that some applied mathematician or mathematical physicist has managed to master. At each level of description, therefore, the reader will be introduced only to those concepts and tools that seem helpful for understanding the behaviors of real neurons or neural systems.

It will be seen in the course of this book that useful mathematical formulations are quite different at chemical and biological levels of description. Thus, the vibrations of individual proteins are governed by Newton's laws of motion, wherein energy is conserved and time can be reversed without introducing qualitative changes. At the level of a nerve impulse, on the other hand, laws of nonlinear diffusion guide relevant phenomena. As in most realms of biology, the energy of an impulse is not conserved by its dynamics, and time has a definite arrow, flowing—as we all come to know—out of the past and toward the future. It is hoped that an appreciation of these deep differences between molecular and neural levels of description will prepare the reader for additional surprises that may appear at global levels of the brain's dynamics.

The first chapter provides an introduction relating current research to the history of neuroscience over the past two centuries. Carried on in subsequent chapters, such recognition of the early contributions is helpful, I believe, both for obtaining a balanced view of present activities and to avoid reinventing the wheel. Following this historical introduction, the chapters are divided into two main parts.

Comprising Chapters 2 through 9, the first part deals with the dynamics of individual neurons, considering the ways they converse with one another via chemical synapses, gap junctions, and external current loops and showing how intrinsic membrane proteins give nerve fibers an active character. Starting with the properties of active membranes, the emergence of an impulse is then described, noting its mathematical character as an *attractor* in the dynamics of the nerve, thereby implying *all-or-nothing* behavior and a *threshold* level for ignition. From these twin properties, the interactive dynamics of impulses at the branching regions of dendritic and axonal fibers are suggested, which may provide bases for information processing within individual neurons.

From an understandable eagerness to get on with the investigation of neural systems, the study of individual neurons has been somewhat neglected by the neuroscience community. Yet much interesting research remains to be done at the level of the nerve cell, some of which may eventually be useful for comprehending the intricate dynamics of a brain.

The remainder of the book (Chapters 10 through 12) presents various mathematical formulations for the brain's global dynamics. These theories are of two general classes. The first is characterized by the classical stimulus–response concepts, which were popular among behavioral psychologists in the middle of the twentieth century, and involve threads of causal implication that flow directly from inputs to outputs. Although not rich enough to model the behaviors of real brains, such theories are useful for the design of computing machines that can learn to recognize patterns and may describe brain activities in limited regions.

The second class of global brain theories invokes closed internal loops of causality, significantly augmenting difficulties of analysis while allowing a far more realistic spectrum of behaviors, including global nonlinear dynamics, propagating waves of information, and attentiveness. From this perspective, the brain is recognized as much more than a biological computer that merely analyzes input data and presents logical conclusions to some cognizant client. Among other differences with present day computers, a human brain operating within a human culture is an *open system* from which new levels of functional abstraction may emerge when convenient or necessary. While organizing itself in the context of experience, it seems, the brain manages to become its own client.

Based as they are on vastly oversimplified models of individual neurons, these various mathematical descriptions of neural systems can offer only nebulous approximations to the behaviors of real brains. Nonetheless, they provide conservative benchmarks from which to regard the intricate problems faced by young researchers entering into the realms of neuroscience.

In formulating the dynamics of neural systems, a central idea is Donald Hebb's classic concept of a *cell assembly,* according to which functionally significant entities in the brain are not limited to individual neurons. From Hebb's perspective, a more significant level of the brain's dynamics involves emerging groups of cells acting in concert, thereby gaining robustness. As with the nerve impulse, each cell assembly is an attractor in the dynamics of the brain, again implying all-or-nothing and threshold behavior. From these properties stems a fruitful hypothesis: the brain is organized into a *nonlinear hierarchy of neuronal assemblies* with ever-increasing intricacy upon ascending the scale.

To provide an overall survey of Hebb's point of view, the first and last chapters have been written to be read together as parts of a single essay. In a second reading, Chapters 2, 3, 9, and 11 can be added to appreciate the broad outlines of neural structures and cell assemblies. More mathematical concepts are discussed in the remaining chapters, with technical details in the appendices.

Broadly speaking, the book is intended as a guide to the directions in which scientific research may be heading as we leave the "century of physics," where relationships among causes and effects are largely sorted out and move into the "century of biology," where causes and effects are densely interwoven and ontological puzzles abound. In the biological realms, we must learn to deal with *organisms*—forms of matter having their own agendas—a challenge that may require modifications in the practice of science. Such ideas are, I believe, particularly important for appreciating the course of future research in neuroscience, a key consideration for all entering the field.

I hope that the perspectives presented here will provide useful guides for those responding to the most exciting and formidable challenge facing contemporary science: understanding the human mind.

Tucson, Arizona   ALWYN SCOTT
2002

Acknowledgments

It is a pleasure to acknowledge the many insights and ideas received over the past four decades from teachers, colleagues, students, and friends at MIT's Research Laboratory of Electronics, the Laboratorio di Cibernetica and the Stazione Zoologica in Naples, the University of Wisconsin Neuroscience Program, the Los Alamos Center for Nonlinear Studies, the Technical University of Denmark, the University of Arizona Programs in Neuroscience and Applied Mathematics, and the UA Center for Consciousness Studies. Without this enlightenment, this book would not have been imagined, never mind written.

Financial support over the years has been provided by the National Library of Medicine, the National Science Foundation, the European Molecular Biology Organization, and the Fetzer Institute, and the following have looked through portions of the manuscript, making helpful comments, criticisms, and corrections: Stephane Binczak, Tom Christensen, Jonathan Coles, Chris Eilbeck, Erik Fransén, Walter Freeman, Chris Jones, Eric Kennedy, Robert Lakatos, Anders Lansner, Roman Poznanski, Leslie Tolbert, Logan Trujillo, and Burton Voorhees. For this generous assistance, I am deeply grateful.

Tucson, Arizona
2002

ALWYN SCOTT

Contents

List of Figures

1
A Short History of Neuroscience

Although attempts to understand the physical bases for mental processes go back to the early Greek and Egyptian civilizations, modern electrophysiology began with the late eighteenth-century investigations by Luigi Galvani on the sciatic nerve–muscle preparation of the frog [8]. In 1791, this Italian physician reported that the muscle would twitch when the nerve was stimulated by a bimetallic contact and also by atmospheric electricity. Thus, Galvani proposed three types of electricity—chemical, atmospheric, and animal—with the latter being different from the two others, but his compatriot Alessandro Volta disagreed. In the attempt to show that Galvani's animal electricity was identical to that produced by bimetallic currents, Volta invented the battery, thereby launching the science of electricity in the historically convenient year of 1800. All of these early experiments were carefully repeated by the German physicist Frederick von Humboldt, confirming both Volta's view that the various forms of electricity are closely related and Galvani's observation that animal electricity has qualitatively distinctive features. Let us consider these differences.

1.1 Dynamics of a Nerve Impulse

Perhaps the most singular property of animal electricity is its *speed of propagation* along a nerve fiber, which in the first half of the nineteenth century was believed to be extremely large. In 1850, experimental measurements of the velocity of a nerve impulse were reported by young Hermann Helmholtz, who went on to become recognized as one of the great German physicists of the nineteenth century. (Among other accomplishments, Helmholtz was among the first to recognize the principle of *energy conservation*, about which we will learn more in this book.)

Using Galvani's classic frog preparation and Emil Du Bois-Reymond's newly invented ballistic galvanometer to observe intervals of time, Helmholtz measured impulse speed by carefully recording the instants at which the muscle twitched for stimulations introduced at two different locations along the attached nerve [23]. Dividing the difference in locations of stimulation by the corresponding difference in twitching times, he obtained a surprisingly low propagation speed of about 27 meters per second (m/s). Interestingly, this value is what thousands of electrophysiology students now find each year using sophisticated electronic measuring equipment that did not become available to neuroscientists until the middle of the twentieth century.

But how, Helmholtz pondered, was this small velocity of propagation explained? What is the underlying physical chemistry? Does some chemical substance actually move? If so, what? Many orders of magnitude below the speed at which electricity propagates through a copper wire or an ionic solution, his empirical observation remained a puzzle throughout the second half of the nineteenth century.

As it turns out, the nerve impulse is now known to be a *wave of activity* quite similar to the flame of a candle or the burning of a dynamite fuse. In such examples, the wave releases energy that is stored in an *active medium*, and the released energy then feeds the dissipative (energy-consuming) aspects of the process.

From this perspective, a wave of activity can be viewed as a *coherent process* [56] represented by the diagram

$$\textbf{Release of energy}$$
$$\downarrow \qquad \uparrow$$
$$\textbf{Dissipation of energy}$$

with each component supporting (or "causing") the other. Such phenomena are examples of *nonlinear diffusion in excitable media*, and the preceding diagram provides an illustration of *positive feedback*. In other words, the nonlinear release of stored energy fuels the dissipative processes, and these processes in turn release stored energy. From the feedback around this *closed causal loop* emerges a new dynamic entity: the nerve impulse.

Of course, candles have been around for a long time, and their scientific importance was emphasized in the 1850s by the brilliant English experimental physicist Michael Faraday in one of his annual "Christmas Lectures" on selected aspects of natural philosophy for young people at London's Royal

Institution [16],[1] but the scientific world is often slow on the uptake. How then did the concept of nonlinear diffusion enter the realms of neuroscience?

At the dawn of the twentieth century, two key contributions to the understanding of nerve conduction appeared. The first of these was the *membrane theory for nerve cells,* suggested by Julius Bernstein in 1902 [4]. According to his hypothesis, the outer covering of a nerve cell is normally impermeable to ionic flow but becomes suddenly permeable during the passage of an impulse. This idea would receive substantial confirmation during the course of the century, leading to the analyses presented in Chapter 5 of this book.

Shortly thereafter, the concept of nonlinear diffusion in an active medium was introduced to the scientific community at the general meeting of the German Society for Applied Physical Chemistry in 1906 [35]. At this meeting, Robert Luther demonstrated a propagating chemical reaction, pointing out that the wave of chemical activity travels at a velocity of about

$$v \sim \sqrt{D/\tau}, \tag{1.1}$$

where τ is the *reaction time* (seconds, s) for the energy releasing process and D is the corresponding *diffusion constant* (in meters squared per second, m^2/s) for ions. Because both diffusion constants and reaction times can vary widely for physical systems, such a wave of activity provides a credible explanation for the modest speed of a nerve impulse. Strangely, although Walther Nernst, the famed German physical chemist, was both present at this meeting and clearly interested in Luther's ideas, this prescient contribution to physical chemistry was neglected for several decades.

Not overlooked by neuroscientists was the extensive work of English electrophysiologist Edgar Douglas Adrian, who proposed the *all-or-nothing* principle of nerve impulse conduction in 1914 at the age of 25 [1]. After earning a medical degree, Adrian went on to pioneer the application of the newly developed vacuum-tube amplifier to the study of nerve dynamics, observing a *refractory period* of diminished excitability that follows the passage of an impulse.

In 1936, the English marine biologist John Zachary ("J.Z.") Young noted that a large cylindrical structure in the common squid is in fact a nerve [69], leading the American biophysicist Kenneth Cole to record the classic oscilloscope photograph of a nerve impulse shown in Figure 1.1 [11]. Taken in 1938, the continuous line in this image shows the dependence of the impulse voltage on time, which is indicated by the black marks at the bottom edge of the figure.[2] Evidently, the impulse voltage rises rapidly to

---

[1]You can find an engraving of Faraday presenting a Christmas Lecture on a recent British twenty-pound note.

[2]In those days, the displacement of the horizontal sweep was not linear in time; thus, experimenters recorded the tips of a 1000 Hertz (Hz) oscillation, making the interval between each mark equal to 1 ms.

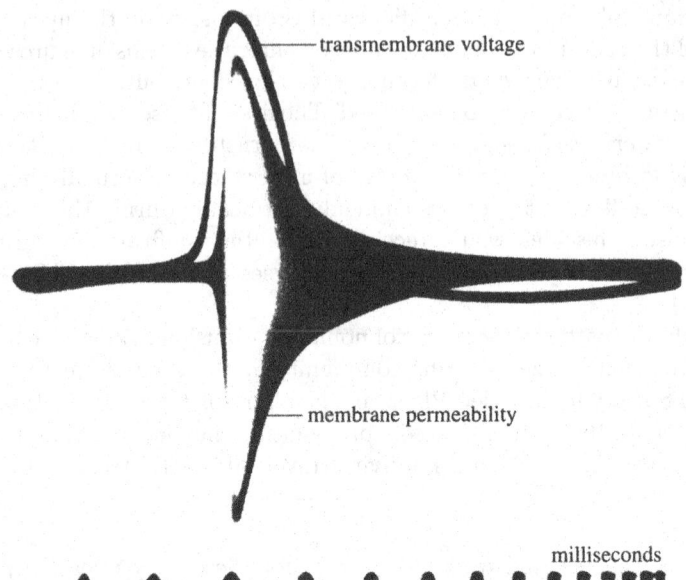

transmembrane voltage

membrane permeability

milliseconds

Figure 1.1. An early oscillogram of the change in membrane conductance (band) and membrane voltage (line) with time during the passage of a nerve impulse on a squid axon. (Time increases to the right, and the marks along the lower edge indicate intervals of 1 ms.) (Courtesy of K.S. Cole.)

a maximum value of about 100 millivolts (mV) in a fraction of a millisecond (ms), and this initial rise is called the *wave front* or *leading edge* of the nerve impulse. The impulse voltage then relaxes more slowly back to its resting level over a time interval of several milliseconds. The broad band also shown in the figure is a measure of changes of membrane permeability (or ionic conductance) from a resting value.[3]

Curiously overlooked by Western scientists was an important paper that also appeared in 1938 by the Soviet scientists Yakov Zeldovich and David Frank-Kamenetsky [71]. Addressing the problem of flame-front propagation, they proposed a simple *nonlinear partial differential equation* (PDE) for nonlinear diffusion in an active medium in which the independent variables were time and distance in the direction of propagation. In this paper, the authors solved their nonlinear PDE for an analytic solution describing a stable traveling wave: the flame front.

As we will see in Chapter 5, this simple equation also predicts both the speed of a nerve impulse on a squid axon and the shape of its leading edge. If these results had been noted by applied mathematicians and be-

---

[3]To measure ionic conductance, an ac bridge was balanced at the resting level of membrane permeability; thus, the width of the band indicates unbalance of the bridge, which stems from the change of permeability during the impulse.

come known to the community of electrophysiologists in the mid-1930s, the collective understanding of nerve impulse dynamics would have been greatly accelerated. Of course, that was a difficult decade for all segments of Russian society, but the almost total neglect of such a key paper is also evidence of a distressing lack of communication among nonlinear scientists, a problem that would continue until well into the 1970s [56].

The next major event in the history of the dynamics of single nerve fibers occurred in 1952 with the publication by English electrophysiologists Alan Hodgkin and Andrew Huxley of a series of papers culminating in their classic "A quantitative description of membrane current and its application to conduction and excitation in nerve" [25]. Taking advantage of wartime developments in electronics, they made careful measurements of individual ionic currents flowing across the membrane of a squid nerve. From these data, they developed a detailed nonlinear partial differential equation describing both the time course of membrane permeability and the nonlinear diffusion of transmembrane voltage along the nerve. Using this formulation—and a feat of numerical analysis that was truly impressive before the digital computer became available—Hodgkin and Huxley were able to reproduce many details of Figure 1.1, including the impulse speed of about 20 m/s, and to predict several other aspects of nerve impulse behavior, including Adrian's all-or-nothing response and the post-impulse refractory period.

In the same year that the H–H work appeared, surprisingly, the English mathematician Alan Turing proposed nonlinear diffusion as a theoretical basis for the appearance of patterns (now called "Turing patterns") in the embryonic development of living organisms, suggesting examples of biological emergence to which the theory might apply [62]. Although Turing's speculations and the Hodgkin–Huxley results were outstanding examples of research in applied mathematics, members of that community took little note of this seminal work until the early 1970s, when nonlinear diffusion in an active medium finally became recognized as an interesting and important form of nonlinear wave propagation [56]. Engineers, on the other hand, were involved in the development of a *neuristor*, or electronic analog of the active nerve fiber, during the 1960s as a novel basis for computer design. This effort led to a fruitful collaboration between the American neuroscientist Richard FitzHugh and the Japanese computer engineer Jin-ichi Nagumo, resulting in a simpler nerve model based on a nonlinear partial differential equation that bears their names [17, 41].

In contrast to the five dynamic variables of the H–H formulation, the FitzHugh–Nagumo model represents a nerve impulse with only three variables: transmembrane voltage, axial ionic current, and a *recovery variable*. Thus, traveling wave solutions are easier to visualize as trajectories in what mathematicians call a *phase space*. Aided by this conceptual simplification and the steady growth of available computing power, applied mathematicians finally began to explore the F–N model as a means for understanding

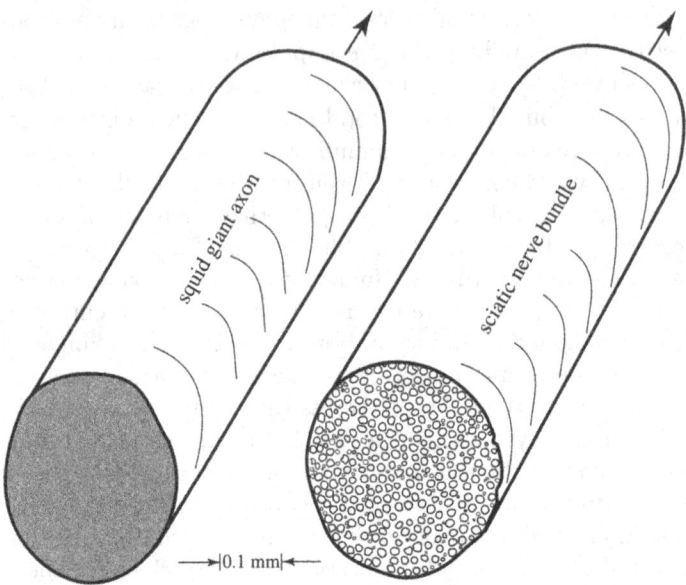

Figure 1.2. Comparison of the squid giant axon (left) and the sciatic nerve bundle controlling the leg muscle of a rabbit (right). There are about 375 myelinated fibers in the rabbit nerve, each conducting an individual train of nerve impulses at up to 80 m/s, about four times faster than the impulse velocity on a squid nerve. (Data from Young [70].)

the qualitative nature of nerve impulse propagation. Continuing throughout the 1970s, this tardy yet essential effort has deepened our understanding of several key phenomena, including all-or-nothing propagation, threshold conditions for nerve impulse formation, impulse stability, impulse response to variations in fiber geometry, decremental conduction, speed of periodic impulse trains, and effects of temperature and narcotization, all of which are considered in this book.

Presently, the propagation of a nerve impulse on a smooth fiber is a well-understood area of mathematical biology, the salient features of which should be appreciated by all serious students of neuroscience.

Interestingly, the sciatic nerve—first studied by Galvani in the late eighteenth century and used as a basic preparation for much subsequent neuroscience research—is not a smooth fiber; in fact, it is not even a single fiber. Like all vertebrate motor nerves, the sciatic nerve is a *bundle* of individual fibers, each carrying a different train of impulses from the spinal cord to a muscle, as was emphasized in a classic image prepared by J.Z. Young from which Figure 1.2 is drawn.

In this figure, we see a squid nerve compared with a rabbit sciatic nerve bundle on the same scale of distance showing that the rabbit nerve has about 375 information channels to one for the squid nerve. Because rabbit fibers also carry impulses up to about four times more rapidly than squid giant axons, the capacity of the rabbit nerve to transmit information is some three orders of magnitude greater than that of the squid nerve for the same area of cross section. How does a myelinated nerve achieve such impressive performance?

Each sciatic nerve fiber is covered almost everywhere by a fatty insulating sheath (called *myelin*), thus allowing only rather widely spaced *active nodes* of membrane to switch (see Figure 7.1). In such a *myelinated nerve* structure, the wave of activity does not progress uniformly along the fibers but jumps from one active node to the next, rather like a falling row of dominos. This process of "saltatory conduction"[4] shares the above-mentioned features of smooth fibers (e.g., all-or-nothing propagation, threshold, refractory period) while introducing new wrinkles. One of these is the unfortunate phenomenon of *failure*, wherein an impulse arriving at one active node is unable to ignite its neighbor.

Although also characterized by a closed causal loop (feedback loop) of the sort

Release of energy

$$\downarrow \qquad \uparrow$$

Dissipation of energy

the speed of saltatory impulse propagation is given by an expression of the form

$$v \sim s/\tau, \tag{1.2}$$

where s is the distance (in meters) between active nodes and τ is the time (seconds) required for the node to respond to a signal from an adjacent node.

Instead of being modeled by a nonlinear partial differential equation (like the squid fiber), a myelinated fiber is described by a set of nonlinear *difference-differential equations* (DDEs). In this formulation, time remains as an independent variable but the space variable is replaced by an integer that counts active nodes along the fiber.

Notwithstanding the problem of impulse failure, the myelinated nerve structure has two functional advantages, which may explain its evolutionary development. The first of these is that the speed of impulse propagation can be raised by increasing the distance between active nodes. This feature permits the fiber diameter to be reduced without decreasing the impulse velocity, as shown in Figure 1.2. The second advantage of a myelinated fiber is that the amount of energy expended in the passage of a nerve impulse is much less than is required for a smooth fiber, an important consideration for living organisms.

---

[4]Coined by Lillie, the term "saltatory" stems from the Italian verb *saltare*, meaning "to jump" [32].

First studied in detail in the late 1940s [54], the nonlinear dynamics of myelinated nerves have been investigated by a variety of physical and biological scientists throughout the ensuing half century [26, 12, 60] and it is currently an area of interest in applied mathematics. Because myelinated nerves are found in many biological organisms and several aspects of saltatory conduction are not yet well-understood, some of the outstanding research problems and opportunities are sketched in this book.

1.2   The Structure of a Nerve Cell

Based on evidence accumulated by the esteemed Spanish physiologist Santiago Ramón y Cajal that nerve cells are independent biological elements, the word *neuron* came into use during the 1890s [50]. Among other abilities, independent biological cells need means for communicating among themselves; thus, the term *synapse* was coined by the English physiologist Charles Scott Sherrington for a channel of interaction or communication between two different neurons. The concept of a neuron is now an essential part of our language, but what does it imply?

From the computer-oriented perspectives of today, a neuron is often described as a simple processor of information, gathering a linear sum of incoming signals from the synapses on the *dendritic trees*, comparing this sum with a threshold, and sending corresponding outgoing signals through the branches of the *axonal tree* (which may be either smooth or myelinated, as described earlier) to influence the behaviors of other neurons or muscle cells. This view is shared by many students of the human brain, raising several questions.

First, do dendritic trees merely gather a linear weighted sum of the incoming signals for presentation to the outgoing axon? If so, the causal implications of the collective input are more readily sorted out, but the behavior of a real neuron may not be so simple. Might not some sort of information processing be implied by the intricate dendritic branching structures?

Second, what happens to the transmitted information as it passes down the main axon? Does this information arrive unmodified at all of the distant (or *distal*) tips of the axonal tree? Or does information processing also occur at branching regions of the axonal tree?

Third, how about the synapses? What are the details of their dynamics? Might they alter their behaviors, providing a basis for memory? Could there also be electrostatic interactions or direct ionic couplings among neurons?

Finally, it is necessary to consider what is implied by "information" in the context of a neural system. Does the term merely indicate the memory capacity that a computer engineer would require to store the transmitted signal? Should not the *meanings* of the messages to other components of the organism be taken into consideration? If so, where are such meanings to be found and how are they to be described? In other words, what language does the neuron speak, and at what level of description is it represented?

An early attempt to answer such questions was published in 1943 by Warren McCulloch—an American psychiatrist who went on to play a leading role in theoretical studies of brain dynamics—and a young mathematician named Walter Pitts. Entitled "A logical calculus of the ideas immanent in nervous activity," this important paper proposed a specific model for the dynamics of a neuron that has become widely known as the *McCulloch–Pitts* (M–P) neuron [37].

In this model, the following assumptions are made. (1) Dendritic trees are assumed to gather a linear weighted sum of the input signals. (2) This sum is compared with a *threshold level* at the *initial segment* of the outgoing axon (see Figure 2.1). (3) If the weighted sum of inputs is greater than this threshold, an outgoing signal is launched on the main trunk of the axonal tree. (4) After a certain time delay, any impulse appearing on the main trunk of the axonal tree is transmitted without failure to all of the distal twigs of this tree. Thus, the only nonlinear feature of the M–P model is a single switch, which is assumed to act at the initial segment of the axonal tree.

Because this M–P neuron is among the simplest models that could be proposed, it has provided a basis for many computer studies of neural systems since the early 1960s. As computer power has grown over the past four decades—doubling every eighteen months, we are told—these studies have become ever more sophisticated, but increasing computer power has had other implications for neural modeling.

As noted previously, the detailed Hodgkin–Huxley description of nerve impulse conduction along the giant axon (or outgoing fiber) of a neuron in the stellate ganglion of the common squid also became available in the early 1950s. Based on independent measurements of parameters characterizing an isolated patch of active membrane, the H–H theory accurately predicts both the speed and the shape of a nerve impulse along with several other aspects of neural behavior. In the context of the Hodgkin–Huxley theory, therefore, each patch of active membrane on a neuron may function like a switch, suggesting far more intricate behavior than is supposed for the McCulloch–Pitts model.

With the growth of available computing power, H–H modeling was extended beyond the confines of a single fiber to provide increasingly sophisticated descriptions of an entire neuron. Thus, two divergent lines of neuroscience research developed, both of which were driven by the increasing numerical capability that the researchers had at their disposal: M–P-based models of so-called "neural networks" and improved models of individual neurons based on the H–H description of a nerve membrane.

By the early 1970s, advances in electron micrography and electrophysiology made it evident to some that the dynamics of real neurons are far
more complex than the McCulloch–Pitts model implies. Thus, in 1972,
U.S. neurologist Steven Waxman proposed the concept of a "multiplex neuron" in which patches of membrane near the branching regions of incoming
(dendritic) and outgoing (axonal) processes of the neuron are viewed as
localities of low safety factor, where propagating nerve impulses can become enhanced or extinguished depending on the presence or absence of
other impulses [63]. In other words, an individual neuron might be able to
perform logical computations at the branching regions of its axonal and
dendritic trees, making it more like an integrated circuit chip than a single
switch.

As reasonable as it seemed to some at the time [55], this expanded view of
the neuron's computational power was far from being universally accepted.
Jerry Lettvin, a noted electrophysiologist in the Electrical and Computer
Engineering Department at MIT, told me (in the spring of 1978) that
when he and his colleagues reported blockages of nerve impulses in the optic nerves of cats and speculated on the possibilities of this phenomenon
for visual information processing [10], his funding (from the National In-
stitutes of Health) was cut off. "When you start doing science again," he
was informed, "we are ready to resume your support."

Why were suggestions of neuronal intricacy so widely ignored in the
1970s? Although one can only speculate, three possibilities come to mind.

(1) Admitting to increased computational power of the individual neuron
compounds the already daunting task of analyzing systems of neurons.

(2) Much of the early research on increased computing power of individual neurons was being done in the former Soviet Union [29], and Western
scientists—particularly those in the United States—tended to ignore or
disparage the fruits of Russian science.

(3) A widespread and uncritical belief in the concept of "all-or-nothing"
propagation left ignored the possibility of impulse failure at branching
regions of axonal or dendritic trees.

An encouraging feature of science, however, is that the truth does eventually become recognized. Like old soldiers, pillars of the establishment "fade
away" as evidence for a new paradigm gathers. What was once considered
flagrant speculation becomes established knowledge, and so it was with our
collective perceptions of the dynamic intricacy of a neuron. Since the end
of the 1980s, the experimental evidence for the impressive computational
power of individual neurons has become compelling, and former heresy is
now recognized as common sense [3, 30, 59].

In the following chapter, we will see one of the features contributing to
the intricacy of a neuron: its dynamics operate at several levels, including those of intrinsic proteins embedded in the active membranes, complex
behavior of individual synapses, nonlinear waves of activity propagating
on smooth or myelinated fibers, impulse failure or enhancement near the

branching regions of dendritic and axonal trees, and the global functioning of the neuron. In subsequent chapters, understanding the behavior of multilevel nonlinear dynamic hierarchies is recognized as a central issue for both mathematical biology and cognitive science.

1.3   Organization of the Brain

Just as there are distinctly different levels of dynamic activity in an individual neuron, the brain itself is organized into several diverse regions with responsibilities for various aspects of behavior. Although our collective understanding of this organization is far from complete, it has advanced well beyond that of Aristotle, who believed that the heart is the seat of human intelligence, and of René Descartes, who proposed in the seventeenth century that the mind was controlled by the pineal gland.

Indeed, much of the above-noted flood of nineteenth-century research in neuroscience was devoted to establishing where the various functions of the brain are located. Notable among these many activities were the discovery by Marie-Jean-Pierre Flourens in 1823 that the cerebellum regulates motor activity, a demonstration by Bartolomeo Panizza in 1855 that the occipital (optic) lobe of the neocortex (or outer layer of the brain) is essential for vision, and Paul Broca's 1861 observation that the left temporal lobe of the neocortex controls our ability to speak. Among others, the results of these efforts contributed to Sherrington's classic *The Integrative Action of the Nervous System*, which contains an early suggestion that neurons act in functional groups rather than as individual units [57].

Where a component of the brain's activity is located is one question, but how it is organized is quite another. What are its principles of design? How do the ten thousand million or more neurons of the brain conspire to guide the behavior of a normal human being?

Such questions were addressed by McCulloch and Pitts in their seminal 1943 paper [37]. In a search for answers, these authors distinguished between two different sorts of neural organization that remain relevant to the present day. The first of their design principles was quaintly termed "nets without circles," implying that no closed loops of causal implication (or feedback loops) were allowed to be embedded within the system. Under this principle, information flows uniformly in one direction from input to output terminals. The second design principle was called "nets with circles," a strategy that takes advantage of closed causal loops situated between sensory inputs and muscular output, which had recently been announced by Rafael Lorente de Nó [33, 34], the last and most brilliant student of Ramón y Cajal.

Interestingly, these observations go back to the mid-1920s but were strongly opposed by Cajal. Cajal urged Lorente not to publish a manuscript describing closed loops of neural activity in the neocortex because such ideas were unacceptable to the neuroscience community and their dissemination would harm Lorente's career. Out of respect for his mentor, therefore, Lorente delayed publication until after Cajal died in 1934 [19]. Why, one asks in retrospect, was the neuroscience community so hesitant to entertain the concept of closed causal loops? Was it that this assumption makes the dynamics of a brain too complicated to understand?

Whatever the answers, the assumption of nets *without* circles greatly simplifies mathematical analysis and is in accord with the strong stimulus–response bias of *behaviorism,* which vitiated North American psychology in the middle of the twentieth century. Following these lines, American psychologist Frank Rosenblatt proposed the *perceptron* in 1958 as a "model for information storage and organization in the brain" [53].

Based on McCulloch–Pitts neurons, the perceptron was conceived as a *layered structure* in which each layer of model neurons receives inputs from the adjacent layer on one side, performs logical computations, and passes on the results of those computations to the adjacent layer on the other side. Allowing the input (or "dendritic") weightings of the model neurons to be appropriately modified during a *learning phase* gives the system an ability to classify patterns that grows with experience. In such systems, the input patterns might be two-dimensional images impressed on a first layer (sometimes called the "retina"), and the output layer could be a digital code classifying various patterns into desired categories such as letters of the alphabet [6, 44].

Although it has been shown that perceptrons can automatically learn to classify incoming patterns, these abilities are limited in several ways [40]. First, the maximum number of patterns recognized can be no more than the number of neural elements from which the system is constructed—an inconvenient restriction in some applications. Second, translations, rotations, and scalings of a learned pattern are perceived as new patterns. Finally, it is difficult for the perceptron to detect patterns that are embedded within other patterns—so-called "patterns in context"—a recognition feat that is rather easily performed by humans.

In the mid-1980s, American physicist Erich Harth and his colleagues showed how the problem of detecting a pattern in context could be solved by allowing internal feedback loops of information processing to form around many different dynamic levels of a neural system. This is the so-called "creative loop," which links the highest and lowest levels of cerebral dynamics [21]. Thus the principle of nets *with* circles is now recognized as an essential aspect of any realistic model of the brain's organization. What does this broader design strategy imply?

We have used closed loops of causal implication in representing the process of nonlinear diffusion through an active medium, which governs the propagation of a nerve impulse along the axon of a neuron. Similarly, a net with circles has an internal structure that may be described by a diagram of the following form.

Information at A

$\downarrow$ $\uparrow$

Information at B

Such a diagram suggests that a new internal pattern (A together with B) *emerges* as a result of the internal feedback loop.

Thus an understanding of feedback is of central significance in describing the behavior of biological organisms, as was recognized and widely promoted by the American mathematician Norbert Wiener in the 1950s [65]. Defining the new research area of *cybernetics* as the "science of communication and control in man and the machine," he emphasized the vital role played by feedback in cognitive systems, but such phenomena may be even more important than Wiener realized. To see this, note that there are two sorts of feedback—positive and negative—both artfully employed in biology, albeit in different ways.

Negative Feedback

A simple example of negative feedback, with which all of us are familiar, is the temperature control on the living room wall. If the room gets too cold, a switch is thrown and turns the furnace on. When the room becomes too warm, the switch relaxes and the heat turns off. By this means, the temperature of the living room is automatically maintained within an acceptable range.

Thus, a control system that is based on negative feedback comprises two basic elements: a sensor, which detects the value of some variable (temperature), and a means for comparing that value with what the system desires (a temperature of about 68° F). Going back to James Watt's eighteenth-century invention of a governor for controlling the speed of his steam engine, this negative feedback principle has long been used in engineering design, but it was put on a sound mathematical basis in the late 1920s in the context of electronic amplifier design.

In the realms of biology, several examples of negative feedback are readily noted.

(1) On hot days, we perspire, and the resulting evaporation cools our body. Conversely, if we are cold, we begin to shiver, and the energy expended in this activity warms our skin.

(2) In bright sunlight—relaxing with a book on a sandy beach at high noon, for example—the irises of our eyes contract, thereby reducing our retinal light intensities. In a dimly lit room, on the other hand, these same irises open wide, allowing more light to enter.

(3) When we need food, the discomfort of hunger encourages us to eat. Having eaten enough, as all know, satiety bids us to stop.

In each of these examples, the effect of negative feedback is to manage (or control) the internal influences of external variables. The cybernetic system or biological organism has certain internal requirements stemming from its wants and needs, and the effect of negative feedback is to prevent external variables (temperature, light intensity, rate of energy consumption, and so on) from pushing the internal variables beyond acceptable bounds.

Positive Feedback
It is easy to confuse positive and negative feedback because both involve closed causal loops of energy or information and are achieved through similar mechanisms, but the two processes are very different.[5] Let us pause, therefore, in our historical survey of neuroscience to get a clear understanding of positive feedback.

In its simpler manifestations, positive feedback does not seem to be a useful mechanism. Consider, for example, a furnace control on the living room wall that an incompetent heating engineer has designed to display positive rather than negative feedback. Thus, if the furnace is off and the room gets too warm, a positive feedback control turns the furnace *on*. If, on the other hand, the furnace is on and the room gets too cold, positive feedback turns the furnace *off*. Clearly, this is not a very satisfactory way to regulate the temperature of a house, and an engineer who arranges the temperature control to operate in such a manner will soon be looking for another line of work.

Whereas a system with negative feedback is essentially *stable*, a positive feedback system is *unstable*. With positive feedback, small deviations from a stationary state are amplified around the closed causal loop, leading to ever larger displacements. Is this ever a useful arrangement?

Think about the conditions on Earth some four thousand million years ago. The Hadean oceans boiled and bubbled like an enormous witches brew, stirred by flashes of lightning under a rain of meteors that spiced the chemical soup. In the midst of this chemical turmoil—evolutionary biologists tell us—certain molecules organized themselves in a manner that allowed *reproduction* and life began. Like our hypothetical positive feedback "control" of the living room temperature, this was a very *unstable* situation. As these protobiological molecules appeared in increasing numbers, they grew at increasing rates because there were ever more molecules available to reproduce themselves. Although an illustration of instability stemming from positive feedback, most would regard this example favorably because we would not exist had life on Earth not managed to get started. How then can the related dynamics be described?

---

[5]Indeed, some use the term "reentry" for positive feedback in neural systems, reserving the term "feedback" for what I term "negative feedback" [15].

If the number of protobiological molecules at a particular time t is indicated by $N(t)$ and the rate at which this number of molecules increases with time is dN/dt, a corresponding positive feedback diagram looks like this:

$$\textbf{Number of molecules } (N)$$
$$\downarrow \qquad \uparrow$$
$$\textbf{Rate of increase of number } (dN/dt)$$

This diagram implies that the rate of increase of the biological molecules is proportional to the number of such molecules that are present.

Another way to describe this situation is with the differential equation $dN/dt = \alpha N$, where α is a constant of proportionality relating the number of molecules and the rate at which this number is increasing with time. This equation has the exponentially growing solution $N(t) = N(0)e^{\alpha t}$, with $N(0)$ indicating the number of protobiological molecules that were present at some initial value of time $(t = 0)$.

The solution of this simple *linear* differential equation implies that the number of protobiological molecules will go on increasing without limit, but this is unrealistic. The number of atoms available to form such molecules is necessarily finite, so the ultimate number of molecules is eventually limited by *nonlinear* effects.

Such a "limit to growth" can be represented by the *logistic* or *Verhulst equation*[6]

$$\frac{dN}{dt} = \alpha N(1 - N/N_0). \tag{1.3}$$

This equation implies that the rate of population increase dN/dt eventually falls to zero when the population $N(t)$ reaches a limiting value of N_0, which is related to the available supply of energy.

A solution for the Verhulst equation that equals $N(0)$ at $t = 0$ and approaches N_0 as $t \to \infty$ turns out to be

$$N(t) = \frac{N_0 N(0) e^{\alpha t}}{N_0 + N(0)(e^{\alpha t} - 1)}, \tag{1.4}$$

and this solution is displayed in Figure 1.3. From this figure, we see that population levels converge toward the steady value of N_0 with increasing time. Thus N_0 is called an *attractor* in the jargon of nonlinear dynamics.

At this point, the reader may be wondering what these discussions of early biology and European population dynamics have to do with the his-

---

[6] Equation (1.3) is named for Belgian mathematician Pierre Francois Verhulst (1804–1849), who derived it in 1846 to describe the growth dynamics of biological populations. He used Equation (1.4) to predict that the upper limit to the population of Belgium would turn out to be 9,400,000. Interestingly, the 1994 population of Belgium was 10,118,000.

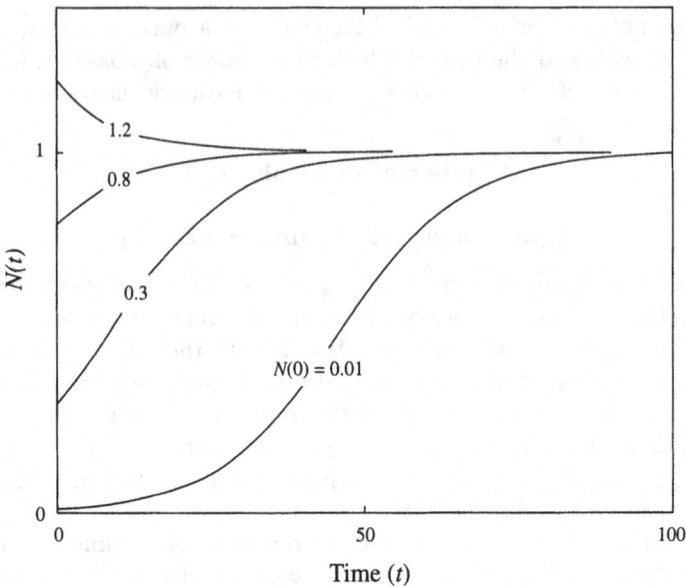

Figure 1.3. Verhulst functions plotted from Equation (1.4), showing the time course of a population $N(t)$ for different initial values $N(0)$. (In these plots, $\alpha = 0.1$ and $N_0 = 1$.)

tory of neuroscience. Why are we wrestling with differential equations and worrying about numbers of organisms? The answer is that the concept of positive feedback is very general, applying to many aspects of neuroscience.

One such application is a description of dynamic events on a nerve fiber when threshold conditions for the ignition of a nerve impulse have been satisfied. To understand these dynamics, think of a candle when the wick has just barely been lit and the fully developed flame has not yet developed. Because the flame is small, its rate of energy loss (dissipation) through the emission of heat and light is less than the rate of energy input from the paraffin; thus the flame grows. As the flame becomes larger, its rate of dissipation increases, eventually coming into balance with the rate at which energy is released from the candle. When this balance is established, a candle flame—like the present population of Belgium—reaches a stable size, neither waxing nor waning.

A similar phenomenon occurs when a nerve fiber is stimulated just above the threshold level at which it ignites. The transmembrane voltage contin-ues to grow until it reaches that of the full impulse—shown for a squid nerve in Figure 1.1—for which energy is released by the impulse at the same rate that it is being consumed. Where else in neuroscience might the dynamics of positive feedback appear?

Toward the end of the nineteenth century, the U.S. psychologist William James suggested that positive feedback loops incorporating facial muscles might be implicated in the phenomenon of emotion [28], but few of his contemporaries picked up on the idea. Inspired by Lorente de Nó's observations of closed causal loops in the neocortex [33, 34], however, Canadian psychologist Donald Hebb in 1949 published a book entitled *Organization of Behavior*, which is among the most significant contributions to neuroscience of the twentieth century [22]. Aiming to "bridge the long gap between the facts of neurology and those of psychology," Hebb proposed that the functional entities in the dynamics of a brain are not confined to individual neurons but include interconnected and interacting groups of neurons. These he called *cell assemblies*, defined as follows.

> Any frequently repeated, particular stimulation will lead to the slow development of a "cell-assembly," a diffuse structure comprising cells . . . capable of acting briefly as a closed system, delivering facilitation to other such systems and usually having a specific motor facilitation. A series of such events constitutes a "phase sequence"—the thought process. Each assembly may be aroused by a preceding assembly, by a sensory event, or—normally—by both. The central facilitation from one of these activities on the next is the prototype of "attention."

To get in mind what Hebb implied by the term "cell assembly," let us turn to a social metaphor. Compare the human brain to a community in which the individual neurons are represented by its citizens. In this metaphor (which is a useful hypothesis in ethnology), the functional realities are not limited to individual citizens but may include groups of people acting together in the pursuit of common goals, with each group termed a "social assembly." A particular individual might be a member of several social assemblies, such as a motorcycle club, a political association, a soccer team, a church, a labor union, a bowling league, a theater group, the parent-teacher's association, a mountaineering club, a weight-watching group, bird watchers, the junior league, and so on.

What is it that allows these various groupings to act as function realities? The members of a particular assembly know each other and share common lists of addresses and telephone numbers, allowing an organization to activate its own members—or even the members of like-minded assemblies—should an appropriate occasion arise. The various activities and interactions of such groups, one can argue, comprise the dynamics of the community; thus, one must understand their structures and the relationships among them in order to read a daily newspaper with comprehension.

The memberships of like-minded social assemblies are interconnected because each has a few overlapping participants. Thus, for example, members of the mountaineering club might join with the hikers and the bird watchers to encourage farmers to resist the development of a tacky theme park near their fields. Or the parents in a school might enlist the support of teachers and the police department to reduce the number of firearms available to children. Just as a particular individual could be both a farmer and a birder, a single nerve cell, in the context of Hebb's cell-assembly theory, would be able to participate in several different neural assemblies of the brain, an observation that greatly increases estimates of how many assemblies there can be.

In Hebb's picture, as noted earlier, the primary functional entities of the brain are assemblies of nerve cells rather than the individual neurons out of which assemblies emerge. Defined by its interconnections, a particular cell assembly forms a sort of "three-dimensional fishnet" involving perhaps hundreds or thousands of neurons extending over much of the brain and linking visual, auditory, and conceptual aspects of a particular perception. Like a candle or an individual neuron, an assembly of neurons "ignites" when a sufficient number of its constituent neurons become active. In the jargon of nonlinear dynamics, each neuronal assembly is also an attractor, displaying the interrelated properties of threshold and all-or-nothing response that characterize individual neurons.

Upon ignition, the activity level of a cell assembly—like the flame of a candle or the impulse on a nerve fiber—will begin to grow exponentially through the action of positive feedback. Eventually, this growth of activity will become limited by the maximum rate at which its constituent neurons can fire, and a constant level will be established, as is suggested by the growth curves in Figure 1.3.

Thus, we see that the process of initial exponential growth followed by ultimate limitation to a stable amplitude—which was used by Verhulst to predict the course of population dynamics in his native Belgium—has a very wide range of applications. With this metaphor, we have at hand a qualitative and potentially quantitative model for the relevant dynamics whenever something new emerges: the flame of a candle or a nerve impulse, life on Earth or a new biological species, a city, or a song.

Over the half century since Hebb's work first appeared, several related contributions to the theory of brain dynamics in "nets with circles" have been presented, including matrix representations of neural interconnections [9], statistical models [2, 13], wave analyses [5, 46, 66], and studies based on "attractor theory" stemming from phase-space analysis of nonlinear systems [27], all of which are discussed in subsequent chapters. Thus, theoretical perspectives have continued to expand as the corresponding experimental knowledge about real neurons and brains has accrued.

With respect to Hebb's cell-assembly theory, there have been some computer simulations [18, 51, 52, 58, 64], psychological experiments [24, 49, 61], and theoretical extensions of the basic ideas [7, 20, 31, 39, 47, 48], and Edelman has used the assembly concept as a basis for his "neural Darwinism" [15], but all of these supporting data are indirect. Is there clear experimental evidence for a cell assembly? Will electrophysiologists ever be able to show that a small and widely dispersed fraction of the brain's neurons are acting as a functional unit? Until a few years ago, I would have supposed this goal to lie beyond the reach of feasibility.

Although it is challenging—to say the least—to obtain recordings from a significant fraction of the neurons in one of Hebb's three-dimensional fishnets, some progress is being made. On simpler species, it is now possible to introduce voltage-sensitive dyes into the neurons and record the dynamics of their global activity on arrays of silicon photodiodes. In this manner, the simultaneous activity of several hundred neurons in the abdominal ganglion of a mollusk (*Aplysia*) has been indirectly observed [68].

As an example of direct voltage measurements, neuroscientists at the University of Arizona are now routinely recording from several dozen extracellular electrodes in the hippocampus of the rat, with each electrode indicating the activity of several neurons for a total of 100 or more individual signals [67]. Interestingly, these multiple recordings can be made while the animal is undergoing psychological testing, and several other laboratories are reporting equally impressive feats [14, 38, 43]. Among the more sophisticated of such devices is the Utah electrode array (UEA), which comprises 100 external electrodes mounted in an area of one square millimeter [36, 45].

Studies of dynamic similarities (called *correlations*) among the individual recordings in these multielectrode experiments are beginning to provide direct empirical evidence for the reality of cell assemblies, leading some to anticipate that Hebb's paradigm will become an established feature of the neuroscience landscape in the next few years [42]. In Chapter 11, we review these developments.

Because most presently available theoretical work on brain dynamics is based on single-switch (essentially McCulloch–Pitts) models of the individual neurons, the real brain is expected to be far more intricate than current formulations suggest. Nonetheless, it is important for mathematically oriented neuroscientists to become familiar with these various theories, because they provide baselines for future representations of the brain's dynamics.

References

[1] ED Adrian, The all-or-none principle in nerve, *J. Physiol. (London)* 47 (1914) 460–474.

[2] S Amari, Characteristics of randomly interconnected threshold-element networks and network systems, *Proc. IEEE* 59 (1971) 35–47.

[3] M Arbib (ed), *The Handbook of Brain Theory and Neural Networks,* MIT Press, Cambridge, MA, 1995.

[4] J Bernstein, Untersuchungen zur Thermodynamik der bioelektrischen Ströme, *Arch. ges. Physiol.* 92 (1902) 521–562.

[5] RL Beurle, Properties of a mass of cells capable of regenerating pulses, *Philos. Trans. R. Soc. (London)* 240A (1956) 55–94.

[6] HD Block, A model for brain functioning, *Rev. Mod. Phys.* 34 (1962) 123–135.

[7] V Braitenberg, Cell assemblies in the visual cortex. In *Theoretical Approaches to Complex Systems*, Springer-Verlag, Berlin, 1978.

[8] MAB Brazier, *A History of the Electrical Activity of the Brain*, Pitman, London, 1961.

[9] ER Caianiello, Outline of a theory of thought-processes and thinking machines, *J. theoret. Biol.* 1 (1961) 204–235.

[10] SH Chung, SA Raymond and JY Lettvin, Multiple meaning in single visual units, *Brain Behav. Evol.* 3 (1970) 72–101.

[11] KS Cole and HJ Curtis, Electrical impedance of nerve during activity, *Nature* 142 (1938) 209.

[12] KS Cole, *Membranes, Ions and Impulses*, University of California Press, Berkeley, 1968.

[13] JD Cowan, A statistical mechanics of nervous activity. In *Some Mathematical Questions in Biology*, American Mathematical Society, Providence, 1970.

[14] SA Deadwyler, T Burn, and RE Hampson, Hippocampal ensemble activity during spatial delayed-nonmatch-to-sample performance in rats, *J. Neurosci.* 16 (1996) 354–372.

[15] GM Edelman, *Bright Air, Brilliant Fire: On the Matter of the Mind*, Basic Books, New York, 1992.

[16] M Faraday, *Faraday's Chemical History of a Candle*. Chicago Review Press, Chicago, 1988. (Republication of *A Course of Six Lectures on the Chemical History of a Candle*, which first appeared in 1861.)

[17] R FitzHugh, Impulses and physiological states in theoretical models of nerve membrane, *Biophys. J.* 1 (1961) 445–466.

[18] E Fransén, *Biophysical Simulation of Cortical Associative Memory*, Doctoral thesis, Royal Institute of Technology, Stockholm, 1996.

[19] WJ Freeman, Premises in neurophysiological studies of learning. In *Neurobiology of Learning and Memory*, G Lynch, JL McGaugh, and NM Weinberger (eds), Guilford Press, New York, 1984.

[20] PH Greene, On looking for neural networks and "cell assemblies" that underlie behavior, *Bull. Math. Biophys.* 24 (1962) 247–275 and 395–411.

[21] E Harth, *The Creative Loop: How the Brain Makes a Mind*, Addison–Wesley, Reading, MA, 1993.

[22] DO Hebb, *Organization of Behavior: A Neuropsychological Theory*, John Wiley & Sons, New York, 1949.

[23] H Helmholtz, Messungen über den zeitlichen Verlauf der Zuckung animalischer Muskeln und die Fortpflanzungsgeschwindigkeit der Reizung in den Nerven, *Arch. Anat. Physiol.* (1850) 276–364.

[24] W Heron, The pathology of boredom, *Sci. Am.* January 1957.

[25] AL Hodgkin and AF Huxley, A quantitative description of membrane current and its application to conduction and excitation in nerve, *J. Physiol. (London)* 117 (1952) 500–544.

[26] AL Hodgkin, *The Conduction of the Nervous Impulse,* Liverpool University Press, Liverpool, 1964.

[27] JJ Hopfield, Neural networks and physical systems with emergent collective computational abilities, *Proc. Natl. Acad. Sci. USA* 79 (1982) 2554–2558.

[28] W James, What is an emotion? In *Psychology Classics: A Series of Reprints and Translations,* K. Dunlap (ed), Wilkins & Wilkins, Baltimore, 1922 (reprinted from *Mind,* 1884, 188–205).

[29] BI Khodorov, *The Problem of Excitability,* Plenum Press, New York, 1974.

[30] C Koch, *Biophysics of Computation: Information Processing in Single Neurons,* Oxford University Press, New York, 1999.

[31] CR Legéndy, On the scheme by which the human brain stores information, *Math. Biosci.* 1 (1967) 555–597.

[32] RS Lillie, Factors affecting transmission and recovery in the passive iron wire nerve model, *J. Gen. Physiol.* 7 (1925) 473–507.

[33] R Lorente de Nó, Studies on the structure of the cerebral cortex: I. The area entorhinalis, *J. Psychol. Neurol.* 45 (1934) 381-438.

[34] R Lorente de Nó, Analysis of the activity of the chains of internuncial neurons, *J. Neurophysiol.* 1 (1938) 207–244.

[35] R Luther, Räumliche Fortpflanzung chemischer Reaktionen. *Z. Elektrochem.* 12(32) (1906) 596–600. (English translation in *J. Chem. Ed.* 64 (1987) 740–742.)

[36] EM Maynard, CT Nordhausen, and RA Normann, The Utah intracortical electrode array: A recording structure for potential brain–computer interfaces, *Electroencephalogr. Clin. Neurophysiol.* 102 (1997) 228–239.

[37] WS McCulloch and WH Pitts, A logical calculus of the ideas immanent in nervous activity, *Bull. Math. Biophys.* 5 (1943) 115–133.

[38] TJ McHugh, KI Blum, JZ Tsien, S Tonegawa, and MA Wilson, Impaired hippocampal representation of space in CA1-specific NMDAR1 knockout mice, *Cell* 87 (1996) 1339–1349.

[39] PM Milner, The cell assembly: Mark II, *Psychol. Rev.* 64 (1957) 242–252.

[40] M Minsky and S Papert, *Perceptrons,* MIT Press, Cambridge, MA, 1969.

[41] J Nagumo, S Arimoto, and S Yoshizawa, An active impulse transmission line simulating nerve axon, *Proc. IRE* 50 (1962) 2061–2070.

[42] MAL Nicolelis, EE Fanselow, and AA Ghazanfar, Hebb's dream: The resurgence of cell assemblies, *Neuron* 19 (1997) 219–221.

[43] MAL Nicolelis, AA Ghazanfar, BM Faggin, S Votaw, and LMO Oliveira, Reconstructing the engram: Simultaneous, multisite, many single neuron recordings, *Neuron* 18 (1997) 529–537.

[44] NJ Nilsson, *Learning Machines: Foundations of Trainable Pattern-Classifying Systems,* Morgan Kaufmann, San Mateo, CA, 1990.

[45] CT Nordhausen, EM Maynard, and RA Normann, Single unit recording capabilities of a 100 microelectrode array, *Brain Res.* 726 (1996) 129–140.

[46] PL Nuñez, The brain wave equation: A model for the EEG, *Math. Biosci.* 21 (1974) 279–297.

[47] G Palm, Toward a theory of cell assemblies, *Biol. Cybern.* 39 (1981) 181–194.

[48] G Palm, Cell assemblies, coherence, and corticohippocampal interplay, *Hippocampus* 3 (1993) 219–226.

[49] RM Pritchard, W Heron, and DO Hebb, Visual perception approached by the method of stabilized images, *Can. J. Psychol.* 14 (1960) 67–77.

[50] S Ramón y Cajal, Structure et connexions des neurons, *Arch. Fisiol.* 5 (1908) 1–25.

[51] A Rapoport, "Ignition" phenomena in random nets, *Bull. Math. Biophys.* 14 (1952) 35–44.

[52] N Rochester, JH Holland, LH Haibt, and WL Duda, Tests on a cell assembly theory of the action of a brain using a large digital computer, *Trans. IRE Inf. Theory* IT–2 (1956) 80–93.

[53] F Rosenblatt, The Perceptron: A probabilistic model for information storage and organization in the brain, *Psychol. Rev.* 65 (1958) 298–311.

[54] WAH Rushton, A theory of the effects of fibre size in medullated nerve, *J. Physiol. (London)* 115 (1951) 101–122.

[55] AC Scott, The electrophysics of a nerve fiber, *Rev. Mod. Phys.* 47 (1975) 487–533.

[56] AC Scott, *Nonlinear Science: Emergence and Dynamics of Coherent Structures,* Oxford University Press, Oxford, 1999.

[57] CS Sherrington, *The Integrative Action of the Nervous System,* Yale University Press, New Haven, 1906.

[58] DR Smith and CH Davidson, Maintained activity in neural nets, *J. Assoc. Comput. Mach.* 9 (1962) 268–279.

[59] G Stuart, N Spruston, and M Häusser, *Dendrites,* Oxford University Press, Oxford, 1999.

[60] I Tasaki, *Physiology and Electrochemistry of Nerve Fibers,* Academic Press, New York, 1982.

[61] WR Thompson and W Heron, The effects of restricting experience on the problem-solving capacity of dogs, *Can. J. Psychol.* 8 (1954) 17–31.

[62] AM Turing, The chemical basis of morphogenesis, *Philos. Trans. R. Soc. London* B237 (1952) 37–72.

[63] SG Waxman, Regional differentiation of the axon: A review with special reference to the concept of the multiplex neuron, *Brain Res.* 47 (1972) 269–288.

[64] H White, The formation of cell assemblies, *Bull. Math. Biophys.* 23 (1961) 43–53.

[65] N Wiener, *Cybernetics,* The Technology Press, Cambridge, MA, 1961.

[66] HR Wilson and JD Cowan, Excitatory and inhibitory interactions in localized populations of neurons, *Biophys. J.* 12 (1972) 1–24.

[67] MA Wilson and BL McNaughton, Dynamics of the hippocampal ensemble code for space, *Science* 261 (1993) 1055–1058.

[68] JY Wu, LB Cohen, and CX Falk, Neuronal activity during different behaviors in *Aplysia*: A distributed organization? *Science* 263 (1994) 820–823.

[69] JZ Young, Structure of nerve fibers and synapses in some invertebrates, *Cold Spring Harbor Symp. Quant. Biol.* 4 (1936) 1–6.

[70] JZ Young, *Doubt and Certainty in Science,* Oxford University Press, Oxford, 1951.

[71] YaB Zeldovich and DA Frank-Kamenetsky, K teorii ravnomernogo rasprostranenia plameni, *Dokl. Akad. Nauk SSSR* 19 (1938) 693–697.

2
Structure of a Neuron

If the human brain is likened to a jungle—and many experimental neuroscientists would find the metaphor apt—the neurons are its flora. Some are large trees; others are small plants. Some are vines, extending their influence over long distances; yet others are more like shrubs, dominating local regions. So it is with many sorts of nerve cells. In the presence of such variability, one wonders what can be said in general about the nature of a neuron.

Expanding on the brief introduction to neurons in the previous chapter, the aim here is to provide a trail guide through the first part of this book, wherein various aspects of neural structure are considered in greater detail. To this end, some basic facts about nerve cells are presented and a few useful formulations are introduced.

2.1 A Generic Neuron

As a basis for organizing Chapters 3 through 9, a cartoon of a neuron is sketched in Figure 2.1, suggesting some of the roles played by various components of these busy cells as they gather and process incoming information for presentation to their output terminals. While regarding this figure, the reader should note how the intricate dynamics of a nerve cell become hierarchically organized into a functional unit.

Among the salient features of typical neurons are the following.

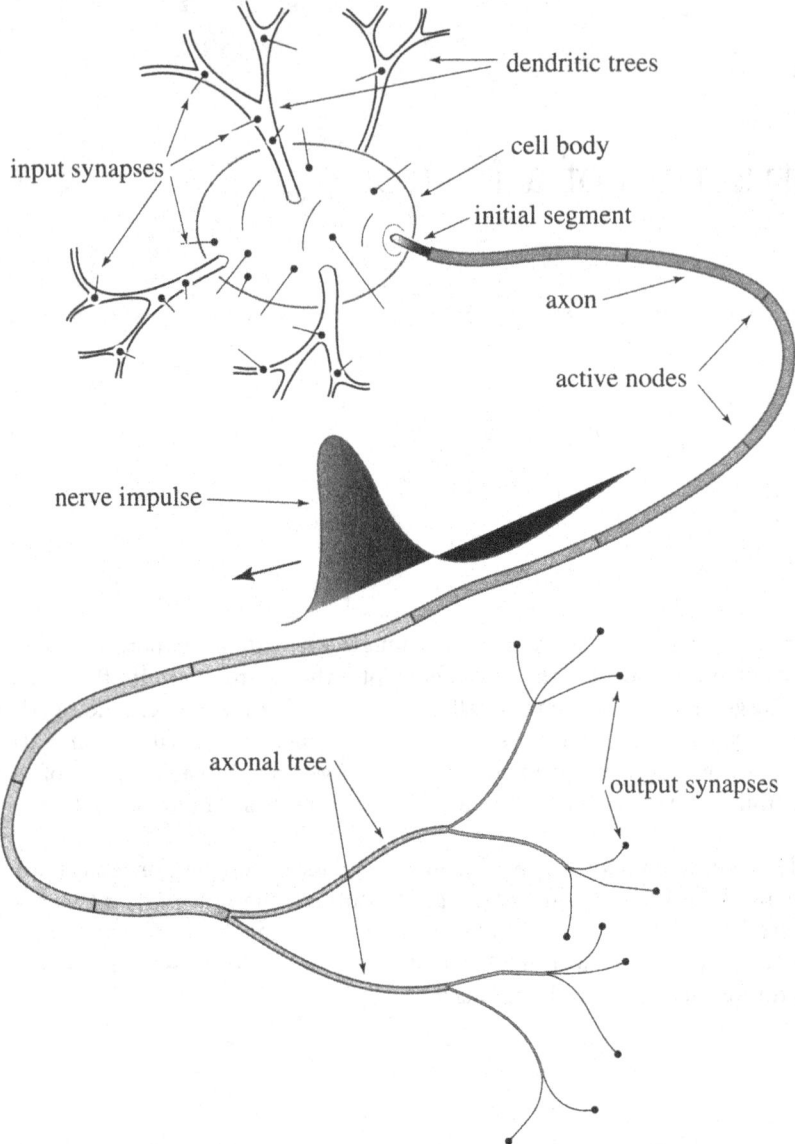

Figure 2.1. A cartoon of a typical nerve cell, or *neuron*, showing *dendrites* that gather incoming information from input *synapses* and an *axon* carrying outgoing signals through the branches of the axonal tree to other neurons or muscles.

- *Axons*: The *axon*, or outgoing channel of a neuron carries information away from the *cell body* and toward the output terminals. As indicated in Figure 1.2 of the previous chapter, an axon may be a relatively large fiber, such as the squid giant axon, or one of the many smaller fibers of a nerve bundle. Squid axons are uniform structures modeled by nonlinear partial differential equations (PDEs), and their dynamics are described in Chapters 4 through 6. Motor nerves of vertebrates, on the other hand, comprise bundles of fibers that are insulated (by a fatty material called *myelin*) except at rather widely spaced *active nodes* where nerve membrane is exposed and the switching action occurs. These axons are said to be "myelinated" and are modeled by nonlinear difference-differential equations (DDEs), as discussed in Chapter 7. Interestingly, motor nerve axons of large mammals can be several meters long, putting them among the most extended of single-cell processes.

- *The nerve impulse*: Propagating along the outgoing axon is an information-laden train of *nerve impulses*, which are solutions of a nonlinear PDE in the case of smooth axons or of a nonlinear DDE for myelinated fibers. As we saw in the previous chapter, each of these impulses expresses a dynamic balance between the release and dissipation of energy, and a major aim of the first half of this book is to offer the reader both analytic and intuitive understandings of this nonlinear wave phenomenon.

- *Initial segment*: The outgoing impulses on an axon are often launched by the switching at the hypersensitive *initial segment* of an axon, which is located near the cell body; thus, it is not necessary for the entire cell body to fire in order to ignite an axonal impulse.

- *Axonal tree*: The axon of a neuron eventually branches (or *bifurcates*) in a tree-like manner, allowing the impulse train on the trunk to be directed toward a variety of locations on muscle cells or other neurons. The nature of impulse propagation through branching regions of axonal trees is studied in Section 9.4 and is important for appreciating what a neuron is about.

- *Dendritic trees*: Likewise, on the input side of a neuron, the dendrites are tree-shaped, with their trunks directed toward the cell body. As indicated in Figure 2.1, this structure allows many input signals to influence the transmembrane potential of the initial segment. Again, possibilities for impulse blockage at the branching regions of dendritic trees are taken up in Section 9.3. (For some views of real dendrites, look ahead to Figures 9.1 and 9.5.)

- *Synapses*: Input signals to a neuron are received through several thousand *synapses*, which inject pulses of ionic current (or electric charge) into the dendrites and cell body in response to signals from other neurons or sensory cells. These synapses can be either chemical or electrical in nature, and chemical synapses can be either passive or active, responding to changes in the transmembrane voltage.

- *Cell membrane*: The *membrane* (or fatty outer covering of a cell) is where the nonlinear switching action of a neuron takes place. In the following chapter, it is shown that the total transmembrane current comprises three components: *displacement current* through the membrane capacitance, *conduction current* of ionic charge in response to the transmembrane electric field, and *diffusion current* of ionic charge in response to transmembrane differences in ionic concentration. An appreciation of the natures of and interplay among these three current components is necessary for understanding the nature of membrane switching.

- *Intrinsic membrane proteins*: Finally, membrane switching is governed by the presence of *intrinsic membrane proteins*, which facilitate the transport of ionic charges across the membrane by altering their physical configurations under the influence of changes in transmembrane voltage. This is the level of description at which a branch of physical chemistry called *molecular dynamics* enters into the description of a neuron [22].

Thus, we see that a neuron is an intricate dynamic system, operating at several levels of activity, including the biomolecular dynamics of intrinsic membrane proteins, the switching behaviors of active membranes, impulse propagation on incoming and outgoing fibers, and the overall functioning of the entire cell.

From a broader perspective, the typical nerve cell provides us with an example of a *nonlinear dynamic hierarchy*, the investigation of which is a central theme of this book. One characteristic of such hierarchies is that the nature of physical reality varies at different levels of description. Understanding this point is, in my view, so important that the following section is devoted to comparing two levels of neural dynamics.

2.2 Two Levels of Neural Dynamics

To begin at the beginning, a neuron is made of atoms, raising the questions: Can its global behavior be reduced to that of its constituent atoms? Or should a neuron be viewed as "more than the sum of its parts"?

In an attempt to answer such questions, we consider the nature of neural dynamics at two different levels of description: the dynamics of biological molecules and of the nerve impulse.

2.2.1 Newtonian Dynamics of Molecules

At a fundamental level, biological organisms are composed of molecules, which in turn are made from atoms. One of the most important classes of biological molecules are the *proteins,* comprising valence-bonded chains (or *polymers*) of amino acids with primary structures (amino acid sequences) determined by the genetic codes of DNA. How can the dynamics of such biological molecules be described?

Consider a molecule that is composed of N atoms. From both experimental measurements (of boiling and freezing points, elastic constants, resonant frequencies, bond lengths, and so on) and numerical computations based upon quantum theory, the *forces* among the atoms of a molecule are found. These interatomic forces can be derived from *potential energy* functions of the form [22]

$$PE = U(x_1, y_1, z_1, x_2, y_2, z_2, \cdots, x_N, y_N, z_N),$$

where $x_j, y_j,$ and z_j are position coordinates of the jth atom. (The potential energy depends on the interatomic distances because the valence electrons—acting as a sort of electronic glue—rearrange themselves as the relative atomic positions change.)

Recognizing that interatomic forces cause atoms to move, we can ask how those forces are related to potential energy. If the positions of the atoms change in such a manner that the potential energy function is reduced, energy becomes available to encourage the change. This change in potential energy is interpreted as a force; thus,

$$\text{force} \equiv \frac{\text{decrease in potential energy}}{\text{change in atomic position}} = -\frac{\text{increase in potential energy}}{\text{change in atomic position}}.$$

In other words, the concepts of force and potential energy are intimately interrelated, both expressing the same aspect of physical reality: the tendency of a stable system to find its natural shape. Potential energy is a global formulation of this tendency, whereas the concept of force enters into a corresponding local statement.

Returning to our molecule, the force acting on the jth atom in (say) the y-direction is

$$-\frac{\partial U}{\partial y_j},$$

and such interatomic forces can cause internal motion of the molecule, but how is this motion calculated? Isaac Newton's second law of motion tells us that the force on a certain atom in a particular direction is equal to its mass times the acceleration of that atom in that direction.

Thus, in a molecule composed of N atoms, there are $3N$ equations of motion of the general form

$$M_j \frac{d^2 y_j}{dt^2} = -\frac{\partial U}{\partial y_j},$$

where mass times acceleration on the left-hand side is equated to force on the right.

If the atoms are at rest and located where the potential energy U takes a minimum value (so that all of its derivatives of U with respect to the atomic coordinates are equal to zero), then Newton's law implies no motion. Under these conditions, the molecule has assumed its natural shape, which may be seen in a chemistry text or obtained on a computer file from the Protein Data Bank [2].

A molecule with all of its atoms at rest is, however, in a very special condition, to be found only at a temperature of absolute zero. Interestingly, if the atoms are allowed to move, the sum of the *kinetic energy*

$$KE \equiv \frac{1}{2} \sum_{j=1}^{N} M_j \left(\left(\frac{dx_j}{dt} \right)^2 + \left(\frac{dy_j}{dt} \right)^2 + \left(\frac{dz_j}{dt} \right)^2 \right)$$

and the potential energy is a constant of the motion. To see this, note that the time derivative of the total energy

$$\frac{d}{dt}(KE + PE)$$

comprises a sum of $3N$ terms of the form

$$\left(M_j \frac{d^2 y_j}{dt^2} + \frac{\partial U}{\partial y_j} \right) \frac{dy_j}{dt},$$

each of which is zero from an application of Newton's second law for a particular atom in a certain direction. Thus,

$$KE + PE = constant$$

for all unforced motions of the system.[1] Why is this interesting?

Energy conservation is an important analytic property of molecular systems for at least three reasons.

(1) Because it effectively reduces the number of dependent variables, the existence of a constant of the motion is sometimes helpful in finding analytic solutions for a vibrating molecule.

(2) Since Newton's equations involve *second* derivatives with respect to time, the direction of time can be reversed without altering the qualitative behavior of the system. Thus, given precise knowledge of the state of a molecule (all of the atomic coordinates and their rates of change with time), it is possible both to *pre*dict the state of the system at some time in the future and also to *retro*dict the state of the system at some time in the past. In other words, the qualitative character of energy-conserving dynamics is preserved under time reversal.

Planetary motion, for which Newton originally developed his dynamics, provides a familiar example of time-reversal symmetry. From present-day

---

[1] Note that $KE + PE$ is a function of $6N$ variables: the $3N$ atomic coordinates and their corresponding velocities. In the argot of nonlinear dynamics, these $6N$ variables define the *phase space* of the system.

measurements of the positions and speeds of the planets in the solar system, astronomers can compute where they (the planets, not the astronomers) will be several hundred years in the future and with equal precision where they were several hundred years in the past.

(3) A third consequence of energy conservation is that it allows one to construct a quantum description of atomic or molecular dynamics. Thus, Erwin Schrödinger's famous wave equation, upon which quantum theory is based, requires conservation of energy as its starting point [33].

Although energy is conserved and time is bidirectional at the molecular and atomic levels of description, only a brief observation of the world about us is required to recognize that these convenient properties do not hold throughout the realms of biology. Knowledge of the past history of a living organism often tells us little of its future, and quantum effects are unimportant in biological dynamics. This is because living creatures are essentially different from energy-conserving systems, which have been the primary focus of physical science during the twentieth century. Instead of conserving it, biological organisms necessarily *consume* energy in the course of their daily activities, just as do the flame of a candle and a nerve impulse.

How then might we describe the behavior of a nerve impulse?

2.2.2 Nonlinear Diffusion of a Nerve Impulse

In the first chapter of this book, we were introduced to the giant axon of the squid, a sketch of which is shown in Figure 1.2. Anticipating the more detailed discussions of Chapters 4 through 6, we now sketch a dynamic description of this nerve.

If the fiber is inactive, one finds a *resting voltage* across the membrane, with the electrical potential inside the nerve about 65 mV negative with respect to the outside potential. Let us define the change in transmembrane voltage from the resting value as $V(x, t)$, with x indicating *where* along the fiber the voltage difference is being measured and t indicating *when*.

In Section 4.4, it will be shown in detail that the nonlinear partial differential equation that governs the spatial and temporal evolution of the transmembrane voltage is the *nonlinear diffusion equation*

$$\frac{1}{rc}\frac{\partial^2 V}{\partial x^2} - \frac{\partial V}{\partial t} = \frac{j_{\text{ion}}}{c}. \tag{2.1}$$

In this equation, the parameters are defined as follows: (1) r is the longitudinal resistance per unit length of the fiber, measured in units of ohms per centimeter. (2) c is the membrane capacitance per unit length of the fiber, measured in units of farads (F) per centimeter. (3) j_{ion} is the ionic current flowing across the membrane (from inside to outside) per unit length of the fiber, measured in units of amperes (A) per centimeter. It is this transmembrane ionic current that introduces nonlinearity into the dynamics, and nonlinear transmembrane current in turn is governed by the behavior of intrinsic proteins embedded in the cell membrane.

To appreciate constraints on the behavior of such nonlinear diffusion equations, let us first assume that the transmembrane ionic current (j_{ion}) is equal to zero (no intrinsic membrane proteins). Equation (2.1) then reduces to the linear diffusion equation

$$D\frac{\partial^2 V}{\partial x^2} - \frac{\partial V}{\partial t} = 0 , \qquad (2.2)$$

where $D = 1/rc$ is a diffusion constant in squared centimeters per second.[2]

Although it is possible to use a linear diffusion equation to project forward in time (predict), this equation cannot project backward (retrodict) beyond some finite limit. As an example, note that an exact solution of Equation (2.2) is

$$V(x,t) = \frac{e^{-x^2/4Dt}}{2\sqrt{\pi D t}} , \qquad (2.3)$$

as can be confirmed by substitution. (A plot of this function is presented in Figure 9.2.) Although Equation (2.3) can be evaluated for all positive times, negative values of time are problematic. Thus, for $t < 0$, the expression in Equation (2.3) becomes imaginary, implying that the values it provides are physically meaningless.[3]

In contrast to Newton's equations—which can be used to determine whatever transpired in the past just as accurately as they can predict the future—the linear diffusion equation tells us nothing about the past behavior before some finite value of time. This property is shared by Equation (2.1) because, as one attempts to project backward in time, the transmembrane ionic current (j_{ion}/c) on the right-hand side can become negligible with respect to peak values of the diffusion terms ($D\partial^2 V/\partial x^2$ and $\partial V/\partial t$) on the left-hand side of the equation, so the properties of Equation (2.3) also restrict the nonlinear dynamic behavior.[4]

---

[2]For a squid nerve, $D = 333$ cm^2/s. To put this number into a physical context, assume a time interval Δt of 1 ms. Then, $\sqrt{D\Delta t}$ is approximately 0.58 cm, indicating that linear disturbances on a squid nerve would diffuse about that distance in a millisecond.

[3]As $t \to 0$ from positive values, the solution in Equation (2.3) approaches a *delta function*, for which $V(x,0) = 0$ except where $x = 0$, and

$$\int_{-\varepsilon}^{+\varepsilon} V(x,0)dx = 1$$

for any $\varepsilon > 0$. Although not a true function because it is undefined at $x = 0$, the delta "function" is useful for many scientific and engineering calculations. See [41] for a detailed discussion of the underlying theory.

[4]From a more general perspective, energy-conserving systems can be expressed in Lagrangian and Hamiltonian formulations (see Appendix A), ensuring that phase-space volumes are preserved by the dynamics. Thus, a given element of the $6N$-dimensional

Thus dynamics at the level of a nerve axon or dendrite are fundamentally different from the energy-conserving dynamics of vibrating molecules or the planets in the solar system. As noted earlier, an astronomer can make careful measurements of the positions and speeds of the planets and then tell us where they were (say) a millennium ago. Neural dynamics, on the other hand, are constrained by an *arrow of time*, which is basic to biology. Similarly, it is *not* possible to use measurements on the flame of a candle—however accurate they might be—to learn when it was lit.

This is not to suggest that Equation (2.1) is of no value. Indeed, as we will see in subsequent chapters, this nonlinear diffusion equation can be viewed as the *fundamental equation of neuroscience,* leading to a variety of useful results that include the following.

(1) With sufficiently accurate measurements of how j_{ion} depends on the transmembrane voltage, Equation (2.1) can be used to compute all of the experimental measurements in Figure 1.1 (time courses of the transmembrane voltage and the membrane permeability) in addition to the propagation speed of a squid nerve impulse [14]. We will study such calculations in Chapters 4 through 6.

(2) If Equation (2.1) is modified by allowing r, c, and j_{ion} to be functions of x, it becomes possible to predict how a nerve impulse will propagate on a tapered fiber or through a varicosity (or local enlargement) of the fiber [37]. These phenomena will be taken up in Chapter 9.

(3) In Section 5.5, we will see how this nonlinear diffusion equation can also be used to compute the amount of *threshold charge* needed to launch an impulse on a smooth fiber.

(4) The existence of a threshold is important because branching regions of the dendritic trees are then recognized as locations of low *safety factor,* at which propagating nerve impulses may be extinguished. In other words, as we will see in Chapter 9, the patch of active membrane near a branching region may act as a local switch, providing a means for logical computations throughout the dendritic trees [19, 35, 43, 46]. Related phenomena have been observed at the branching regions of axonal trees [5].

2.2.3 A Qualitative Comparison

Why is an entire section of this chapter devoted to comparing the dynamic behavior of a vibrating molecule with that of a nerve impulse? Because the differences at these two levels of dynamic description illustrate the difficulties that arise when one attempts to reduce the behavior of a neuron to the dynamics of its constituent molecules, and these difficulties are sometimes

---

phase space ($3N$ position coordinates and $3N$ velocity components) might—like a piece of taffy—be stretched or contracted in various directions, but its total volume remains constant. For diffusion equations, on the other hand, phase-space volumes are not preserved under the dynamic flow, allowing a direction of time to be defined.

overlooked by physical scientists who venture into biology [36], it seems appropriate to conclude with a summary of qualitative differences between molecular dynamics (MD) and nonlinear diffusion (ND).

- MD conserve energy, whereas nothing is conserved under ND. (See Appendix A for a mathematical definition of energy conservation.)

- Time is bidirectional under MD, whereas the ND time has a definite "arrow," or preferred direction of flow. Thus, the qualitative nature of time is essentially different at these two levels of neural dynamics.[5]

- In MD, there is a continual interchange (or periodic sloshing back and forth) between kinetic and potential energies, whereas ND is characterized by stable attractors (e.g., a candle flame or a nerve impulse) toward which the system may rapidly evolve.

- MD forces acting on individual masses add linearly, allowing chains of causality to be traced. Under ND, on the other hand, nonlinear interactions among several nerve impulses propagating on typical axonal and dendritic trees can lead to very intricate behavior for which relationships among causes and effects become difficult to sort out.

- MD systems are *closed* in the sense that all aspects of the dynamics are included in the original statement of Newton's laws. ND in neural networks, on the other hand, leads to *open* systems, which may form new levels of functional reality when necessary or convenient for the organism involved.

- Although quantum theory can be formulated for energy-conserving systems such as MD, the ND equation for a nerve impulse does not conserve energy and therefore has no quantum description.[6]

Physical scientists sometimes assert—seemingly as a matter of faith—that whatever transpires at the level of (say) a nerve impulse or a living cell can be reduced to a description that is based on the interactive motions of the constituent atoms, but the preceding considerations challenge the

---

[5]For an engaging appraisal of the nature of time as an independent variable in different realms of science, two books by J.T. Fraser are recommended [7, 8].

[6]Some contend that energy is conserved throughout the universe, implying a quantum wave function of the universe that necessarily includes a quantum description of all nerve impulses. Although such unconfirmed theoretical speculation may be correct, it is not helpful for electrophysiologists who deal with empirical studies of the dynamic behaviors of individual impulses.

Further doubt on the feasibility of a quantum description for a nerve impulse is cast by estimates of the *decoherence times* over which quantum predictions in nerves remain valid and after which "quantum entanglements" are destroyed. These times are estimated to be about ten orders of magnitude below the relevant time scale of miliseconds for neural dynamics, effectively eliminating the possibility of observing quantum effects in a laboratory of electrophysiology [44].

feasibility of such a program. We will return to this question in Chapter 12, but for now it is interesting to consider the opinion of Schrödinger, whose little book *What Is Life?* helped to launch the flourishing field of molecular biology. To the question "Is life based on the laws of physics?," he replied [34]:

> From all we have learned about the structure of living matter, we must be prepared to find it working in a manner that cannot be reduced to the ordinary laws of physics. And that is not on the ground that there is any new force directing the behaviour of the single atoms within a living organism, or because the laws of chemistry do not apply, but because life at the cellular level is more ornate, more elaborate than anything we have yet attempted in physics.

Having appreciated certain difficulties in developing a reductive formulation for the neuron, we now consider a few "ornate elaborations" of its structure.

2.3 Synapses

Like all living organisms, nerve cells do not exist in isolation but communicate among themselves, primarily through specialized contacts called *synapses*. These contacts may be chemical or electrical in nature, showing different qualitative behaviors. Let us look at some of the details.

2.3.1 Chemical Synapses

One of the synapses indicated in Figure 2.1 is sketched in Figure 2.2, where it should be noted that these diagrams are merely schematic, showing the broad outlines of structures that exhibit many variations among real neurons [6, 16, 23].

From Figure 2.2(a), we see that the the enlarged end bulb of an axon stores a number of *vesicles* (some 30–40 nm in diameter), each containing psychoactive chemicals (called *neurotransmitters*) that can alter the ionic conductivity of the postsynaptic membrane. The postsynaptic membrane, sometimes forming a *dendritic spine*, is separated from the presynaptic membrane by a *synaptic cleft* having a width of about 20 nm.

We will see in Chapter 4 that an increase in sodium ion permeability tends to depolarize the postsynaptic membrane, encouraging the development of an action potential, whereas increasing the potassium ion permeability hyperpolarizes the membrane, thereby inhibiting impulse formation. Thus, synapses can be either *excitatory* or *inhibitory* in nature.

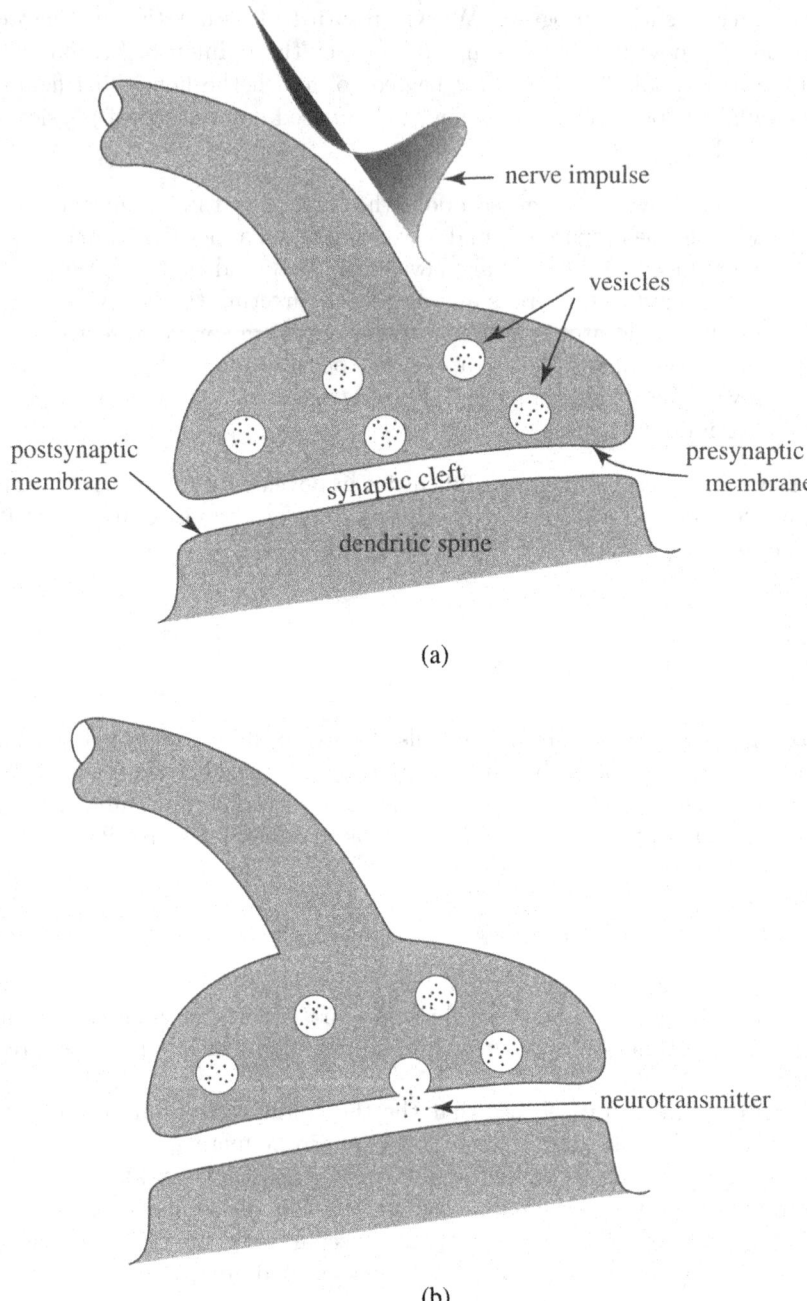

Figure 2.2. Sketches of a chemical synapse. (a) A nerve impulse arrives at the synapse, inducing a vesicle to fuse with the presynaptic membrane. (b) The process of exocytosis, wherein a vesicle is releasing its neurotransmitter molecules into the synaptic cleft. (The drawings are not to scale.)

When a nerve impulse arrives at a synapse, a rather complicated series of events transpires. First, the transmembrane voltage of the impulse causes an inward flow of calcium ions. In a complex manner, these calcium ions then induce some of the vesicles to fuse with the presynaptic membrane. Upon fusing with the presynaptic membrane, a vesicle then spills its neurotransmitter molecules into the synaptic cleft—a process called *exocytosis*. Next, as suggested by Figure 2.2(b), the neurotransmitter molecules diffuse across the cleft, influencing the conductivity of the postsynaptic membrane. Finally, the dendritic spine may act as a postsynaptic switch, modifying the global character of dendritic response to the input signals [26, 29].

Anticipating results of the following chapter, current of a particular ion (I_{ion}) flows across the postsynaptic membrane as determined by an expression of the form

$$I_{ion} = G(t)(V - V_{ion}).\tag{2.4}$$

In this equation, $G(t)$ is the postsynaptic membrane conductivity for that particular ion, and V_{ion} is the voltage of a fictitious "membrane battery," indicating the tendency for those ions to diffuse into or out of the postsynaptic membrane.[7] Generally, V differs from V_{ion}, so I_{ion} follows $G(t)$, which is determined by the foregoing sequence of events.

If the postsynaptic membrane becomes depolarized (making the transmembrane voltage V more positive than its resting value), the synapse is said to induce an "excitatory postsynaptic potential" (EPSP). In typical neocortical neurons, the EPSP appears with an initial delay of about 0.3 ms—the time required for liberation of a vesicle of neurotransmitter and its subsequent diffusion across the synaptic cleft. After this delay, the EPSP rises to a maximum in about 1 ms and then decays in about 10 ms. Hyperpolarization of the postsynaptic membrane, on the other hand, induces an "inhibitory postsynaptic potential" (IPSP) with a longer initial delay ($\sim$ 1.5 ms) and a somewhat longer decay time ($\sim$ 10–20 ms).

From a mathematical perspective, a chemical synapse (CS) exhibits the following characteristics.

- Causality in a CS is *unidirectional* because the synaptic dynamics (e.g., calcium release, vesicle discharge) influence the postsynaptic membrane, but changes in the postsynaptic potential have little or no effect on presynaptic dynamics.

- Because its response time can be no less than the time required for the neurotransmitter molecules to diffuse across the synaptic cleft, the CS is a relatively slow means of neural interaction, requiring a millisecond or two for signal transmission.

---

[7]More precisely, V_{ion}, the equilibrium (or Nernst) value of the postsynaptic transmembrane potential (V), at which no current flows regardless of the postsynaptic membrane permeability.

- The behavior of a CS is *strongly nonlinear,* involving an intricate sequence of biochemical and bioelectrical effects,. In other words, $G(t)$ in Equation (2.4) depends in intricate ways on the dynamics of several intermediate variables, including the postsynaptic transmembrane potential V. In some cases, the relatively isolated membrane of the dendritic spine can switch (as described in Chapter 4) to launch a dendritic signal.

- In addition to being strongly nonlinear, the behavior of a chemical synapse is also *stochastic* [17]. Thus, if there are n sites for vesicle fusion and exocytosis on the presynaptic membrane and each site has a probability p for neurotransmitter release, then the probability of exactly k vesicles releasing their molecules is[8]

$$P(n, k) = \frac{n!}{(n - k)!k!} p^k (1 - p)^{n-k} , \qquad (2.5)$$

where $k = 0, 1, \ldots, n$.

Where the number of vesicle release sites (n) is small (as in central regions of the brain), this equation implies a highly variable incidence of exocytosis under the same experimental parameters [30, 45]. Thus, for example, $P(1, 1) = p$, and p can take values ranging between 0.1 and 0.9. Although the reason for this wide variation is not understood, it may be related to statistical variations in presynaptic calcium ion concentrations [21].

Although real chemical synapses are intricate dynamic systems, they are often modeled by rather simple phenomenological expressions that capture essential aspects of their dynamics.

Passive Chemical Synapses.
In a passive synapse, the conductance function $G(t)$ in Equation (2.4) is a prescribed function of time, a convenient analytic form being [21, 25]

$$G(t) \propto te^{-t/\tau} . \qquad (2.6)$$

Here, τ establishes the time scale of the conductance change, and the proportionality constant determines the maximum increase of conductance.

If the synapse is excitatory, the membrane battery pushes positive ions (usually sodium or calcium ions) into the postsynaptic membrane, raising its voltage from the resting value. For inhibitory synapses, on the other hand, the postsynaptic membrane voltage is lowered from its resting value. This can be accomplished either by pumping positively charged potassium ions out of the postsynaptic cell or negatively charged chloride ions into it.

---

[8]This expression can be obtained by computing $[(1 - p) + p]^n$ and interpreting the kth term as $P(n, k)$.

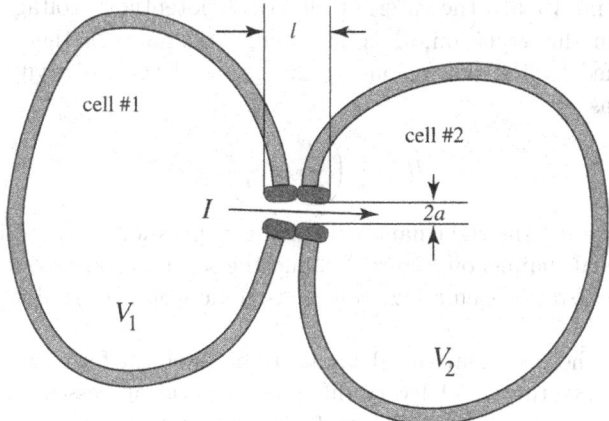

Figure 2.3. A gap junction or electrical synapse.

Active Chemical Synapses.
If the postsynaptic membrane conductance depends on the transmembrane
voltage V, the synapse is said to be *active*. This situation can come about,
for example, if ion channels are blocked at the resting voltage (say, by
magnesium ions), becoming unblocked as the membrane voltage changes.
For active excitatory synapses, an approximate model assumes [21, 25]

$$G(t,V) \propto \frac{e^{-t/\tau_1} - e^{-t/\tau_2}}{1 + Ke^{-\gamma V}}, \tag{2.7}$$

where $\tau_1 \gg \tau_2$ gives a rapid rise ($\sim \tau_2$) and a slow decay ($\sim \tau_1$).

2.3.2   Gap Junctions

Since the mid-1970s, it has been evident from electron microscope images
of dendritic fields that neurons may communicate through direct electrical
connections, which allow ionic current to flow straight out of one neuron and
into another through channels that are defined by the dimensions of intrinsic membrane proteins [32]. Such electrical synapses are called *gap junctions*
because they require interconnecting openings in the cell membrane.

Gap junctions are well-described by *Ohm's law*, which states that current
flows in proportion to the difference in electric potential across the gap.
A task of the experimentalist, however, is to determine the constant of
proportionality, called the *gap resistance*.

Using the notation from Figure 2.3, the current I flowing from cell #1
into cell #2 is given by

$$I = \frac{V_1 - V_2}{R}, \tag{2.8}$$

where V_1 and V_2 are the internal electrical potentials (voltages) of the two cells. In this equation, R is the ohmic resistance of the gap, which is determined by the dimensions of the gap and the resistivity ρ of the cytoplasm as

$$R = \frac{\rho}{a}\left(\frac{l}{\pi a} + \frac{1}{2}\right). \tag{2.9}$$

The first term on the right-hand side of this expression is the resistance of the cylindrical channel on Figure 2.3, and the second term accounts for the *spreading resistance* stemming from constriction of current flow near the channel [40].[9]

Although there is a superficial similarity between the forms of Equations (2.4) and (2.8), they model very different physical processes. We will see in the next chapter how Equation (2.4) represents the algebraic sum of two current components of a particular ion: drift current in response to a voltage gradient and diffusion current in response to a concentration gradient. Equation (2.8), on the other hand, accounts for the drift current of *all* charge carriers in the cytoplasm. It follows that the nature of a gap junction (GJ) is very different from that of a chemical synapse. In particular:

- A GJ is *bidirectional*, allowing current to flow as easily from cell #1 into cell #2 as from cell #2 into cell #1.

- The time required for current to flow in a GJ after a voltage difference between the cells has been established is very fast. This response time is essentially the collision time for cytoplasmic charge carriers, which is several orders of magnitude shorter than the time scales recorded in a laboratory of electrophysiology.

- Because Equation (2.4) is a true expression of Ohm's law, the relation between current I and voltage difference $(V_1 - V_2)$ for a GJ is *linear* over a wide range, readily including the voltages appearing in normal nerves.

- Because a large number of charge carriers are involved in the flow of current through a GJ, the phenomenon is *deterministic*.[10]

---

[9]To get a feeling for the magnitudes involved in Equation (2.9), assume that $\rho = 35.4$ ohm-cm (the value measured by Hodgkin and Huxley for the cytoplasm of a squid axon), $l = 15$ nm and $a = 3$ nm. Then, the corresponding resistance (R) of a single gap junction would be about 250 megohm.

[10]This statement is not quite true because the number of cytoplasmic charge carriers—while large—is not infinite. If the average number of charge carriers flowing through the gap is N, then there will be a relative uncertainty in the current given by $1/\sqrt{N}$, leading to a percentage error of $\pm 100/\sqrt{N}$. Called *shot noise* by electrical engineers, this statistical fluctuation causes the hiss of an untuned radio and is mathematically identical to the errors faced by political pollsters.

2.4   Neural Models

The aim of this section is to consider how one might construct mathematical models for a nerve cell. As we have seen in the foregoing discussion, this is not a straightforward task because a neuron—like all living cells— is an intricate dynamic system that must be described on several levels, and these descriptions may involve qualitatively different properties of the independent variable: time.

To appreciate the magnitude of the task, note that the levels of dynamic activity in a typical neuron include those in the following hierarchical diagram.

Neuron
Branching regions of fibers
Axonal and dendritic fibers
Synapses
Patches of nerve membrane
Biomolecular dynamics

Each of these several levels of description presents a substantial challenge to science, requiring dedicated efforts by many skillful and highly trained experimentalists. Each level is described by some nonlinear dynamics out of which new phenomena can emerge, but these formulations differ at different levels of description. How can all of the resulting information be put together into a reliable representation of a real neuron?

Over the past six decades, some answers to this question have been suggested.

2.4.1   The McCulloch–Pitts (M–P) Neuron

In their classic paper on the dynamics of a brain, McCulloch and Pitts assumed the most simple model for an individual neuron. Briefly, they supposed that the dendritic trees (see Figure 9.1) gathered a linear weighted sum of the incoming synaptic signals and compared this sum with a threshold level at the base (initial segment) of the axonal tree. In this model, if the sum of input signals exceeds the threshold, then an impulse is launched on the main trunk of the axonal tree. Once launched, the impulse travels out to the first axonal branching, where two impulses are generated that proceed down both secondary fibers. This process continues until all synapses at the tips of the tree have received impulses, with none being lost at axonal branchings.

In other words, the M–P premises were the following.

- The activity of a neuron is an all-or-nothing process.

- A certain weighted sum of the synaptic input signals must exceed a threshold level in order for a neuron to fire.

- The only significant time delays are due to synapses.

Formally, these assumptions can be expressed as the input–output relationship

$$V_j(t+\tau) = H\left(\sum_{k=1}^{N} \alpha_{jk} V_k(t) - \theta_j\right), \tag{2.10}$$

where $H(\cdot)$ is the Heaviside step function with properties

$$H(x) = \begin{cases} 1 & \text{for} \quad x \geq 0 \quad \text{and} \\ \\ 0 & \text{for} \quad x < 0. \end{cases}$$

In this formulation, a model brain is represented by N neurons that are joined by an $N \times N$ interconnection matrix

$$A = [\alpha_{jk}],$$

with each element indicating how the firing of the kth neuron influences the tendency of the jth neuron to fire. In particular, if the weighted sum of dendritic inputs

$$\sum_{k=1}^{N} \alpha_{jk} V_k$$

is greater than the threshold θ_j of the jth neuron, then (after a synaptic time delay τ) the model neuron will ignite its axon.

There are at least two reasons why McCulloch and Pitts chose such a simple model for their basic neuron. First, not as much was known about the dynamic properties of real neurons in 1943 as now, and it seemed prudent to avoid speculation about more exotic possibilities for neuronal behavior. Second, and perhaps more important, these authors were interested in using their model neuron as a basis for a theory of how the brain works, and a simpler model for the neuron eased this more ambitious task.[11]

Since the late 1950s, several variations of the M–P neuron have been suggested for computer-based models of a brain, with diverse means for adjusting the (synaptic) interconnection weights (α_{jk}) during the course of neural activity [4, 13, 15, 31]. Under the names "linear threshold unit" (LTU) or "threshold logic unit" (TLU), similar concepts were employed by the engineering community as a basis for designing machines that could

---

[11] An excellent survey of simple neuron models, which are useful for network studies, has been published by Gerstner [10]. Although such representations might seem too simple to be useful for the description of individual cells, Keat et al. [18] have demonstrated that the parameters of Gerstner's "spike response model" [11, 12] can be adjusted to predict the outputs of retinal ganglion cells in salamander, rabbit, and cat.

learn to recognize certain classes of patterns [27, 28, 47]. (The proliferation of names for the same concept reflects attempts to establish proprietary positions.) Such "brain models" will be taken up in Chapter 10, but we are presently concerned with the modeling of individual neurons.

How have M–P models fared over the past half century?

2.4.2   The Multiplex Neuron

During the 1960s and 1970s, the increasing availability and numerical power of electronic computers had two divergent influences on neuroscience research. First, computers made it feasible to study ever larger numbers of interacting neurons. In these investigations, understandably, priority was given to modeling as many neurons as possible, requiring that representations of individual neurons be kept simple. Thus the M–P model neuron and its variants became established as the basis for "neural network theory" in which the lack of biological realism embodied in the neuron models often failed to be questioned. These same computers, on the other hand, allowed individual neurons to become increasingly better understood through ever more realistic mathematical models.

By the early 1970s, improved understanding of the possibilities for membrane dynamics coupled with detailed electron micrograms of real neural structures led neurologist Steve Waxman to propose the *multiplex neuron* as a more realistic neural model [46]. (The term "multiplex" is borrowed from communications engineering, where it implies the ability of a circuit to deal with several messages at the same time.) A significant feature of this representation is that it allows for nonlinear interactions among impulses near branching regions of dendritic and axonal trees [19, 35]. As is discussed in detail in Chapter 9, impulses can be lost at branching regions, permitting computations to occur within the dendritic and axonal trees and not merely at the initial segment. Why is this difference important?

Like the furnace control on your living room wall, an M–P neuron can be compared to a single switch, responding to a linear weighted sum of input signals. This assumption of linearity has profound implications for the nature of causality in the model because linear interactions allow an assortment of causes to be sorted into independent threads—a simplification not generally possible for nonlinear interactions. Nonlinear interactions comprise myriad varieties, confounding the threads of causality and making it difficult to discern who did what to whom [3].

This is an important philosophical point to which we will return in Chapter 12, but for now we note that nonlinear interactions among causes (neural input signals) allow far greater intricacy of the overall dynamics, including the phenomenon of *emergence* [37]. In other words, the whole neuron may be more than the sum of its parts.

2.4.3 Real Neurons?

Interestingly, experimental evidence suggests that even the multiplex neuron may not be sufficiently flexible to capture the properties of real neurons [1, 9, 25, 42, 43]. What phenomena are being left out of the picture?

As Christof Koch has proposed, a biological neuron may operate more like an analog computer than a digital one, converting the incoming streams of information into spatially and temporally distributed variables. Subsequent to this transformation, he speculates [20]:

> Information is then processed in the analog domain, using a number of linear and nonlinear operations (multiplication, saturation, amplification, thresholding) implemented in the dendritic cable structure and augmented by voltage-dependent membrane and synaptic conductances. The resulting signal is then converted back into digital pulses and conveyed to the following neurons.

In his book *Biophysics of Computation* [21], this same author expands on such ideas, arguing that "dendrites can indeed be very powerful, nontraditional computational devices, implementing a number of continuous operations."

In attempting to evaluate such speculations, it is important to remember that the description of a chemical synapse presented in Section 2.3.1 provides only a sketch of the design possibilities open to the processes of evolution. There are several different neurotransmitters under genetic control, including amino acids, biogenic amines (such as dopamine, noradrenaline, and serotonin), and neuropeptides (smallish proteins), and these neurotransmitters influence a variety of postsynaptic membrane receptors [21]. Amino acids (which are the building blocks of proteins) give rise to relatively rapid (on the order of a few milliseconds) postsynaptic potentials that can be either excitatory or inhibitory. Biogenic amines act on longer time scales (hundreds of milliseconds to seconds), whereas the effects of neuropeptides can persist for minutes, opening significant avenues for interplay between body chemistry and the mind.

For their part, the currents flowing through gap junctions present opportunities for intricate *dendrodendritic interactions,* which are not sketched in Figure 2.1 [38, 39]. As Schmitt, Dev, and Smith suggested in the 1970s, evidence from the electron microscope was giving rise to a "quiet revolution" in our understanding of neural circuitry [32]. In their words:

> The new view of the neuron, based primarily on recent electron microscope evidence and supported by intracellular electrical recording, holds that the dendrite, far from being only a passive receptor surface, may also be presynaptic, transmitting information to other neurons through dendrodendritic synapses. Such neurons may simultaneously be the site of many elec-

trotonic current pathways, involving components as small as dendrite membrane patches or individual dendrites.

In their book *Dendrites*, Stuart, Spruston, and Häusser have assembled a fascinating collection of essays on the structure, morphogenesis, biochemistry, electrochemical behavior, functional and structural plasticity, and functional significance of dendrites, confirming the early vision of Schmitt, Dev, and Smith [43].

Yet another sort of subtle electrical interaction arises when impulses are propagating on closely situated parallel fibers, as in the sciatic nerve bundle shown in Figure 1.2. In this case, the external current loop associated with one impulse can influence the propagation of the other impulse (and vice versa), leading to the phenomenon of *impulse synchronization*, in which two or more impulses become coupled together, traveling at exactly the same speed. This effect (called *ephaptic*) suggests interesting mathematical studies that are presented in Chapter 8.

Thus the behaviors displayed by real neurons are far from fully understood, leaving much research to be carried out at this primary level of the brain's dynamics. As the following chapters will show, this work is of interest to the mathematically oriented neuroscientist.

2.5 Recapitulation

The broad aim of this chapter is to present an overview of individual nerve cells that provides a realistic perspective on neural intricacy. To this end, a general neuron is sketched in Figure 2.1, and its major components are defined and described.

It is emphasized that the dynamic behavior of biomolecules—intrinsic membrane proteins, for example, which permit active nerve membranes to switch—is distinctly different from the dynamics of impulse propagation on a nerve fiber. In the Newtonian dynamics of biomolecules, energy is conserved and time reversal is allowed, whereas energy is not conserved and there is an arrow of time governing the nonlinear diffusion processes from which nerve impulses emerge. Recognition of the profound differences between these two levels of neural organization suggests some of the difficulties facing those who would reduce a neuron's behavior to the dynamics of its constituent molecules.

Two sorts of neural interactions are then introduced—chemical synapses and direct electrical communication through gap junctions—that exhibit different physical behaviors. Thus, chemical synapses are unidirectional, relatively slow, strongly nonlinear, and stochastic, whereas gap junctions are bidirectional, very fast, linear, and deterministic.

In preparation for subsequent discussions of nerve models, the dynamics of a typical neuron are parsed into several hierarchical levels of its global nonlinear activity. As the simplest such model, the McCulloch–Pitts (M–P) neuron (proposed in 1943) is described in which a linear weighted sum of input signals is compared with a threshold level in order to decide whether or not the outputs will become active. This M–P model is compared with Waxman's more intricate "multiplex neuron" (proposed in 1972), which features opportunities for substantial information processing in the branching structures of the dendritic and axonal trees. In other words, the M–P model is equivalent to a single switch, whereas the multiplex neuron—like an integrated circuit—comprises many switches. Finally, Waxman's model is evaluated in the light of more recent observations of the behavior of real neurons, revealing them to be vastly intricate dynamic systems yet only dimly understood. It is an exciting time to be a neuroscientist.

References

[1] Y Amitai, A Friedman, B Connors, and M Gutnik, Regenerative electrical activity in apical dendrites of pyramidal cells in the neocortex, *Cereb. Cortex* 3 (1993) 26–38.

[2] HM Berman, J Westbrook, Z Feng, G Gilliland, TN Bhat, H Weissig, IN Shindyalov, and PE Bourne, The Protein Data Bank, *Nucleic Acids Res.* 28 (2000) 235–242. (See www.rcsb.org/pdb/)

[3] M Bunge, *Causality and Modern Science*, third edition, Dover, New York, 1979.

[4] ER Caianiello, Outline of a theory of thought processes and thinking machines, *J. theoret. Biol.* 1 (1961) 204–235.

[5] SH Chung, SA Raymond, and JY Lettvin, Multiple meaning in single visual units, *Brain Behav. Evol.* 3 (1970) 72–101.

[6] JC Eccles, *The Physiology of Synapses*, Springer-Verlag, Berlin, 1964.

[7] JT Fraser, *The Genesis and Evolution of Time*, Harvester Press, Brighton, England, 1982.

[8] JT Fraser, *Of Time, Passion, and Knowledge: Reflections on the Strategy of Existence*, second edition, Princeton University Press, Princeton, 1990.

[9] F Gabbiani, W Metzner, R Wessel, and C Koch, From stimulus encoding to feature extraction in weakly electric fish, *Nature* 384 (1996) 564–567.

[10] W Gerstner, Spiking neurons. In *Pulsed Neural Networks*, S Maass and CM Bishop (eds), MIT Press, Cambridge, MA, 1999.

[11] W Gerstner and JL van Hemmen, Associative memory in a network of 'spiking' neurons, *Network* 3 (1992) 139–164.

[12] W Gerstner, R Ritz, and JL van Hemmen, A biologically motivated and analytically soluble model of collective oscillations in the cortex: I. Theory of weak locking, *Biol. Cybern.* 68 (1993) 363–374.

[13] S Grossberg, Adaptive pattern classification and universal recoding I: Parallel development of coding of neural feature detectors, *Biol. Cybern.* 23 (1976) 121–134.

[14] AL Hodgkin and AF Huxley, A quantitative description of membrane current and its application to conduction and excitation in nerve, *J. Physiol. (London)* 117 (1952) 500–544.

[15] JJ Hopfield, Neural networks and physical systems with emerging collective computational abilities, *Proc. Natl. Acad. Sci. USA* 79 (1982) 2554–2558.

[16] B Katz, *Nerve, Muscle and Synapse,* McGraw–Hill, New York, 1966.

[17] B Katz, *The Release of Neural Transmitter Substances,* Liverpool University Press, Liverpool, 1969.

[18] J Keat, P Reinagel, RC Reid, and M Meister, Predicting every spike: A model for the responses of visual neurons, *Neuron* 30 (2001) 803–817.

[19] BI Khodorov, *The Problem of Excitability,* Plenum Press, New York, 1974.

[20] C Koch, Computation and the single neuron, *Nature* 385 (1997) 207–210.

[21] C Koch, *Biophysics of Computation: Information Processing in Single Neurons,* Oxford University Press, New York, 1999.

[22] JA McCammon and SC Harvey, *Dynamics of Proteins and Nucleic Acids,* Cambridge University Press, Cambridge, 1987.

[23] H McLennan, *Synaptic Transmission,* Saunders, Philadelphia, 1970.

[24] WS McCulloch and WH Pitts, A logical calculus of the ideas immanent in nervous activity, *Bull. Math. Biophys.* 5 (1943) 115–133.

[25] BW Mel, Synaptic integration in an excitable dendritic tree, *J. Neurophysiol.* 70 (1993) 1086–1101.

[26] JP Miller, W Rall, and J Rinzel, Synaptic amplification by active membranes in dendritic spines, *Brain Res.* 325 (1985) 325–330.

[27] M Minsky and S Papert, *Perceptrons,* MIT Press, Cambridge, MA, 1969.

[28] NJ Nilsson, *Learning Machines,* second edition, Morgan Kaufmann, San Mateo, CA, 1990.

[29] DH Perkel and DJ Perkel, Dendritic spines: Role of active membrane in modulating synaptic efficiency, *Brain Res.* 325 (1985) 331–335.

[30] S Redman, Quantal analysis of synaptic potentials in neurons of the central nervous system, *Physiol. Rev.* 70 (1990) 165–198.

[31] F Rosenblatt, *Principles of Neurodynamics,* Spartan Books, New York, 1959.

[32] FO Schmitt, P Dev, and BH Smith, Electrotonic processing of information by brain cells, *Science* 193 (1976) 114–120.

[33] E Schrödinger, Quantisierung als Eigenwerteproblem, *Ann. Phys.* 79 (1926) 361–376.

[34] E Schrödinger, *What Is Life?* Cambridge University Press, Cambridge, 1944 (republished in 1967).

[35] AC Scott, The electrophysics of a nerve fiber, *Rev. Mod. Phys.* 47 (1975) 487–533.

[36] AC Scott, *Stairway to the Mind,* Springer-Verlag, New York, 1995.

[37] AC Scott, *Nonlinear Science: Emergence and Dynamics of Coherent Structures,* Oxford University Press, Oxford, 1999.

[38] GM Shepherd, The neuron doctrine: A revision of functional concepts, *Yale J. Biol. Med.* 45 (1972) 584–599.

[39] GM Shepherd, The dendritic spine: A multifunctional unit, *J. Neurophysiol.* 75 (1996) 2197–2210.

[40] W Shockley, *Electrons and Holes in Semiconductors*, Van Nostrand, New York, 1950.

[41] I Stackgold, *Green's Functions and Boundary Value Problems*, John Wiley & Sons, New York, 1974.

[42] G Stuart and B Sakmann, Action propagation of somatic action potentials into neocortical pyramidal cell dendrites, *Nature* 367 (1994) 69–72.

[43] G Stuart, N Spruston, and M Häusser, *Dendrites,* Oxford University Press, Oxford, 1999.

[44] M Tegmark, The importance of quantum decoherence in brain processes, *Phys. Rev. E* 61 (2000) 4194–4206.

[45] B Walmsley, FR Edwards, and DJ Tracey, The probabilistic nature of synaptic transmission at a mammalian excitatory central synapse, *J. Neurosci.* 7 (1987) 1027–1048.

[46] SG Waxman, Regional differentiation of the axon: A review with special reference to the concept of the multiplex neuron, *Brain Res.* 47 (1972) 269–288.

[47] B Widrow and JB Angell, Reliable, trainable networks for computing and control, *Aerospace Eng.* (September, 1962) 78–123.

3
Nerve Membranes

Who has not wondered at a soap bubble as it floats into the sky on a spring morning? From what magic does this perfect sphere emerge? How might its form be related to the teeming confusion of a handful of soap suds gathered in the evening bath? Or to the regular hexagons of a honeycomb? Are there mathematical principles guiding the development of these structures? Is there a science of *morphogenesis*? Might such a study help us to comprehend the development of biological forms, thereby inferring aspects of their functions? Such questions are the seeds of science.

Living organisms are composed of cells—one for a bacterium and many billions of neurons for each of our brains—and every cell has a characteristic shape. Although a discussion of how these shapes come about is beyond the scope of this book, the energetics of cell membranes play a significant role, the understanding of which leads to an appreciation of the physical natures of electrical capacitance and transmembrane ionic currents.[1]

From one perspective, membrane energetics are related to those of chemical molecules, as was briefly discussed in the preceding chapter. In that case, you will recall, intramolecular forces arise from the action of an electronic cloud that nestles among the atomic nuclei. Variations of the potential energy of this cloud stemming from changes in the relative positions of the

---

[1]For the general reader wishing to understand the relations between physical principles and the natural shapes of biological organisms, there is no better introduction than the book *On Growth and Form* [19] by Scottish classicist, mathematician, and biologist D'Arcy Wentworth Thompson, some comments from which are included in the front matter of this book.

nuclei can be interpreted as forces acting on the atoms. When the atoms are positioned so that these forces cancel, a stationary structure of the molecule is attained; thus, the energy of the interatomic electronic cloud takes a minimum value at the stable structure of the molecule.

We are aware of this phenomenon of energy minimization from many familiar examples. After falling from the sky, rainwater streams downhill and settles in the characteristic shape of a pond, where its total energy is lowest. Our planet assumes a spherical shape because this reduces the total gravitational energy to its smallest value, and similarly the spherical shape of a soap bubble reduces its surface energy to a minimum.

How do such considerations influence the properties of neural membranes? What forces and types of energy are involved?

3.1   Lipid Bilayers

To understand the energetic nature of biological membranes, note that they are composed largely of fatty (lipid) molecules that have the general structure shown in Figure 3.1(a) [3]. This cigar-shaped biomolecule is distinguished by the charged *head group* at one end, which is the source of an electric field. Building on a previous demonstration by Lord Rayleigh that oil films on the surface of water can become monomolecular, Irving Langmuir [5] showed that this monolayer appears as in Figure 3.1(b), with the head groups of the lipids uniformly directed toward the water surface. Why is this so?

Whenever the head group of a lipid molecule is exposed to air, its charge creates an electric field, requiring the expenditure of energy. If a positively charged head group is close to water, on the other hand, its electric field is largely canceled, reducing the amount of electric field energy that is generated by the head charge.

Water achieves this because its constituent molecules are electric dipoles that rotate themselves such that their negative ends are directed toward the positive head group, thereby canceling most of the electric field. Thus, an individual lipid molecule can substantially decrease its associated electric field energy merely by orienting itself so that its charged end is located as close as possible to water, just as rock on a steep hillside can lower its gravitational energy by rolling down into a valley. For a monomolecular lipid layer on the surface of water, the energetically preferred structure is as shown in Figure 3.1(b).

As described by Isaac Newton [13], you can observe a related effect for soap films in the following manner. First, take a few inches of wire (a paper clip will do, although smaller wire works better) and bend the end into a loop with a diameter of a millimeter or so. Then mix some liquid dish soap with warmish water, and dip the loop to form a soap film. Under a

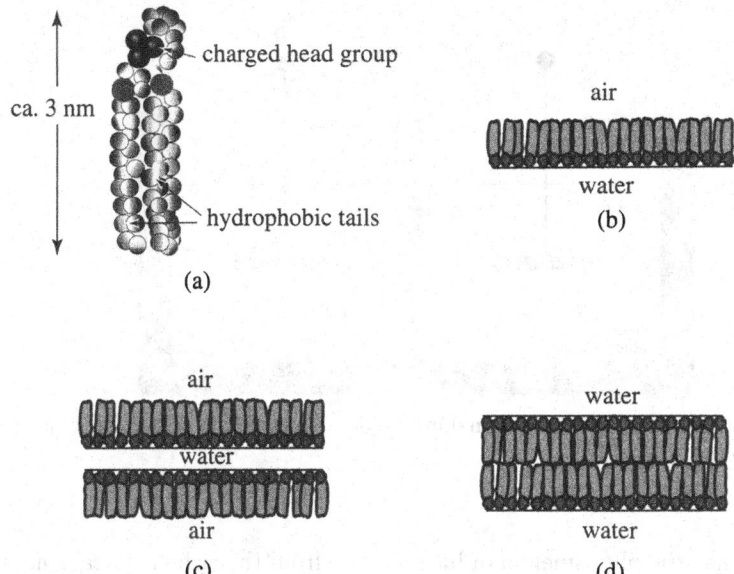

Figure 3.1. (a) A lipid (fatty) molecule (redrawn from Goodsell [3]). (b) A monomolecular lipid layer on water surface. (c) A bimolecular soap film. (d) A lipid bilayer.

bright light, you will at first observe the colored interference bands of the film that are familiar from childhood observations of soap bubbles. These color bands indicate that the film thickness is of the order of a wavelength of visible light ($\sim$ 4000 Å, or 400 nm) [1]. If you watch the film for a few minutes, however, it undergoes a dramatic change. Without breaking, the film becomes almost completely reflectionless, which indicates that its thickness has suddenly reduced to a value well below the wavelength of visible light, causing it to appear *black*. You are now observing a *bimolecular soap film* with the structure shown in Figure 3.1(c). (Within this film, a thin layer of water remains that attracts their charged head groups.)

Because the membrane of a biological cell is totally immersed in water, an energetically favorable structure is the lipid bilayer film, shown in Figure 3.1(d), and extended films can assume a variety of interesting geometries. If the film is a closed surface, for example, its natural form will be a sphere because that shape minimizes total energy, just as for soap bubbles.[2] Re-

---

[2]Collections of bubbles are yet more intricate. The next time you are washing up, you might take a careful look at a handful of soapsuds under a good light, noting that interior divisions tend toward a fourteen-sided figure, called a *tetrakaidekahedron* by Lord Kelvin. Just as the hexagon fills a two-dimensional area with a minimum boundary, this 14-gon is a space-filling shape with minimum wall area [19].

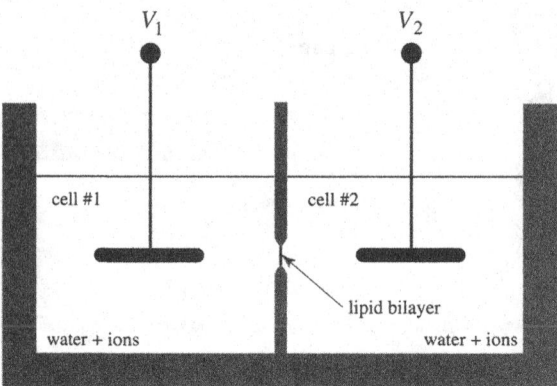

Figure 3.2. An experiment for making physical measurements on artificial cell membranes.

calling that the phenomenon of life emerged from the rich chemical soup of the Hadean seas some four billion years ago [8], one wonders whether the spontaneous formation of lipid bilayer compartments might have played a role. How difficult is it for such films to form?

In the early 1960s, Paul Mueller, Donald Rudin, Ti Tien, and William Wescott showed that a biological lipid bilayer can be reconstructed using the apparatus shown in Figure 3.2 [9, 14, 20]. In this experiment, a vessel is arranged with two chambers (cell #1 and cell #2), each of which is held at different electric potentials (V_1 and V_2) and filled with aqueous solutions containing different ionic concentrations. Additionally, there is a small hole between the two chambers that can be covered with a lipid bilayer using a technique similar to the formation of a bilayer soap film. Touching the hole with a camel's hair brush that has been dipped in lipid will first cause the formation of a relatively thick film, which soon collapses into the energetically more favorable bilayer of Figure 3.1(d).[3]

Using such an experimental apparatus, Mueller et al. [9], among others, were able to perform a number of physical experiments on lipid bilayers, including visual observation of the bilayer formation, measurement of electrical conductivity per unit area of a pure lipid bilayer, measurement of capacitance per unit area of a lipid bilayer, measurement of resting potential across the lipid bilayer as a function of ionic concentration, and observation of the influence of intrinsic (embedded) membrane proteins on membrane conductivity. Some fruits of these studies are the following [4].

---

[3] A short film showing the formation of a lipid bilayer is available on the Internet at www.msu.edu/user/ottova/soap_bubble.html.

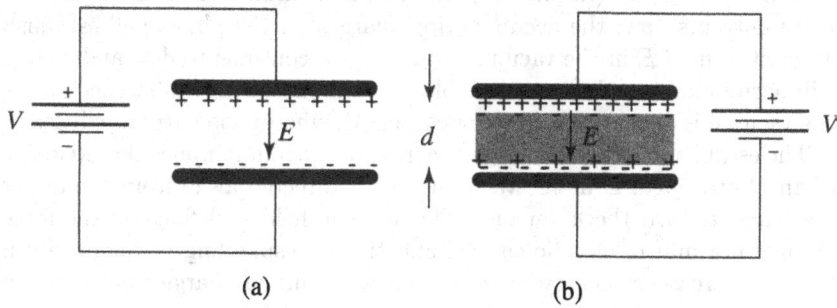

Figure 3.3. (a) A capacitor in a vacuum. (b) A capacitor that is filled with a material substance.

- The electrical capacitance of a lipid bilayer is about 1 microfarad (μF) per square centimeter.

- The electrical conductivity (or ionic permeability) of a pure lipid bilayer is very small, corresponding to that of a good insulator such as quartz.

- Membrane permeability is very sensitive to the presence of intrinsic proteins. If certain proteins are dissolved in the lipid bilayer, membrane conductivity increases by several orders of magnitude.

- With a proper choice of embedded membrane proteins, the switching action of a nerve membrane can be reproduced [10].

Because these observations are relevant to studies of the nerve, let us consider them in greater detail.

3.2   Membrane Capacitance

As we will see in the following chapter, the electrical capacitance of a nerve membrane plays a key role in the dynamics of its switching; thus, it is important for neuroscientists to understand what a capacitor is and the nature of the electric charge that it stores.

Consider first the vacuum capacitor shown in Figure 3.3(a) in which two parallel conducting plates of area A are separated by distance d. The plates are insulated from each other by a vacuum, so how does electric current manage to flow into the capacitor on the upper wire and out of it on the lower wire? Does electric current actually flow through the vacuum?

To answer such questions, let us connect a battery of voltage V across an uncharged capacitor with its positive (negative) terminal to the upper (lower) wire. Initially, current will flow into (out of) the upper (lower) plate, leaving it positively (negatively) charged, as is indicated in the figure. As these currents flow, the accumulating charges on the plates will establish an electric field E in the vacuum. Current will continue to flow and charge will accumulate on the capacitor plates until the voltage difference across the vacuum is equal to the battery voltage V, whereupon current will cease.

The establishment of a voltage V across the vacuum implies the existence of an *electric field E* in the vacuum that is directed away from the upper plate and toward the lower one. This electric field is defined as the force acting on a unit of electric charge, and the corresponding voltage is given by the decrease in energy experienced by a unit of charge upon moving from the lower to the upper plate. As was noted in the previous chapter, the concepts of force and energy are related by an expression of the form

$$\text{force} \equiv \frac{\text{decrease in energy}}{\text{change in position}},$$

where the direction of the force corresponds to the change in position.

Because electric field (E) is defined as a force per unit charge, and voltage (V) is energy per unit of electrical charge, it follows that the electric field between the capacitor plates is

$$E = \frac{V}{d}. \tag{3.1}$$

After this electric field is established and the flow of current has ceased, there will remain a constant positive charge $(+Q)$ on the upper plate and a constant negative charge $(-Q)$ on the lower plate. If the battery is removed, these charges and the corresponding electric field energy will remain stored in the capacitor. How much charge will be stored?

If V is measured in volts and the charge Q is measured in coulombs (C), it is a basic law of electricity that

$$\frac{Q}{A} = \varepsilon_0 E = \varepsilon_0 \frac{V}{d},$$

where A is the area of a capacitor plate and ε_0 is a fundamental constant describing the electrical properties of the vacuum. Thus, the charge on a vacuum capacitor is related to the voltage across it by the relation

$$Q = \frac{\varepsilon_0 A}{d} V = \tilde{C} V.$$

In this relation,

$$\tilde{C} = \frac{\varepsilon_0 A}{d}$$

is called the *capacitance*, which is measured in *farads*, or coulombs per volt.[4] Finally, ε_0, called the *electric permittivity of the vacuum*, has an empirical value of 8.84×10^{-12} farads per meter (F/m).

Not surprisingly, no electric current flows through the vacuum. Whenever the voltage across the capacitor plates is changed, electric current flows into one plate and out of the other, thereby readjusting the amount of electric charge that is stored.

What happens if we slip a material insulator (mica, wax, or metallic oxide) between the capacitor plates, as indicated in Figure 3.3(b)? (Although there may be a leakage of electric current through the insulator, let us assume that this effect is small enough to be neglected.) We can again measure the charge flowing into the upper plate, which turns out to be always larger than for the vacuum capacitor. Thus,

$$\frac{Q}{V} \equiv \tilde{C} = \frac{\kappa \varepsilon_0 A}{d} ,$$

where $\kappa > 1$ is called the *relative dielectric constant* describing the insulating material. Evidently the value of κ is unity for a vacuum, and it is of the order of 3 to 4 for ordinary insulating materials (e.g., glass, wax, wood, plastics). (Interestingly, the relative dielectric constant of water is about 80, because each molecule of this important biological fluid is an electrical dipole that can rotate. Energetically speaking, it is this large value of κ that reduces the electric field in the neighborhood of charged lipid head groups shown in Figure 3.2.)

The reason that $\kappa > 1$ for a material insulator is because the electric field $E = V/d$ acts on the electric charges (electrons and atomic nuclei) comprising the material, pulling negative charge slightly out of the upper surface. Similarly, positive charges are pulled slightly out of the lower surface of the insulating material.

In other words, there will be a negative *bound charge* on the upper surface of the insulator and a positive bound charge on the lower surface, as indicated in Figure 3.3(b). This *bound surface charge* cancels some of the *free charge* (Q) on the plates, allowing Q to be larger for the same value of voltage across the capacitor plates.

With constant voltage, the current through an ideal capacitor is zero; only when the voltage varies with time does stored charge change and current flow. The amount of this current (I) is then

$$I \equiv \frac{dQ}{dt} = \tilde{C}\frac{dV}{dt} ,$$

which is the basic description of a capacitor as an electric circuit element.

---

[4]Be careful not to confuse the abbreviation for coulomb (C) with the symbol (C) for capacitance.

3.3   Transmembrane Ionic Currents

Consider again the experimental apparatus of Figure 3.2 with the following assumptions: sodium and potassium salts have been dissolved in the water, appropriate intrinsic proteins are embedded in the lipid bilayer, and a steady voltage is impressed across the electrodes. Under these circumstances, a steady current is observed to flow between the two terminals (into one and out of the other), indicating a constant transport of ions across the lipid bilayer membrane. For each ion, this steady ionic current has two distinct components: a *conduction current* responding to the voltage difference across the membrane and a *diffusion current* responding to the difference in ionic concentrations across the membrane. Let us examine these two components in detail.

3.3.1   Conduction Current

Assuming that a steady voltage is maintained across a lipid bilayer, the average electric field within that membrane will be given by Equation (3.1), where d is the membrane thickness. Because this electric field exerts a force on the electric charges of ions, it can cause an ionic current to flow, which is called the *conduction current*. How is conduction current calculated?

Suppose that the membrane is permeable to sodium ions, each carrying a charge of $+q$, where q has the magnitude of the electronic charge; thus,

$$q = 1.602 \times 10^{-19} \text{ C}.$$

In general, each of these ions exhibits a stochastic motion stemming from the irregular manner in which it is randomly pushed and pulled about by other thermally agitated components of the system. When a voltage is applied and an electric field E is present inside the membrane, this field adds a small *drift velocity* in the direction of E to the random thermal motion. If the drift velocity is small compared with thermal motion, its magnitude is proportional to the electric field, and we can write

$$v_{\text{drift}} = \mu_{\text{Na}} E,$$

where μ_{Na} is a constant of proportionality called the *mobility* of the sodium ion.

Conduction current density (J_c) is defined as the ionic charge times the average of this drift current per unit area (perpendicular to the direction of the electric field). Thus, it can be written

$$J_c = q v_{\text{drift}}[\text{Na}^+] = q \mu_{\text{Na}}[\text{Na}^+] E,$$

where $[\text{Na}^+]$ is the concentration of the sodium ions in ions per unit volume.

Because $E \propto V$, the conduction current density is expected to be given by an expression of the form

$$J_c = \tilde{G}_{\text{Na}} V,$$

where $\tilde{G}_{Na}$ is a *sodium ion conductivity per unit area* in units of mhos per unit area.[5]

The form of this equation, however, does not imply that transmembrane conduction current is proportional to transmembrane voltage. As we have seen in the previous section, ionic current flow through a lipid bilayer is greatly facilitated by the presence of *intrinsic* proteins, which are embedded in the membrane. Even at small values of transmembrane voltage, the electric field within a membrane can be quite large, causing the membrane proteins to change shape and altering the values of μ_{Na} and $\tilde{G}_{Na}$. We will see that this effect is of central importance in understanding how a patch of nerve membrane manages to act as an electric switch.

3.3.2   Diffusion Current

In addition to the conduction current flowing across a membrane, there is also a *diffusion current*. Just as the conduction current is caused by an electric field (or change in the electric potential with distance), the diffusion current stems from spatial changes in the ionic concentrations.

With D_{Na} defined as the diffusion constant for sodium ions, an expression for the diffusion current density is

$$J_d = -qD_{Na}\frac{d[Na^+]}{dx} .$$

In this equation, the minus sign appears because ions diffuse in the direction for which their concentration is diminishing.[6]

In thinking about diffusion current, keep in mind that it is an independent component of transmembrane current. Just as the conduction current can be set to zero by making the voltage difference across the membrane zero, the diffusion current for a particular ion becomes zero whenever the concentrations of that ion are the same on both sides of the membrane.

In other words, J_c and J_d in a cell membrane can be independently adjusted by the electrophysiologist. Yet, for each ion, the diffusion constant (which determines the diffusion current) is related in a fundamental way to the mobility (which fixes the drift velocity of the corresponding conduction current).

---

[5] The unit of resistance (volts/amperes) is the ohm, whereas the unit of conductance (amperes/volts) has two names. For some, it is whimsically called the *mho*, which is "ohm" spelled backward. For others, this same unit of conductance is called the *siemens* (S). In this book, the terms "ohm" and "mho" are used.

[6] Note that this diffusion constant for sodium ions is totally unrelated to the diffusion constant for voltage on a squid axon, which was mentioned in Equation (2.2) of the previous chapter.

3.3.3 Einstein's Relation

In the year 1905, while struggling to finish a problematic doctoral thesis and beginning what would become a difficult marriage, a young Swiss patent clerk named Albert Einstein found the time and mental energy to publish three short papers on different subjects, each with revolutionary implications for twentieth-century physics.

The first of these papers—as is widely known—presented his special theory of relativity, fundamentally altering Newtonian mechanics, and the second made basic contributions to the quantum theory of electromagnetic radiation, for which he was to receive the Nobel Prize. Interesting as these two works are, it is the third short paper—investigating the stochastic behavior of a small particle suspended in a liquid, called *Brownian motion*[7]—that concerns us here. Why was Einstein worrying about Brownian motion along with everything else on his plate?

The atomic theory of matter was not universally accepted a century ago because no one had actually seen an atom. In this context, therefore, Einstein suggested that the random movement of a small particle in aqueous suspension could be taken as evidence for collisions with randomly moving molecules of water [2], and he was able to find a simple relationship between the diffusion constant and the mobility of such a randomly moving particle. This relation is important for formulating the dynamics of a nerve membrane.

To see how Einstein derived his relationship, return to Figure 3.2 and assume that only a single ionic species (say sodium ions) is able to penetrate the bilayer membrane, no external voltage is applied to the electrodes, and the system is in *thermal equilibrium.*

What is meant by "thermal equilibrium"? In simple terms, this phrase implies that there is no way of knowing the direction of time. Given a detailed movie of a system in thermal equilibrium, in other words, it is not possible to tell whether the film is being run forward or backward. More formally, this is called the principle of *detailed balance,* which states that every local process and its reverse proceed at the same rate.[8]

In the context of Figure 3.2, thermal equilibrium implies that the conduction current J_c of the sodium ions is everywhere exactly canceled by the diffusion current J_d. Thus,

$$J_c + J_d = -q\mu_{Na}[Na^+]\frac{dV}{dx} - qD_{Na}\frac{d[Na^+]}{dx} = 0, \qquad (3.2)$$

---

[7]First observed by the English botanist Robert Brown in 1827.

[8]This type of temporal reversibility is quite different from that related to conservation of energy mentioned in the previous chapter. The present instance is a special assumption made in statistical analyses of systems with many degrees of freedom. An essential feature of biological organisms is the violation of this assumption.

where $E = -dV/dx$. This can be integrated to obtain

$$\frac{[\text{Na}^+]_2}{[\text{Na}^+]_1} = \exp\left[-\frac{\mu_{\text{Na}}}{D_{\text{Na}}}(V_2 - V_1)\right], \tag{3.3}$$

where the subscripts (1 and 2) refer to different sides of the lipid bilayer. Let us consider the physical meaning of this equation.

Equation (3.3) tells us that if the voltage is higher on side #2 of the lipid bilayer, then the thermal equilibrium concentration of sodium ions will be lower on that side. In other words, there are fewer sodium ions where their energy is higher and more where it is lower. From a physical perspective, this is to be expected, just as a mountaineer finds the density of atmospheric molecules to be lower at high altitudes, where their gravitational energy is greater.

This phenomenon was well known to Walther Nernst and Max Planck at the end of the nineteenth century [12, 15] and can be stated as follows. In thermal equilibrium, the relative probabilities of finding an energy-conserving system in two states of different energies are given by the ratio

$$\frac{P_2}{P_1} = \exp\left[-\frac{\Delta U}{kT}\right], \tag{3.4}$$

where ΔU is the increase in energy of state #2 over that of state #1, T is absolute temperature, and

$$k = 1.38 \times 10^{-23} \text{ joules per degree kelvin}$$

is the *Boltzmann constant*. (From a different perspective, this constant tells us the average amount of thermal energy that a system stores in each degree of freedom at thermal equilibrium.)

Noting that the energy difference of an ion on two sides of a membrane is given by

$$\Delta U = q(V_2 - V_1),$$

it follows from a comparison of Equations (3.3) and (3.4) that for any charged particle undergoing Brownian motion the diffusion constant and the mobility are related by

$$D = \frac{kT}{q}\mu. \tag{3.5}$$

Equation (3.5), called the *Einstein relation*, plays an important role not only in the study of electrolytes and in electrophysiology but also in semiconductor electronics, where it establishes a connection between drift and diffusion currents of minority carriers in the base regions of bipolar transistors [16].

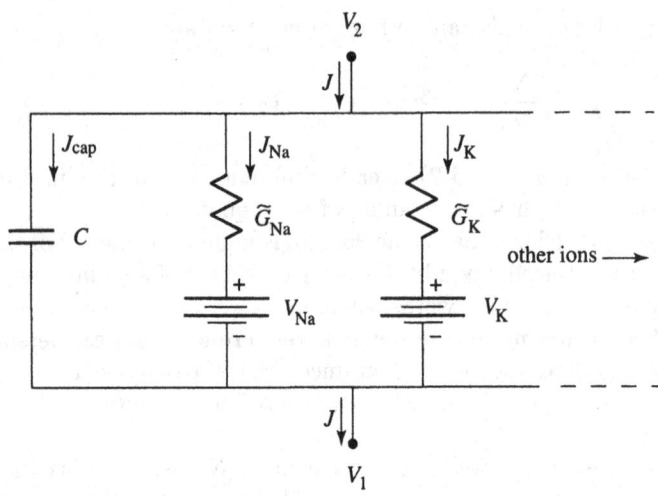

Figure 3.4. An electric circuit model for a unit area of the lipid bilayer membrane shown in Figure 3.2.

3.4   A Membrane Model

We are now in a position to assemble an electrical model for a lipid bilayer membrane that is permeable to an arbitrary number of ionic species, taking account of the following current components.

First, there is the capacitive component, which from Section 3.2 contributes a current density of

$$J_{\text{cap}} = C\frac{dV}{dt},$$

where $C = \kappa\varepsilon_0/d$ is the capacitance per unit area of the bilayer. This current is represented as the left-hand branch in Figure 3.4, where in the context of Figure 3.2

$$V \equiv V_2 - V_1.$$

In addition to the capacitive current, there is also an ionic current for each species of ion that is able to pass through the membrane. Let us first consider the sodium ion current, which is represented as the second branch (counting from the left) in Figure 3.4.

From the previous section, sodium ion current consists of two independent components: conduction current (which flows in response to the voltage difference across the membrane) and diffusion current (which responds to the difference of sodium ion concentrations on the two sides of the membrane). Although they can be independently adjusted, these two components are linked by the Einstein relation between mobility and diffusion constant for each ion.

To establish this connection, refer to Equation (3.2), which expresses the sum of conduction and diffusion currents under the condition of thermal equilibrium. Inverting Equation (3.3) and using the Einstein relation (Equation (3.5)), the total transmembrane sodium current (drift plus diffusion) is zero for

$$V_2 - V_1 = V = \frac{kT}{q} \ln \left(\frac{[\text{Na}^+]_1}{[\text{Na}^+]_2} \right).$$

Thus, the steady-state current can be written as

$$J_{\text{Na}}(V) = \tilde{G}_{\text{Na}} \left[V - \frac{kT}{q} \ln \left(\frac{[\text{Na}^+]_1}{[\text{Na}^+]_2} \right) \right], \qquad (3.6)$$

where the bracketed factor is the *electrochemical potential* across the membrane, which comprises an electromotive component (the first term) and a diffusion component (the second term).

Calculation of $\tilde{G}_{\text{Na}}$ is a daunting task requiring detailed knowledge of (or assumptions about) the several obstacles that an ion might encounter on its way across the membrane, including potential barriers, steric barriers, making and breaking of valence bonds, viscous effects, and so on. Empirically, however, $\tilde{G}_{\text{Na}}$ can be defined as

$$\tilde{G}_{\text{Na}} \equiv \frac{J_{\text{Na}}(V)}{V - (kT/q) \ln \left([\text{Na}^+]_1 / [\text{Na}^+]_2 \right)}. \qquad (3.7)$$

The first term on the right-hand side of Equation (3.6) is the conduction current (responding to the potential difference V), whereas the second term is the diffusion current (responding to the difference in ionic concentrations across the membrane). Through the definition of $\tilde{G}_{\text{Na}}$, these two components are linked by the Einstein relation, but what do we know about $\tilde{G}_{\text{Na}}$?

First, the artificial membrane experiments (discussed in Section 3.1) strongly suggest that ions move through the membrane via pores or channels of some sort that are formed by embedded proteins. Further evidence that sodium current passes through protein pores is the fact that it can be selectively blocked by small amounts of tetrodotoxin (TTX), a water-soluble paralytic poison found in the tissues of the Japanese puffer fish *Spheroides rubrides*.

Second, as noted earlier, the externally imposed electric field ($E = V/d$) within the membrane can be quite large. If, for example, a potential difference of $V = 50$ mV is impressed across a typical bilayer membrane of thickness $d = 10$ nm (100 Å), the resulting electric field is 50,000 V/cm. Thus, $\tilde{G}_{\text{Na}}$ is expected to depend strongly upon the value of V because the configurations (shapes) of embedded membrane proteins may change with the forces induced by this large electric field inside the bilayer. Protein configuration changes subsequently alter the sodium ion mobility (μ_{Na}) and the corresponding diffusion constant, $D_{\text{Na}} = (kT/q)\mu_{\text{Na}}$.

Finally, $\tilde{G}_{Na}$ is also expected to be a function of the sodium ion concentrations, $[Na^+]_1$ and $[Na^+]_2$. In general, $\tilde{G}_{Na}$ should increase as the ionic concentrations increase because there will be more charge carriers within the membrane. Conversely, as $[Na^+]_1$ and $[Na^+]_2$ approach zero, $\tilde{G}_{Na}$ must also go to zero because there will be no sodium ions available to carry the current. (In the following chapter, we will see how this property can be used to separate various components of nerve membrane current.) If the sodium ion concentrations remain at their normal values, however, $\tilde{G}_{Na}$ in Equation (3.6) can be viewed as a function of V alone.

As represented in Figure 3.4, therefore, Equation (3.6) can be written as

$$J_{Na}(V) = \tilde{G}_{Na}(V)(V - V_{Na}),$$

where

$$J_c = \tilde{G}_{Na}(V)V$$

is the conduction component of the sodium ion current that flows downward when the voltage at terminal #2 is more positive than the voltage at terminal #1. The last term in this equation is the sodium diffusion current, which is conveniently represented as

$$J_d = -\tilde{G}_{Na}(V)V_{Na},$$

where

$$V_{Na} \equiv \frac{kT}{q} \ln\left(\frac{[Na^+]_1}{[Na^+]_2}\right)$$

is a battery representing the *diffusion* (or *Nernst*) potential. The negative sign before this last term implies that the diffusion component flows upward if the sodium concentration in chamber #1 is higher than in chamber #2.[9]

Additional ionic currents are readily brought into the picture. Potassium current, which is selectively blocked by small concentrations of tetraethylammonium (TEA), can be modeled as

$$J_K = \tilde{G}_K(V)(V - V_K), \tag{3.8}$$

where

$$V_K \equiv \frac{kT}{q} \ln\left(\frac{[K^+]_1}{[K^+]_2}\right),$$

and included as another branch in the circuit representation of Figure 3.4.

Similarly, the transmembrane current carried by doubly charged calcium ions (Ca^{++}) can be represented as

$$J_{Ca} = \tilde{G}_{Ca}(V)(V - V_{Ca}),$$

---

[9]In using Equation (3.6) for interpreting empirical data, it is convenient to remember that $kT/q = 25.2$ mV at a temperature of 293 K, or 20°C, and 26.7 mV at 310 K, or 37°C.

where

$$V_{Ca} \equiv \frac{kT}{2q} \ln \left(\frac{[Ca^{++}]_1}{[Ca^{++}]_2} \right),$$

with the "2" in the denominator accounting for the fact that each ion carries a double charge.

In a general case, therefore, the total ionic current through a membrane will be given by an expression of the form

$$J = C\frac{dV}{dt} + J_{Na} + J_K + J_{Ca} + \cdots + \text{etc.} \tag{3.9}$$

Bear in mind that nothing has yet been said about the dynamics of the ionic conductances $\tilde{G}_{Na}$, $\tilde{G}_K$, $\tilde{G}_{Ca}$, and so on. This question will be taken up in the following chapter.

3.5 Resting Potential and the Sodium–Potassium Pump

Suppose that the terminals in the experiment of Figure 3.2 are left open and V is allowed to relax to a steady voltage called the *resting potential* (V_R) of the membrane. To compute V_R, both J and dV/dt are set equal to zero in Equation (3.9), implying that the sum of the ionic currents must also be zero.

If the bilayer is permeable only to sodium and potassium ions—as is approximately so for the squid axon membrane—it follows directly from Equations (3.6), (3.8), and $J_{Na} + J_K = 0$ that

$$V_R = \frac{kT}{q} \left(\frac{\tilde{G}_{Na} \ln \left([Na^+]_1/[Na^+]_2\right) + \tilde{G}_K \ln \left([K^+]_1/[K^+]_2\right)}{\tilde{G}_{Na} + \tilde{G}_K} \right). \tag{3.10}$$

The generalization of this result to an arbitrary number of ionic current components is straightforward. (In the following chapter, we will see that it is convenient to measure all voltages with respect to the resting potential.)

From Equation (3.10), an interesting fact emerges. If the resting value of $\tilde{G}_K$ is much larger than the resting value of $\tilde{G}_{Na}$, then the resting voltage will be close to the potassium diffusion voltage. In other words,

$$V_R \to \frac{kT}{q} \ln \left(\frac{[K^+]_1}{[K^+]_2} \right)$$

as $\tilde{G}_K/\tilde{G}_{Na} \to \infty$. For the active membrane of a squid axon, this is approximately the case.

Measurements on a normal squid nerve also show that the sodium ion concentration outside the axon is much higher than its inside value, whereas the outside level of potassium ion concentration $[K^+]_o$ is lower than the inside level $[K^+]_i$. Thus,

$$V_K = \frac{kT}{q} \ln \left(\frac{[K^+]_o}{[K^+]_i} \right) \approx -77 \text{ mV},$$

implying that $[K^+]_o/[K^+]_i \approx 0.047$.

Normal squid nerve has a resting potential

$$V_R \approx -65 \text{ mV},$$

indicating that positively charged ions have lower energy inside than outside. At the resting potential, therefore, the potassium conduction current is directed from outside to inside, whereas the potassium diffusion current is from inside to outside, so the two components largely cancel, as is expected for the resting potential lying close to the potassium diffusion potential.

For sodium ions, on the other hand,

$$V_{Na} = \frac{kT}{q} \ln \left(\frac{[Na^+]_o}{[Na^+]_i} \right) \approx +50 \text{ mV},$$

which corresponds to $[Na^+]_o/[Na^+]_i \approx 7.3$. In this case, both the conduction current and the diffusion components of the total sodium current are directed inward when $V = V_R$. How is such a dynamic imbalance maintained?

First, it is necessary that the value of the sodium conductance $(\tilde{G}_{Na})$ be very small at the resting voltage, which is consistent with the observation that $\tilde{G}_{Na} \ll \tilde{G}_K$, but merely making sodium conductance small will not maintain the dynamic imbalance of sodium ions over a long period of time. Even a small inward flow of sodium ions would eventually reduce the large ratio of outside to inside concentrations. To maintain the ratio, some mechanism is needed to transport sodium ions across the membrane, from inside to outside, and potassium ions from outside to inside. This mechanism turns out to be provided by yet another intrinsic membrane protein, which is informally known as the *sodium–potassium pump*.

More precisely called "Na/K-ATPase," this enzyme was first isolated in the 1950s by the Danish biochemist Jens Christian Skou and his colleagues from the membranes of the nerves from the legs of some 25,000 crabs [17, 18]. Using the biological energy available from hydrolysis of adenosine-triphosphate (ATP) to adenosine-diphosphate (ADP), Na/K-ATPase pumps sodium ions out of nerve cells as it pumps potassium ions in [6, 7, 11]. The resulting resting potential provides the motive force that drives neural activity.

3.6   Recapitulation

The emphasis in this chapter has been on developing a physical understanding of nerve membranes and their subsequent dynamics; thus the energetics of lipid bilayer formation was introduced in the context of soap films. Both the nature of charge storage in an electric capacitor and an experimental arrangement to study ionic current flows through artificial lipid bilayers were then considered in some detail.

Building on this physical picture, the nature of conduction and diffusion components of transmembrane current were described and related through a derivation of the Einstein relation, famously linking the mobility and diffusion constant for each ionic species. To provide a basis for discussions of nerve impulse propagation in the following chapter, a membrane model was developed for an arbitrary number of ionic species that relates the resting potential of a nerve membrane to ionic permeabilities. Finally, the sodium–potassium pump was mentioned. This vital enzyme uses biological energy to establish a resting potential across real nerve membranes, thereby supplying the energy needed for nerve impulse propagation.

References

[1] CV Boys, *Soap Bubbles: Their Colors and Forces Which Mold Them*, Dover, New York, 1959.

[2] A Einstein, Über die von der molekularkinetischen Theorie der Wärme geforderte Bewegung von in ruhenden Flüssigkeiten suspendierten Teilchen, *Ann. Phys.* 17 (1905) 549–560.

[3] DS Goodsell, *The Machinery of Life*, Springer-Verlag, New York, 1993.

[4] MK Jain, *The Bimolecular Lipid Membrane: A System*, Van Nostrand Reinhold, New York, 1972.

[5] I Langmuir, The constitution and fundamental properties of solids and liquids, *J. Am. Chem. Soc.* 39 (1917) 1848–1906.

[6] JB Lingrel, Na,K-ATPase: Isoform structure, function, and expression, *J. Bioenerg. Biomembr.* 24 (1992) 263–270.

[7] S Lutsenko and JH Kaplan, Organisation of P-type ATPases: Significance of structural diversity, *Biochemistry* 34 (1996) 15607–15613.

[8] L Margulis and D Sagan, *What Is Life?* Simon & Schuster, New York, 1995.

[9] P Mueller, DO Rudin, HT Tien, and WC Westcott, Reconstitution of cell membrane structure *in vitro* and its transformation into an excitable system, *Nature* 194 (1962) 979–980.

[10] P Mueller and DO Rudin, Action potentials induced in bimolecular lipid membranes, *Nature* 217 (1968) 713–719.

[11] JV Møller, B Juul, and M le Maire, Structural organisation, ion transport, and energy transduction of P-type ATPases, *Biochim. Biophys. Acta* 1286 (1996) 1–51.

[12] W Nernst, Die elektromotorische Wirksamkeit der Ionen, *Z. Phys. Chem.* 4 (1889) 129–181.

[13] I Newton, *Optiks*, Dover, New York, 1952. (Based on the fourth edition, London, 1730.)

[14] AL Ottova, T Martynski, A Wardak, and HT Tien, Self-assembling BLMs on solid support. In *Molecular & Biomolecular Electronics*, RR Birge (ed), Advances in Chemistry Series #240, American Chemical Society, Washington, DC, 1994.

[15] M Planck, Über die Erregung von Elektricität und Wärme in Elektrolyten, *Ann. Phys. Chem.* 39 (1890) 161–186.

[16] W Shockley, *Electrons and Holes in Semiconductors*, Van Nostrand, New York, 1950.

[17] JC Skou, The influence of some cations on an adenosine triphosphatase from peripheral nerves, *Biochim. Biophys. Acta* 23 (1957) 394–401.

[18] JC Skou and M Esmann, The Na,K-ATPase, *J. Bioenerg. and Biomembr.* 24 (1992) 249–261.

[19] DW Thompson, *On Growth and Form*, Cambridge University Press, Cambridge, 1961.

[20] HT Tien and AL Ottova, The lipid bilayer concept and its experimental realization: From soap bubbles, the kitchen sink, to bilayer lipid membranes, *J. Membr. Sci.* 189 (2001) 83–117.

4
The Hodgkin–Huxley (H–H) Axon

Amid the devastation of the Second World War, technology prospered. In the course of developing radar systems for detecting aircraft and for control of responding fire, substantial progress was made in the design of electronic pulse amplifiers and generators and of cathode-ray tubes for the visual display of such pulses. At the end of 1945, this knowledge became available for applications to peaceful pursuits.

One application of the new technology was a careful study of the membrane dynamics of the giant axon of the squid (*Loligo*) (see Figure 1.2), which was carried out by Alan Hodgkin and Andrew Huxley in the early 1950s [16, 17, 18, 19]. Although obtained for a specific nerve in a particular species, the concepts emerging from this work have guided much subsequent research in electrophysiology. Thus, students of neuroscience should be familiar with the key features of Hodgkin–Huxley (H–H) theory. In this chapter, we will consider some essential aspects of this formulation of nerve impulse dynamics, paying particular attention to the relationships between mathematical models and biological reality.

4.1 Space and Voltage Clamping

In carrying out their work, Hodgkin and Huxley used the techniques of *space clamping* and *voltage clamping* to obtain a dynamic characterization of a localized patch of squid membrane. Although these two phrases sound alike, they describe quite different experimental techniques and must be individually understood.

Space Clamping

Geometrically, a squid nerve is a long cylindrical tube of ion-conducting axoplasm encased in a lipid bilayer membrane.[1] Variations in transmembrane voltage with distance along the structure are expected as a result of longitudinal current flowing through the resistive axoplasm, and such variations make it difficult to measure the specific (per unit area) ionic permeability of the membrane as a function of transmembrane voltage. Because of the relatively large cross section of a squid nerve, this difficulty can be overcome by inserting a conducting wire longitudinally through the nerve to serve as a terminal for the internal voltage. Under such circumstances, longitudinal variations in voltage are eliminated and the transmembrane voltage is everywhere held constant—or "space-clamped"—at the same value.

Voltage Clamping

The term "voltage clamping" implies the use of a *negative feedback amplifier* to set the potential difference across a nerve membrane at a desired value [7]. This is important because the key for understanding the dynamics of ion currents through a squid nerve is to achieve precise control of the transmembrane voltage, holding it at fixed values in the face of changes in other experimental variables such as ionic currents, membrane permeability, electrode resistance, temperature, and the like.

The advantages of using negative feedback are twofold: the input resistance of the amplifier is high, making it easy to set the desired voltage level, and the output resistance is very low, keeping the output (membrane) voltage fixed in the presence of experimental variations in the axon. In other words, the negative feedback amplifier output looks like an *ideal voltage source* (a battery with zero internal resistance, the level of which can be easily adjusted by an experimenter).

With space clamping, the total current per unit area flowing through the membrane is given by an expression of the form

$$\frac{\text{Total membrane current}}{\text{Membrane area}} = J_{\text{ion}} + C\frac{dV}{dt}, \qquad (4.1)$$

where the first term on the right-hand side represents the transmembrane ionic current per unit area through the membrane, and the second term is displacement current per unit area through the capacitance of the lipid bilayer. The current flowing into one terminal of the membrane capacitance measures the change in free charge of (say) positive ions collecting near one side of the membrane. The capacitive current flowing out of the other terminal is then the change in negative ionic charge near the other side of the membrane.

---

[1] For those who wish to learn more about squid experiments, the booklet by Arnold et al. [3] is an excellent place to begin. Also, Cole's book [7] has many experimental details and a lovely color photograph of the squid.

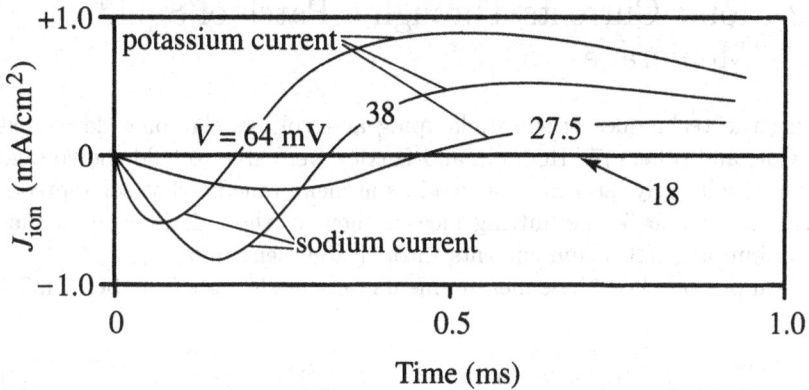

Figure 4.1. One of Cole's earliest measurements of squid membrane current under both space and voltage clamping. (Drawn from data in [7].)

As we learned in Chapter 3, the capacitance per unit area (C) of a typical nerve cell membrane is about 10^{-6} F/cm^2. A farad has the units of coulombs per volt or ampere-seconds per volt, so a typical nerve membrane with a voltage that is changing at a rate of about 100 mV per millisecond (as on the leading edge of the squid nerve impulse shown in Figure 1.1) will carry a displacement (or capacitive) current of 100 μA/cm^2. Under space-clamped conditions, this displacement current—the last term in Equation (4.1)—will be measured on an external current meter in addition to the ionic current that passes through the membrane.

If the membrane is also clamped at a constant voltage (V), on the other hand, the capacitive current is zero because

$$\frac{dV}{dt} = 0$$

and only the ionic current through the membrane is measured. Figure 4.1 shows results from one of the earliest such measurements, which was recorded by Kenneth Cole in 1947 [7].

When the voltage inside the axon was made 18 mV positive with respect to the resting value, no ionic current was observed. At 27.5 mV, however, the ionic current initially flowed into the axon, responding to the fact that sodium ion concentration is larger outside than inside. After a half millisecond, this current turned outward, responding to the opposite concentration ratio of potassium ions. At larger values of the space-clamped voltage, the onset of sodium current was more rapid and the subsequent potassium current larger.

4.2   Ionic Currents Through a Patch of Squid Membrane

Using the techniques of space clamping and voltage clamping developed by Cole and others [7], Hodgkin and Huxley were able to hold the voltage across a relatively large area of squid axon membrane fixed at some predetermined voltage V, permitting measurement of the individual dynamics of sodium and potassium currents through the membrane.

To appreciate how these measurements were made, note from Section 3.4 that

$$\text{sodium current} = \tilde{G}_{\text{Na}}(V, t)(V - V_{\text{Na}}) \tag{4.2}$$

and

$$\text{potassium current} = \tilde{G}_{\text{K}}(V, t)(V - V_{\text{K}}), \tag{4.3}$$

where V_{Na} (V_{K}) is the voltage at which the sum of the conduction and diffusion components of sodium (potassium) current cancel each other, and the corresponding membrane conductances are $\tilde{G}_{\text{Na}}$ and $\tilde{G}_{\text{K}}$.

In the previous chapter, V was defined as the total voltage difference across the cell membrane; thus, $V \equiv V_2 - V_1$ in Figures 3.2 and 3.4. Both experimentally and analytically, however, Hodgkin and Huxley found it more convenient to define voltages with respect to the resting potential of the membrane, which is defined in Equation (3.10). Here and henceforth in this book, therefore, all components of transmembrane voltage (V, V_{Na}, and V_{K}) are measured with respect to the resting potential, $V_{\text{R}} \approx -65$ mV. In the notation of the present chapter, therefore,

$$V_{\text{Na}} \;=\; 25 \log \left(\frac{[\text{Na}^+]_o}{[\text{Na}^+]_i} \right) + 65,$$

$$V_{\text{K}} \;=\; 25 \log \left(\frac{[\text{K}^+]_o}{[\text{K}^+]_i} \right) + 65,$$

$$V(\text{present chapter}) \;=\; V(\text{previous chapter}) + 65.$$

In other words, all membrane voltages in this and subsequent chapters are $-V_{\text{R}} \approx +65$ mV larger than the corresponding voltages in Chapter 3. Under these new definitions, the sodium current remains zero whenever $V = V_{\text{Na}}$, and the potassium current is zero for $V = V_{\text{K}}$.

For typical squid nerves, Hodgkin and Huxley found that

$$V_{\text{Na}} = +109 \pm 11\% \text{ mV}$$

and

$$V_{\text{K}} = -11 \pm 14\% \text{ mV},$$

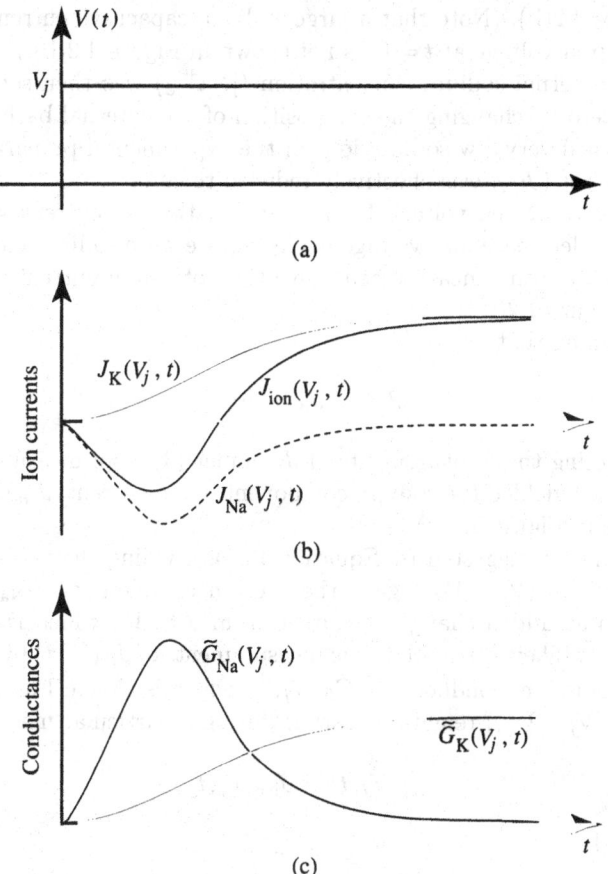

Figure 4.2. Figures related to the Hodgkin–Huxley determination of membrane conductances. (a) The applied voltage as a function of time. (b) Measurements of total ionic current and potassium current, from which sodium current can be calculated. (c) Sodium and potassium conductances at $V = V_j$ as functions of time. (See the text for details.)

where these diffusion potentials depend on the ratios of outside to inside ion concentrations.

To measure the individual (sodium plus potassium) components of membrane conductivity, Hodgkin and Huxley proceeded as follows [16].

(1) As indicated in Figure 4.2(a), the space-clamped membrane voltage was suddenly changed from the resting value ($V = 0$) at $t = 0$ to V_j, where

$$V_{Na} > V_j > 0,$$

and held there under voltage clamping. At this voltage, the total ion current, $J_{ion}(V_j, t)$, through the membrane was measured as a function of time, as in Figure 4.2(b). (Note that a large spike of capacitive current, caused by the jump in voltage at $t = 0$, is not shown in Figure 4.2(b).)

(2) The external sodium concentration ($[Na^+]_o$) was then set approximately to zero by changing the composition of the external bath. Because there remained very few sodium ions in the experimental preparation, the sodium current (J_{Na}) was effectively reduced to zero.

(3) Using the same voltage level as in (i), the ion current was again measured under space and voltage clamping. Because sodium current had been eliminated, this measurement gave the potassium current, $J_K(V_j, t)$, shown in Figure 4.2(b).

(4) Assuming that

$$J_{ion} = J_K + J_{Na}$$

and subtracting the measurements of J_K under (3) from measurements of J_{ion} under (1) yielded the sodium component of ion current, $J_{Na}(V_j, t)$, the dashed line in Figure 4.2(b).

(5) Finally, as suggested by Equation (3.7), dividing the measurements of $J_{Na}(V_j, t)$ by $(V_j - V_{Na})$ gave the sodium conductivity $\tilde{G}_{Na}(V_J, t)$ in response to a sudden change of voltage from 0 to V_j, which is shown in Figure 4.2(c). Likewise, dividing the measurements of $J_K(V_j, t)$ by $(V_j - V_K)$ gave the potassium conductivity $\tilde{G}_K(V_j, t)$, which is also in Figure 4.2(c).

Because $(V_j - V_{Na})$ remains constant during a particular measurement,

$$\tilde{G}_{Na}(V_J, t) \propto J_{Na}(V_j, t),$$

and similarly

$$\tilde{G}_K(V_J, t) \propto J_K(V_j, t),$$

but $\tilde{G}_{Na}$ has the opposite sign from J_{Na} because $(V_j - V_{Na})$ is negative.

Performing this measurement for a dozen different values of V_j, Hodgkin and Huxley observed that the sodium conductance (or permeability) initially rose from zero to a maximum value in a time of less than one millisecond and then decayed back to zero in a few milliseconds. Going further, they showed that the dynamics of sodium ion current at voltage V could be modeled by the formula

$$J_{Na}(V) = G_{Na}m^3(V, t)h(V, t)(V - V_{Na}), \tag{4.4}$$

where G_{Na} is a maximum sodium conductance (or permeability) per unit area, m is a "sodium turn-on" variable, and h is a "sodium turn-off" variable. Similarly, the dynamics of potassium ion current at voltage V were represented as

$$J_K(V) = G_K n^4(V, t)(V - V_K),$$

Table 4.1. Parameter values measured by Hodgkin–Huxley for the giant axon of the squid [19].

Parameter	Mean	Range	Standard	Units
C	0.91	(0.8—1.5)	1.0	$\mu\mathrm{F/cm^2}$
G_{Na}	120	(65—260)	120	$\mathrm{mmhos/cm^2}$
G_{K}	34	(26—49)	36	$\mathrm{mmhos/cm^2}$
G_{L}	0.26	(0.13—0.5)	0.3	$\mathrm{mmhos/cm^2}$
V_{Na}	+109	(95—119)	+115	mv
V_{K}	−11	(9—14)	−12	mv
V_{L}	+11	(4—22)	+10.5995	mv

where n is a "potassium turn-on" variable. Thus, $\tilde{G}_{\mathrm{Na}} \equiv G_{\mathrm{Na}}m^3 h$ and $\tilde{G}_{\mathrm{K}} \equiv G_{\mathrm{K}}n^4$, where m, h, and n are all constrained to lie between zero and one.[2]

The total ionic current per unit area across the membrane, as expressed in Equation (4.1), then becomes

$$J_{\mathrm{ion}} = G_{\mathrm{Na}}m^3 h(V - V_{\mathrm{Na}}) + G_{\mathrm{K}}n^4(V - V_{\mathrm{K}}) + G_L(V - V_L), \qquad (4.5)$$

where m, h, and n are functions of both V and t, and the last term is a small leakage current, accounting for ionic current missed by the direct measurements of sodium and potassium components. From such measurements, Hodgkin and Huxley were able to construct first-order ordinary differential equations governing the dynamics of m, h, and n, which are recorded in Appendix B.

As in Figure 3.4, V_{Na}, V_{K}, and V_{L} can be viewed as ionic batteries that cause diffusion components of ionic currents to flow across the membrane. We will see that these batteries supply energy to a nerve impulse to make up for the energy lost to ohmic dissipation of the circulating ionic currents.

---

[2] As Cole has pointed out, however, the assignment of powers to m and n is somewhat arbitrary [7]. Although several other representations of the dynamics have been suggested, the original Hodgkin–Huxley formulation has become widely accepted in the neuroscience literature [12, 14, 20, 30, 34, 38].

Values for the parameters in Equation (4.5) are given in Table 4.1. Interestingly, these parameter values exhibit normal physiological variations.[3]

The "standard" values in this table lie close to the average values and were selected for reasons to be explained later. In particular, there is no empirical basis for the standard value of +10.5995 mV for V_L. This value was chosen to make $J_{ion} = 0$ when $V = 0$ in Equation (4.5) for the other standard parameters.[4]

That such a formulation is not a special property of the squid axon is shown in Table 7.1 presenting corresponding standard parameters for the sciatic nerve of a frog [7]. In this case, the measurements were made on small active nodes because each nerve fiber is largely covered with an insulating sheath, called myelin, as indicated in Figure 7.1. The effect of this structure is to increase the conduction velocity of an impulse without increasing the nerve diameter, a phenomenon that was briefly discussed in Chapter 1 and will be considered again in Chapter 7. Comparison of the data in Tables 4.1 and 7.1 shows that the properties of nerve membranes from different phyla of the animal kingdom are similar.

To repeat: using the experimental methods sketched above, Hodgkin and Huxley formulated the dynamics of $m(V,t)$, $h(V,t)$, and $n(V,t)$ as the first-order ODEs recorded in Appendix B. Using these equations, it is straightforward to compute how a space-clamped membrane will switch when it is not voltage-clamped.

4.3   Space-Clamped Action Potentials

Suppose that we have a length of squid axon that is space clamped so the voltage is uniform over the entire membrane area. To simplify the arithmetic, take this area to be one square centimeter. Then, from Equation (4.1), the basic equation governing the dynamics of membrane voltage is

$$\frac{dV}{dt} = \frac{I_0(t) - J_{ion}}{C}, \tag{4.6}$$

where $I_0(t)$ is a current injected into the axon by the experimenter and J_{ion} is given in Equation (4.5). A short pulse of this injected current will

---

[3]Such diversity may surprise physical scientists, who usually deal with *homogeneous* classes of objects, in which the members are *identical* (such as electrons, protons, oxygen atoms, benzene molecules, and so on), but the science of biology aims to understand *heterogeneous* classes of living organisms, in which the members are *similar* but not identical. See Walter Elsasser's *Reflections on a Theory of Organisms* [11] for a discussion of this distinction, which we will revisit in Chapters 10 and 12.

[4]This value for V_L is slightly below the value of 10.613 mV given by Hodgkin and Huxley [19].

charge the membrane capacitance, which raises the transmembrane voltage from its resting value ($V = 0$) to a threshold level that allows sodium ion current to flow into the axon and initiate a *space-clamped action potential*.

It turns out that sodium turn-on (mediated by m) is about an order of magnitude faster than potassium turn-on and sodium turn-off (mediated by n and h, respectively). Using the standard values for the H–H parameters from Table 4.1 and referring to the equations for $m(V,t)$, $h(V,t)$, and $n(V,t)$ given in Appendix B, the subsequent dynamics of Equation (4.6) can be described as follows.

(1) At the resting potential ($V = 0$), the sodium conductance is almost zero. This is because (see Appendix B) $m_0(0) = 0.053$ and $h_0(0) = 0.596$ so sodium ion permeability

$$G_{Na}m_0^3(0)h_0(0) = 0.000089\, G_{Na}\,.$$

The potassium ion current is also small at rest because $n_0(0) = 0.318$ so potassium ion permeability

$$G_K n_0^4(0) = 0.0102\, G_K\,.$$

(2) As the membrane voltage is increased from its resting value (by a short pulse of injected current, $I_0(t)$), sodium channels open ($m \to 1$) on a time scale of $\tau_m \sim 0.2$ ms.

(3) This influx of sodium ions brings the membrane voltage to a level approaching $+115$ mV with respect to its resting value. Then $(V - V_{Na}) \approx 0$, implying from Equation (4.2) that the sodium ion current becomes small.

(4) At this voltage, potassium ion permeability turns on ($n \to 1$) as sodium ion permeability turns off ($h \to 0$) on time scales of a few milliseconds.

(5) Because $(V - V_K) \sim +100$ mV, potassium ions flow rapidly out of the axon and carry the membrane voltage back to its resting value on a time scale of a few milliseconds.

To describe these dynamics quantitatively, one can integrate Equation (4.6) together with

$$\frac{dm}{dt} = -\frac{m - m_0(V)}{\tau_m(V)}\,,$$
$$\frac{dh}{dt} = -\frac{h - h_0(V)}{\tau_h(V)}\,, \qquad (4.7)$$
$$\frac{dn}{dt} = -\frac{n - n_0(V)}{\tau_n(V)}\,,$$

from Appendix B.

Hodgkin and Huxley [19] carried through a number of such membrane-switching calculations for a variety of experimental conditions, which demonstrated that their formulation is in quantitative agreement with measurements of $V(t)$ on real squid membranes.

Although the initial H–H computations were laborious in 1950, presently available computers make them convenient. The numerically inclined are referred to Chapter 9 of Hugh Wilson's book *Spikes, Decisions, and Actions,* which comes with a collection of MATLAB codes [39]. These easily used codes are integrated into the text and encourage the reader to study many features of Equation (4.6), including membrane switching, periodic membrane dynamics (bursting) with constant I_0, dependence of bursting frequency on I_0 and various sorts of ionic current, stochastic resonance under the influence of random noise, and subharmonic resonance.

Before leaving the subject of transmembrane ionic dynamics, three items should be underscored.

- Introductory discussions of nerve membrane dynamics sometimes leave the impression that sodium current ceases to flow at the peak of an action potential because "sodium ions rush in," altering the concentration ratio. Straightforward calculations for the Hodgkin–Huxley system, however, show that ionic ratios change by a negligible amount during the passage of a single nerve impulse. In other words, the sodium and potassium batteries are hefty enough to conduct many impulses before needing to be recharged by the sodium–potassium pump. Sodium current ceases to flow at the peak of the switching cycle because sodium diffusion current becomes canceled by oppositely directed conduction current when V is about equal to V_{Na}.

- The equations obtained by Hodgkin and Huxley describe ionic currents that are averaged over many thousands of channels. It is now possible to observe the currents flowing through individual protein channels using the *patch clamp* technique, in which a single channel is fixed on the end of a glass microelectrode.[5] Such measurements indicate that individual channels switch from being fully closed to being fully open, carrying currents on the order of a picoampere for times on the order of a millisecond [6, 37]. Thus, the formulation of Hodgkin and Huxley describes the average probabilities of individual membrane channels being open as a function of time, and it is the temporal variations of these probabilities that are described by the dynamics of m, h, and n in Equations (4.7).

- In neurons of the mammalian central nervous systems, there are currently known to be several dozen types of ionic channels, exhibiting a variety of dynamic behaviors [26]. Although two ionic species (sodium activating and potassium deactivating) may be satisfactory for qualitative studies of the human brain, quantitative analyses often require

---

[5]Electrophysiologists seem fascinated by the term "clamp."

formulations for the ionic current that include more terms than in Equation (4.5).

This last caveat is especially apt for mammalian cell bodies and dendrites, where calcium ions can cause a depolarization of the membrane that results in several types of action potentials. Because it is more difficult to make space- and voltage-clamped measurements on neocortical dendrites than on the giant axon of the squid, the models of these calcium currents are often less reliable than the H–H equations. Present formulations tend to follow the H–H paradigm of Equation (4.4), but the internal calcium ion concentrations can be quite small (tens of nanomoles), which brings intracellular calcium diffusion into the picture [4, 22, 24]. Interactions between sodium and calcium activation currents often lead to complex bursting behavior at the cell body, which the reader can explore numerically using MATLAB codes from Wilson's book [39].

4.4 The "Cable Equation"

At this point, we know how a patch of space-clamped nerve switches; thus, we are ready to consider how this local switching activity propagates along a nerve that is not space-clamped. The relevant parameters in impulse propagation are the following.

- r is the longitudinal resistance per unit length of the fiber, which is usually measured in units of ohms per centimeter. For a cylindrical fiber of radius a, $r = \rho/\pi a^2$, where ρ is the specific resistivity of the cytoplasm in ohm-centimeters.

- c is the membrane capacitance per unit length of the fiber and is measured in units of farads per centimeter. For a cylindrical fiber of radius a, $c = 2\pi a C$, where C is the capacitance per unit area of the membrane in farads per square centimeter.

- j_{ion} is the ionic current flowing across the membrane (from inside to outside) per unit length of the fiber and is measured in units of amperes per centimeter. For a cylindrical fiber of radius a, $j_{\text{ion}} = 2\pi a J_{\text{ion}}$, where J_{ion} is the transmembrane current per unit area, which is shown in Figure 4.1 and given in Equation (4.5).

How are these parameters to be introduced into a theory of nerve impulse propagation? To answer this question, consider the differential circuit diagram of a nerve fiber displayed in Figure 4.3(b). To first order in Δx, Ohm's law implies that

$$V(x,t) - V(x + \Delta x, t) = i(x + \Delta x/2, t)r\Delta x, \qquad (4.8)$$

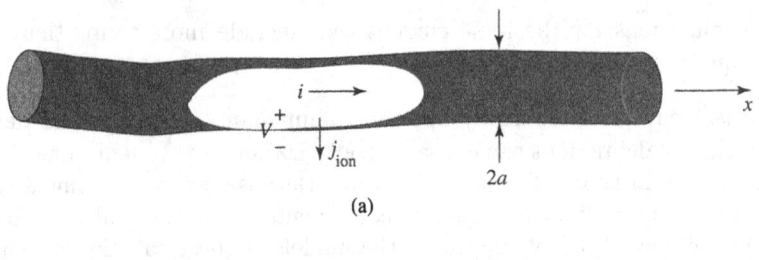

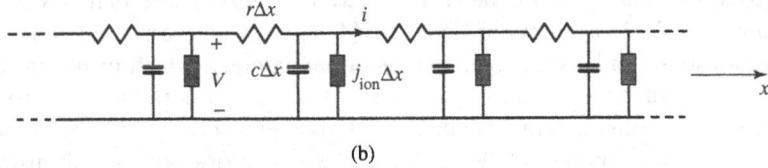

Figure 4.3. (a) Sketch of a squid axon. (b) A corresponding differential circuit diagram that can be used to derive the cable equation for impulse propagation.

where i is the longitudinal (x-directed) current flowing through the nerve. From conservation of electric charge, we also know that to first order in Δx

$$i(x, t) - i(x + \Delta x, t) = \left(c\frac{dV(x + \Delta x/2, t)}{dt} + j_{\text{ion}}(x + \Delta x/2, t) \right) \Delta x. \quad (4.9)$$

Combining these two equations to eliminate i and taking the limit as $\Delta x \to 0$ yields the following *nonlinear diffusion equation*:

$$\frac{1}{rc}\frac{\partial^2 V}{\partial x^2} - \frac{\partial V}{\partial t} = \frac{j_{\text{ion}}}{c}. \quad (4.10)$$

Motivated by familiarity with a related partial differential equation that arose in the analysis of telegraph lines, Equation (4.10) is often called the "cable equation" by electrophysiologists, but this name is misleading. Propagation of dits and dahs over a telegraph line is a linear electromagnetic phenomenon, whereas Equation (4.10) represents *nonlinear* electrostatic diffusion.[6]

From the perspectives of modern nonlinear science, Equation (4.10) is a *nonlinear field equation* out of which emerges an elementary particle of neural activity: the nerve impulse [35]. It is nonlinear because of the nonlinear dependencies of j_{ion} on m, h, and n, which in turn depend nonlinearly on V.

Let us now analyze the cable equation to understand how a nerve impulse emerges from the mathematical structure that we have developed.

---

[6] Using Maxwell's equations, one can take magnetic effects into account in the derivation of Equation (4.10), but the error involved in neglecting this correction is about one part in 10^8 [33].

4.5 Traveling-Wave Solutions of the Hodgkin–Huxley Equations

A solution of Equation (4.10) that represents a nerve impulse is a *traveling wave* for which

$$V(x, t) = \tilde{V}(x - vt).$$

For a traveling wave, the general dependencies on both x and t are constrained by the traveling-wave variable

$$\xi \equiv x - vt,$$

where v is the propagation velocity of the traveling wave.

This traveling-wave assumption implies that the partial derivatives with respect to x and t in Equation (4.10) are related in the following manner:

$$\frac{\partial V}{\partial x} = \frac{d\tilde{V}}{d\xi}\frac{\partial \xi}{\partial x} = \frac{d\tilde{V}}{d\xi}$$

and

$$\frac{\partial V}{\partial t} = \frac{d\tilde{V}}{d\xi}\frac{\partial \xi}{\partial t} = -v\frac{d\tilde{V}}{d\xi}.$$

Thus, Equation (4.10)—a *partial* differential equation (PDE)—is reduced under the traveling-wave assumption to the *ordinary* differential equation (ODE)

$$\frac{1}{rc}\frac{d^2\tilde{V}}{d\xi^2} + v\frac{d\tilde{V}}{d\xi} = \frac{j_{\text{ion}}}{c},\tag{4.11}$$

but there is no free lunch. Because it is an ODE rather than a PDE, Equation (4.11) is easier to solve than Equation (4.10), but it contains less information. We can see this in two ways.

First, an electronic computer could nowadays be employed to integrate Equation (4.10) numerically with a stable traveling-wave solution and its corresponding velocity v emerging from appropriate initial conditions. In Equation (4.11), on the other hand, v is an undetermined parameter. Second, the existence of a numerical solution for Equation (4.10) demonstrates the stability of that solution, whereas Equation (4.11) tells us nothing about the stability of any traveling waves that we might use it to discover.

The existence of a particular traveling-wave velocity can be seen from the energetics of nerve impulse propagation, which were introduced in Chapter 1. Thus, an H–H nerve impulse can be viewed as a coherent process represented by the closed causal loop (or positive feedback diagram)

Release of energy

↓ ↑

Dissipation of energy

with each component supporting (or being the cause of) the other.

Propagation at a fixed speed implies that the rate at which energy is dissipated must equal the rate at which energy is released by the moving impulse. The released energy is provided by the "ionic batteries" in Equation (4.5), with dissipation stemming from the ohmic losses of circulating ion currents.[7]

In other words, in a nerve axon, the energy released by a moving impulse is primarily the electrical field energy that is stored in the membrane capacitance, and the corresponding dissipation stems from ohmic losses of internal and external ionic current flows. We will see in the following chapters that this concept of "power balance" provides insights into the dynamics of many simple nerve models.

4.5.1 Phase-Space Analysis

Constrained by the lack of a digital computer in the early 1950s, Hodgkin and Huxley had no choice but to study Equation (4.11). In fact, the most advanced numerical tool they had was a mechanical adding machine. How then did they proceed?

Written as a first order ODE system, Equation (4.11) involves five dependent variables: V, $W \equiv dV/d\xi$, m, h, and n. From Equation (4.7), this system is

$$
\begin{aligned}
\frac{dV}{d\xi} &\equiv W \,, \\[2mm]
\frac{dW}{d\xi} &= r j_{\text{ion}}(V, m, h, n) - rcvW \,, \\[2mm]
\frac{dm}{d\xi} &= \frac{m - m_0(V)}{v\tau_m(V)} \,, \\[2mm]
\frac{dh}{d\xi} &= \frac{h - h_0(V)}{v\tau_h(V)} \,, \\[2mm]
\frac{dn}{d\xi} &= \frac{n - n_0(V)}{v\tau_n(V)} \,,
\end{aligned}
\qquad (4.12)
$$

a set of equations that determine the course of a solution trajectory in a *phase space* of five dimensions that represents the dependent variables V, W, m, h, and n. The right-hand sides of these equations fix the components

---

[7]Requiring a traveling-wave solution to have a speed fixed by the parameters is a strong constraint on dissipative nonlinear wave systems that does not hold for energy-conserving systems. In *soliton* systems, for example, the speed of a traveling wave is determined by the initial conditions from a continuum of allowed values [35]. Thus a nerve impulse is *not* a soliton.

of a *rate vector*

$$\left(\frac{dV}{d\xi}, \frac{dW}{d\xi}, \frac{dm}{d\xi}, \frac{dh}{d\xi}, \frac{dn}{d\xi} \right)$$

telling how solution trajectories move as functions of the independent variable ξ in the five-dimensional space

$$(V, W, m, h, n).$$

In other words, at each point in the phase space, we could draw a little arrow with components

$$\left(\frac{dV}{d\xi}\Delta\xi, \frac{dW}{d\xi}\Delta\xi, \frac{dm}{d\xi}\Delta\xi, \frac{dh}{d\xi}\Delta\xi, \frac{dn}{d\xi}\Delta\xi \right),$$

which would show where a phase point at (V, W, m, h, n) would move as ξ increases to $\xi + \Delta\xi$.[8]

Equations (4.12) are said to be *autonomous* because the functions on the right-hand sides do not depend on the independent variable ξ. Thus, the lengths and directions of our little arrows are independent of ξ throughout the phase space, providing simplifications for visualizing a solution and finding it.

Understanding solution trajectories of an autonomous system of first-order ODEs is facilitated by first learning where in the phase space the rate vector is equal to zero. Such loci are called *singular points* (SPs), and for Equations (4.12) there is only one SP at

$$(V, W, m, h, n) = (0, 0, 0.05293, 0.59612, 0.31768),$$

which is indicated in Figure 4.4.

Sufficiently near the singular point, Equations (4.12) are linear because the components of the rate vector are very close to zero. Analysis of this linear system reveals trajectories that are directed toward the SP as $\xi \to +\infty$ and also trajectories that are moving away from the SP for large negative values of ξ.

From all of these possibilities, we must select a trajectory corresponding to a nerve impulse, for which $V \to 0$ as $\xi \to \pm\infty$. This is evidently a trajectory that leaves the SP at $\xi = -\infty$ and returns to it at $\xi = +\infty$.

---

[8]These considerations provide a basis for integrating Equations (4.12). Thus starting with a phase point at $(V(\xi), W(\xi), m(\xi), h(\xi), n(\xi))$, one could add

$$\left(\frac{dV}{d\xi}\Delta\xi, \frac{dW}{d\xi}\Delta\xi, \frac{dm}{d\xi}\Delta\xi, \frac{dh}{d\xi}\Delta\xi, \frac{dn}{d\xi}\Delta\xi \right)$$

to get the position of the phase point at $(V(\xi + \Delta\xi), W(\xi + \Delta\xi), m(\xi + \Delta\xi), h(\xi + \Delta\xi), n(\xi + \Delta\xi))$. Continuing this process will determine the trajectory as a function of ξ. This is essentially what Hodgkin and Huxley did many times with a mechanical calculator.

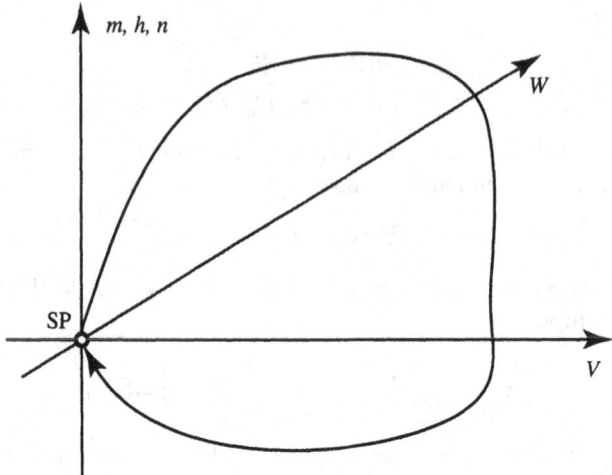

Figure 4.4. Schematic representation of traveling-wave trajectories in the five-dimensional phase space of V, $W \equiv dV/d\xi$, m, h, and n. The assumed traveling-wave speed (v) must be adjusted so that a wave originating from the singular point (SP) as $\xi \to -\infty$ approaches it again as $\xi \to +\infty$.

Such a trajectory that starts at an SP and returns to the same one is called *homoclinic* as opposed to a *heteroclinic* trajectory, which starts at one SP and terminates at another.

In general, of course, a trajectory that leaves an SP would not be expected to return—it is much more likely to cruise off to the far reaches of phase space—but we have one more tool at our disposal. This is the undetermined traveling-wave speed v, which is an adjustable parameter in Equations (4.12). In other words, v can be selected to establish a homoclinic trajectory in the phase space.

Called the "shooting method," the numerical procedure used by Hodgkin and Huxley was as follows. (1) Select a value for v and begin integrating Equations (4.12) along a trajectory that leaves the SP at $\xi = -\infty$. (2) Note how this trajectory moves around the phase space, eventually returning to the vicinity of the SP. (3) Record the smallest distance between the trajectory and the SP. (4) Change the selection of v so that this smallest distance is reduced. (5) Repeat the process until $V(\xi)$ and v are sufficiently well-determined.

4.5.2 Numerical Results

The moment of truth has arrived. We are now ready to consider whether the numerical procedure just described leads to a traveling-wave solution corresponding to experimental observations.

Table 4.2. Standard Hodgkin–Huxley parameters for the giant axon of the squid.

Parameter	Value	Units
ρ	35.4	ohm-cm
a	238	μm
r	2.0×10^4	ohms/cm
c	1.5×10^{-7}	F/cm

In attempting to answer this question, Hodgkin and Huxley were faced with two problems. First, to calculate the traveling-wave solution on a single nerve, it was necessary to carry through many tedious repetitions of the shooting method without the benefit of an electronic computer. Also the parameter values for real nerves vary over a rather wide range, as is seen in the third column of Table 4.1.

To limit the number of parameters considered—and thereby the number of integrations on their mechanical calculator—they selected one axon with an internal (axoplasmic) resistivity (ρ) of 35.4 ohm-cm, a radius (a) of 238 μm, and the "standard" membrane values indicated in Table 4.1 as a candidate for detailed numerical analysis. For this specific axon, the relevant parameters for the nonlinear field equation are given in Table 4.2. Because the action potential on this nerve was measured at a temperature of 18.5°C, the numerical calculations of the impulse were also performed at this temperature.

From a comparison of Figures 1.1 and 4.5, the shape of the traveling-wave solution calculated for this nerve was found to be in good qualitative agreement with experimental observations. Quantitatively, the calculated impulse velocity was 18.8 m/s, whereas the measured speed was 21.2 m/s: again in substantial agreement. Like a present-day celebrity who is "famous for being well known," therefore, this particular nerve has become widely recognized and studied as the *standard Hodgkin–Huxley axon* defined by the parameters in Tables 4.1 and 4.2.

Although it was impractical for Hodgkin and Huxley to carry out the necessary calculations in the early 1950s, there are in fact *two* homoclinic solutions of Equations (4.12) at two different values for the traveling-wave speed, and the corresponding solutions of the PDE system are shown in Figure 4.5. In this figure, the higher-amplitude solution (at $v = 18.8$ m/s) corresponds to the impulse displayed in Figure 1.1 of Chapter 1. This traveling-wave solution is *stable* in the sense that deviations from it relax to zero with increasing time as solutions of the full PDE given by Equation (4.10). (See Sections 5.4 and 6.5.2 and Appendix D for discussions of nerve impulse stability criteria.)

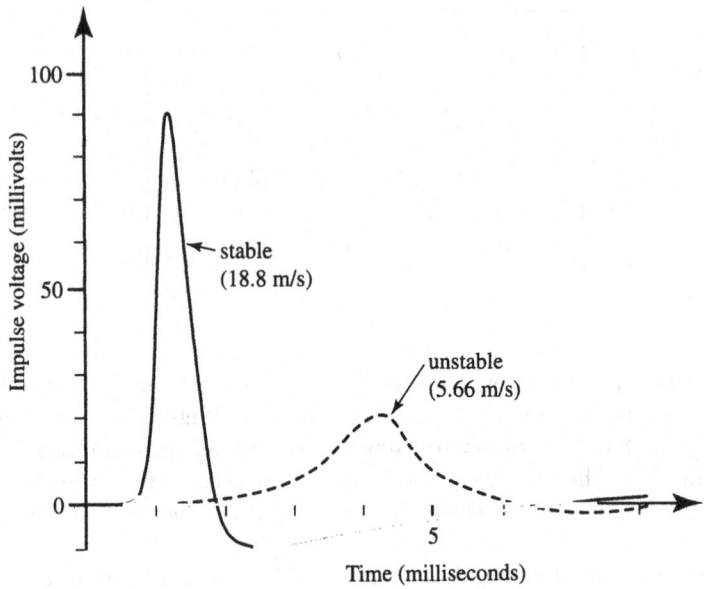

Figure 4.5. A full-sized spike (at $v = 18.8$ m/s) and an unstable threshold impulse (5.66 m/s) for the Hodgkin–Huxley axon at 18.5°C. (Redrawn from Huxley [21].)

The smaller-amplitude traveling wave solution, with a speed of 5.66 m/s, was found by Huxley in 1959 using an electronic computing machine [21]. This solution is *unstable* in the sense that deviations from it diverge with increasing time as solutions of the full PDE given by Equation (4.10). Slightly smaller solutions decay to zero, and slightly larger solutions grow to become the fully developed nerve impulse; thus, this unstable solution defines *threshold conditions* for igniting an impulse.

In the language of modern nonlinear theory, the stable traveling wave of greater amplitude can be viewed as an *attractor* in the solution space of the PDE system of Equation (4.10); thus, solutions lying within a *basin of attraction* converge to the attractor as $t \to +\infty$. The lower-amplitude unstable solution, on the other hand, lies on a *separatrix* dividing an impulse's basin of attraction from that of the null solution.

4.6   Degradation of a Squid Nerve Impulse

By the middle of the 1960s, electronic computing machines had developed to a level where the original Hodgkin–Huxley calculations were fairly

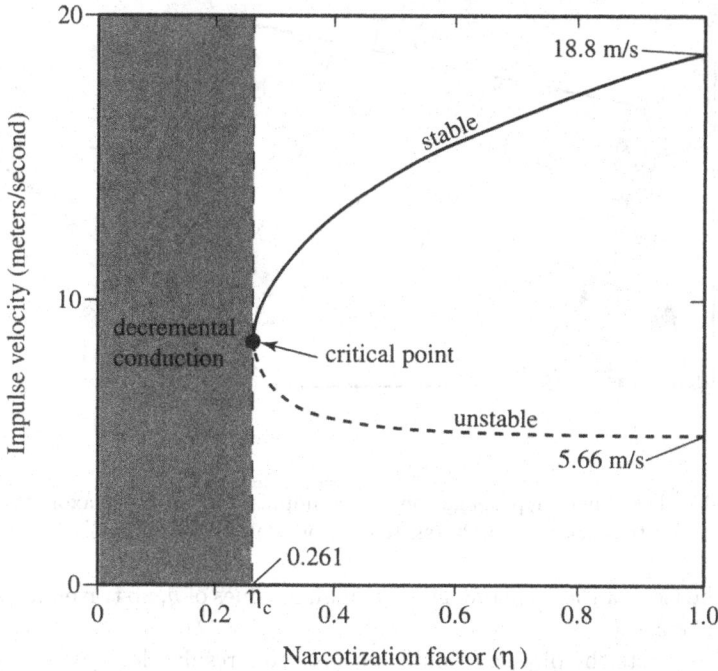

Figure 4.6. Power balance loci showing impulse speeds for the Hodgkin–Huxley equations as a function of a "narcotization factor" (η). (Redrawn from Cooley and Dodge [9].)

straightforward, and additional results began to emerge from numerical studies of their formulation. Among these new investigations, Cooley and Dodge looked at the effects of "narcotizing" a nerve fiber, thereby broadening understanding of how nerves work [9].

To represent the effects of a narcotic agent on a nerve, the standard H–H equations were altered by reducing the values of the maximum sodium and potassium conductances (given in Table 4.1) by a unitless *narcotization factor*

$$\eta < 1 \, .$$

Thus,

$$G_{\mathrm{Na}} \to \eta G_{\mathrm{Na}} \quad \text{and} \quad G_{\mathrm{K}} \to \eta G_{\mathrm{K}}$$

are the values used in numerical computations.

As is seen from Figure 4.6, one qualitative effect of this type of narcotization is to reduce the speed of the larger-amplitude stable traveling-wave solution shown in Figure 4.5. Conversely, the speed of the smaller-amplitude unstable traveling-wave solution is increased. Eventually, a *critical point* is reached at

$$\eta_c = 0.261 \, ,$$

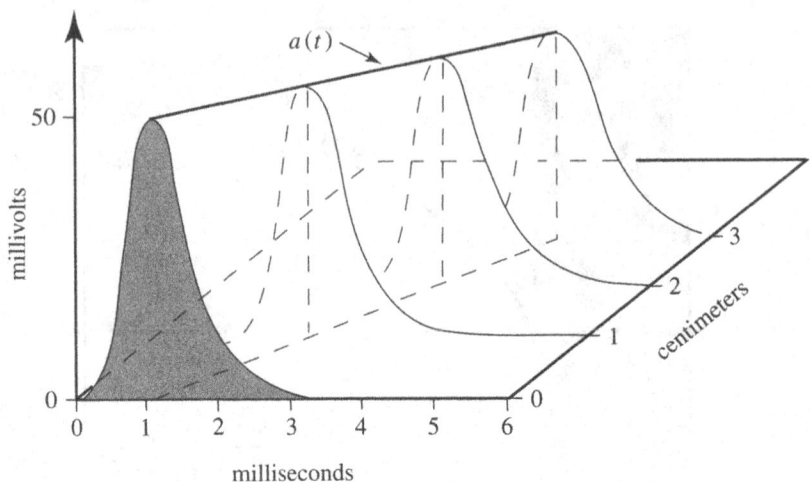

Figure 4.7. Decremental propagation of an impulse on an H–H axon that is narcotized by the factor $\eta = 0.25$ (sketched from data in [9] and [25]).

where the two solutions merge. For yet smaller values of η, no traveling-wave solutions exist.

To appreciate the physical significance of these results, look at the (v, η) parameter plane of Figure 4.6. The curve plotted in this plane shows the loci of parameters where a balance is established between the rate at which energy is generated by the ionic batteries in Equation (4.5) and the rate at which it is dissipated by the ionic currents associated with a nerve impulse.

The upper curve indicates stable traveling-wave solutions, implying that a small change of an impulse solution (either positive or negative) will relax back to zero and restore the original wave. The lower curve indicates unstable traveling waves, implying two different effects. An increase in amplitude of the solution will grow (because energy generation is greater than dissipation) until the total solution reaches the stable solution of the upper curve. If its amplitude is decreased, on the other hand, the impulse will decay (because energy generation is less than dissipation) until it falls to zero. These numerical results provide an explanation for the *all-or-nothing* property of a nerve impulse noted by Adrian in 1914 [2].

Although the concept of all-or-nothing propagation holds for $\eta > \eta_c$, its logical basis evaporates for $\eta < \eta_c$. In this regime, however, one can find *decremental* propagation of a nerve impulse, as is sketched in Figure 4.7 [9, 25]. For such a decremental impulse, the rate at which energy is generated is only slightly less than the rate of dissipation, so the solution relaxes rather slowly to zero. As has been emphasized by Lorente de Nó and Condouris [27], this phenomenon was long overlooked by electrophysiologists who had concentrated their attentions on the properties of standard nerves.

These qualitative conclusions stemming from the computations of Cooley and Dodge are quite general, applying to several other experimental situations in which the ability of a nerve to conduct impulses is degraded. Thus, results resembling those displayed in Figure 4.6 are also observed in the following cases.

- Changes in the internal or external concentrations of sodium or potassium ions can alter the robustness, or *safety factor*, of an impulse. As a specific example of this phenomenon, Adelman and FitzHugh have augmented the Hodgkin–Huxley formulation to study buildup of external potassium ion concentration in a localized "periaxonal space" (30–40 nm across) surrounding a squid giant axon [1]. Widely observed in measurements on squid nerves, this increase of potassium concentration is caused by repeated stimulation of the fiber, which alters the equilibrium potential for potassium ions (V_K in Table 4.1) on time scales of the order of 100 ms. We will meet this phenomenon again in Chapter 9.

- Higher temperatures increase the temporal rates of membrane conductance change, shortening both the sodium turn-on time (τ_m) and the potassium turn-on time (τ_n). Near 18.5°C, decreasing τ_m causes the squid impulse velocity to increase, but around 30°C the decrease in τ_n dominates, causing the impulse speed to decrease and degrading the ability of a nerve to conduct impulses [7, 21].

- Increasing the leakage conductance (G_L) degrades the performance of a nerve. If the H–H value of 0.3 mmho/cm^2 (given in Table 4.1) is increased beyond a critical value of about 8.6 mmho/cm^2, for example, only decremental propagation occurs [7].

- Abrupt changes in the cross-sectional area of the nerve can lead to a failure of impulse propagation [23]. As we will see in Chapter 9, this phenomenon has implications for the possibilities for information processing at the axonal or dendritic branchings of a neuron.

- The propagation of a periodic train of impulses becomes degraded as the interval between individual impulses becomes less than a certain value. This effect occurs because each impulse is then propagating through a *refractory zone* in the wake of the preceding impulse.

4.7 Refractory and Enhancement Zones

The difficulties encountered by a nerve impulse that follows too closely on the heels of another have been studied empirically through *double impulse* measurements [10, 36]. In such experiments, pairs of impulses are launched on single fibers, and the ratio of impulse speeds is measured as a function of the impulse interval (T).

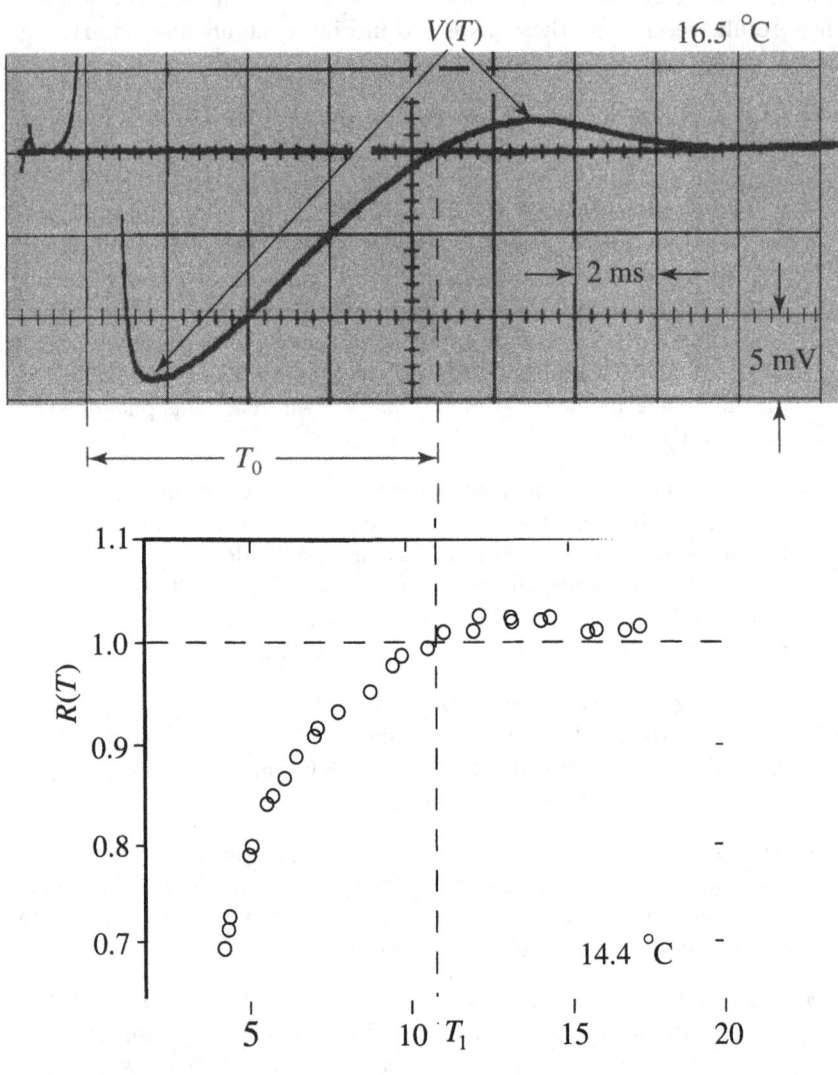

Figure 4.8. Two data sets related to the refractory zone in the wake of a propagating impulse. UPPER: An oscilloscope photograph showing the oscillatory tail of a single action potential $V(t)$. LOWER: Data from a double impulse experiment, where T is the time interval between impulses and $R(T)$ is the ratio of the speeds of the two impulses defined in Equation (4.13). (Note that the crossover points in the two data sets are aligned and the time scales have been adjusted for different temperatures.)

The lower part of Figure 4.8 shows the results of one series of measurements of the ratio

$$R(T) \equiv \frac{\text{speed of second impulse}}{\text{speed of first impulse}} \tag{4.13}$$

for two impulses propagating on a particular giant axon of the squid (*Lologo vulgaris*) collected by the Stazione Zoologica from the Bay of Naples. A significant feature of these data is the impulse interval $T = T_1$ at which $R(T) = 1$.

Expressed in terms of T_1, there are qualitative differences among four regions of the impulse interval.

- *Absolute refractory zone:* For impulse intervals less than

$$T < 0.4 T_1,$$

  it is impossible for the second impulse to propagate; thus, $R(T) = 0$.

- *Relative refractory zone:* For impulse intervals in the range

$$0.4 T_1 < T < T_1,$$

  the second impulse is able to propagate, albeit with diminished speed.

- *Enhancement zone:* In the range

$$T_1 < T < 1.8 T_1,$$

  the second impulse is observed to travel *faster* than the first impulse.

- *Decoupled zone:* If the impulse interval is greater than about $1.8 T_1$, both impulses travel at the same speed, with the second impulse uninfluenced by the first.

The critical time interval T_1 increases as the temperature is lowered, as implied by the Hodgkin–Huxley equations given in Appendix B. From Equation (B.3), the rate constants (τ_m^{-1}, τ_h^{-1}, and τ_n^{-1}) for the sodium turn-on and turn-off variables (m and h) and for the potassium turn-on variable (n) are to be multiplied by the factor

$$\kappa = 3^{(\text{Temp}-6.3)/10} \tag{4.14}$$

if the temperature ("Temp") differs from 6.3°C. Assuming that the timings of refractory effects stem from the dynamics of h and n on the trailing edge of the first impulse, Equation (4.14) implies $T_1 \propto 3^{-\text{Temp}/10}$ in the lower part of Figure 4.8. Analysis of data from 13 freshly prepared squid axons at temperatures between 14.4°C and 20.4°C gives the relationship [36]

$$T_1 = 58.2 \times 3^{-\text{Temp}/10} \text{ ms} \pm 13\%, \tag{4.15}$$

where the percent error indicates rms deviation from the mean.

Although absolute and relative refractory zones are well-established in neuroscience lore, the enhancement zone is less widely recognized. Observed in the frog (*Rana pipiens*) sciatic nerve by Graham as early as 1934 [15] and confirmed by Bullock in giant fibers of the earthworm (*Lumbricus terrestris*) in 1951 [5], enhancement of the second impulse speed can be understood with reference to Figure 4.8.

On the upper part of this figure is reproduced an oscilloscope photograph of the trailing edge of a single squid impulse into which a second impulse would attempt to propagate. On this photograph, T_0 is the time between the point of maximum slope on the leading edge of $V(t)$ (occurring at about 50 mV, this is off the scale of the figure) and the point on the trailing edge where $V(t)$ crosses over from being hyperpolarizing (more negative than the resting value) to depolarizing.

Both T_0 and T_1 have been measured on the above-mentioned cohort of 13 squid axons [36], showing that

$$T_0 \approx T_1$$

to within an experimental error of about ±15%.[9]

To facilitate the reader's comprehension of the data displayed in Figure 4.8, the upper part of the figure has the crossover point between hyperpolarizing and depolarizing voltages aligned with $T = T_1$ on the lower part of the figure. Also because the upper part of the figure is at 16.5°C whereas the lower part was measured at 14.4°C, the time scale on the lower part is contracted by the factor $3^{(14.4-16.5)/10} = 0.794$.[10]

Evidently, the qualitative behaviors of these two measurements are similar in the neighborhood of the crossover points, which suggests that the enhancement zone stems from the depolarizing phase on the trailing edge of the first impulse. In other words, if the leading edge of the second impulse is within the depolarized region in the wake of the first impulse, the second impulse propagates faster than it would on an undisturbed axon, leading to the experimental observation that $R > 1$.

Why is there a depolarizing phase in the wake of the first impulse? Because squid membrane near its resting level is *oscillatory*.

Discovered by Cole and Baker in 1941 [8] and noted by Hodgkin and Huxley in 1952 [19], subthreshold membrane oscillations were investigated

---

[9]Comparing these measurements with Hodgkin–Huxley calculations reveals two disagreements. (1) Under H–H, the time interval

$$T_1 = 36.6 \times 3^{-\text{Temp}/10} \text{ ms},$$

which lies below the limits of Equation (4.15) [29]. (2) Over the temperature range between 14 and 18.5°C, the H–H model indicates that T_1 is about 1.3 ms less than T_0 [31]. George and Silberstein [13] have suggested that periaxonal buildup of potassium ion concentration [1] in the wake of the first impulse can decrease the speed of the second, which tends to increase T_1.

[10]Although the two data sets in Figure 4.8 were taken on different axons, the shape of $V(t)$ in the upper part of the figure is in agreement with detailed measurements of $V(t)$ for the axon of the lower part, from which $R(T)$ was also calculated [36].

numerically by Sabah and Liebovic in 1969 [32] and both experimentally and numerically by Mauro and his colleagues in 1970 [28]. These studies have shown that a patch of squid membrane below threshold has a damped resonance ($Q \sim 3$) at around 50–100 Hz, depending on the temperature [28].

From the perspective of an electrical engineer, a subthreshold patch of membrane looks like a "parallel GLC circuit," where C is the membrane capacitance, G is the subthreshold membrane conductance, and L is a *phenomenological inductance* generated by time delays associated with the dynamics of h and n, the sodium turn-off and potassium turn-on variables. In considering this membrane resonance, it is important to remember that the phenomenological inductance has nothing to do with storage of magnetic field energy, which is negligible in neural dynamics [33].

4.8 Recapitulation

Fundamental for experiments on nerve membrane dynamics are the techniques of *space clamping* (to remove space variations from membrane measurements) and *voltage clamping* (for holding the transmembrane voltage at a preassigned value). Based on these techniques, the Hodgkin–Huxley (H–H) formulation of ionic dynamics was presented, and the related phenomenon of squid axon membrane switching was discussed.

The "cable equation" for propagation along a nerve fiber was then derived and used to describe traveling-wave impulses on the H–H model of a squid giant axon. In the language of nonlinear science, the ODE system for traveling-wave propagation can be viewed in a *phase space* of five dimensions, wherein a nerve impulse is represented by a *homoclinic* trajectory that begins and ends at the same singular point. In the context of the PDE system governing the full dynamics of the nerve, a nerve impulse is an *attractor* with a *basin of attraction*. H–H calculations of impulse shape and speed are shown to be in good agreement with measurements of impulse propagation on squid axons.

Finally, various experimental and physiological influences that degrade the robustness of an impulse were considered, refractory and enhancement zones in the wake of a propagating impulse were described, and the subthreshold resonance of a nerve membrane was noted.

References

[1] WJ Adelman, Jr and R FitzHugh, Solutions of the Hodgkin–Huxley equations modified for potassium accumulation in a periaxonal space, *Fed. Proc. (Fed. Am. Soc. Exp. Biol.)* 34 (1975) 1322–1329.

[2] ED Adrian, The all-or-none principle in nerve, *J. Physiol. (London)* 47 (1914) 460–474.

[3] JM Arnold, WC Summers, DL Gilbert, RS Manalis, NW Daw, and RJ Lasek, *A Guide to the Laboratory Use of the Squid* Loligo pealei, Marine Biological Laboratory, Woods Hole, MA, 1974.

[4] OV Aslanidi, OA Mornev, O Skyggebjerg, P Arkhammer, O Thastrup, MP Sørensen, PL Christiansen, K Conradsen, and AC Scott, Excitation wave propagation as a possible mechanism for signal transmission in pancreatic islets of Langerhans, *Biophys. J.* 80 (2001) 1195–1209.

[5] TH Bullock, Facilitation of conduction rate in nerve fibers, *J. Physiol. (London)* 114 (1951) 89–97.

[6] WA Catterall, Structure and function of voltage-sensitive ion channels, *Science* 242 (1988) 50–61.

[7] KS Cole, *Membranes, Ions and Impulses,* University of California Press, Berkeley, 1968.

[8] KS Cole and RF Baker, Longitudinal impedance of the squid giant axon, *J. Gen. Physiol.* 24 (1941) 771–788.

[9] JW Cooley and FA Dodge, Digital computer solutions for excitation and propagation of the nerve impulse, *Biophys. J.* 6 (1966) 583–599.

[10] F Donati and H Kunov, A model for studying velocity variations in unmyelinated axons, *IEEE Trans. on Biomed. Eng.* BME-23 (1976) 23–28.

[11] WM Elsasser, *Reflections on a Theory of Organisms: Holism in Biology,* The Johns Hopkins University Press, Baltimore, 1998 (first published in 1987).

[12] R FitzHugh, A kinetic model of the conductance changes in nerve membrane, *J. Cell. Comp. Physiol.* 66 (1965) 111–117.

[13] SA George and PT Silberstein, Conduction velocity after effects of spike activity, *Neurosci. Abstr.* 3 (1977) 217.

[14] SS Goldstein and W Rall, Changes of action potential shape and velocity for changing core conductor geometry, *Biophys. J.* 14 (1974) 596–607.

[15] HT Graham, Supernormality, a modification of the recovery process in nerve, *Am. J. Physiol.* 110 (1934) 225–242.

[16] AL Hodgkin and AF Huxley, Currents carried by sodium and potassium ions through the membrane of the giant axon of *Loligo, J. Physiol. (London)* 116 (1952) 449–472.

[17] AL Hodgkin and AF Huxley, The components of membrane conductance in the giant axon of *Loligo, J. Physiol. (London)* 116 (1952) 473–496.

[18] AL Hodgkin and AF Huxley, The dual effect of membrane potential on sodium conductance in the giant axon of *Loligo, J. Physiol. (London)* 116 (1952) 497–506.

[19] AL Hodgkin and AF Huxley, A quantitative description of membrane current and its application to conduction and excitation in nerve, *J. Physiol. (London)* 117 (1952) 500–544.

[20] RC Hoyt, The squid giant axon: Mathematical models, *Biophys. J.* 3 (1963) 399–431.

[21] AF Huxley, Can a nerve propagate a subthreshold disturbance? *J. Physiol. (London)* 148 (1959) 80P–81P.

[22] J Keener and J Sneyd, *Mathematical Physiology*, Springer-Verlag, New York, 1998.

[23] BI Khodorov, *The Problem of Excitability*, Plenum Press, New York, 1974.

[24] C Koch, *Biophysics of Computation: Information Processing in Single Neurons*, Oxford University Press, New York, 1999.

[25] KN Leibovic, *Nervous System Theory: An Introductory Study*, Academic Press, New York, 1972.

[26] RR Llinás, The intrinsic electrophysiological properties of mammalian neurons: Insights into central nervous system function, *Science* 242 (1988) 1654–1664.

[27] R Lorento de Nó and GA Condouris, Decremental conduction in peripheral nerve: Integration of stimuli in the neuron, *Proc. Natl. Acad. Sci. USA* 45 (1959) 593–617.

[28] A Mauro, F Conti, F Dodge, and R Schor, Threshold behavior and phenomenological impedance of the squid giant axon, *J. Gen. Physiol.* 55 (1970) 497–523.

[29] RN Miller and J Rinzel, The dependence of impulse propagation speed on firing frequency and dispersion for the Hodgkin–Huxley model, *Biophys. J.* 34 (1981) 227–259.

[30] W Rall and GM Shepherd, Theoretical reconstruction of field potentials and dendrodendritic synaptic interactions in olfactory bulb, *J. Neurophysiol.* 31 (1968) 884–915.

[31] J Rinzel, private communication, September 1980.

[32] NH Sabah and KN Liebovic, Subthreshold responses of the Hodgkin–Huxley cable model for the squid giant axon, *Biophys. J.* 9 (1969) 1206–1222.

[33] AC Scott, Effect of series inductance of a nerve axon upon its conduction velocity, *Math. Biosci.* 11 (1971) 277–290.

[34] AC Scott, The electrophysics of a nerve fiber, *Rev. Mod. Phys.* 11 (1975) 487–533.

[35] AC Scott, *Nonlinear Science: Emergence and Dynamics of Coherent Structures*, Oxford University Press, Oxford, 1999.

[36] AC Scott and U Vota-Pinardi, Velocity variations on unmyelinated axons, *J. Theor. Neurobiol.* 1 (1982) 150–172.

[37] W Stuhmer, Structure-function studies of voltage-gated ion channels, *Ann. Rev. Biophys. Biophys. Chem.* 20 (1991) 65–78.

[38] J Tille, A new interpretation of the dynamic changes of the potassium conductance in the squid giant axon, *Biophys. J.* 5 (1965) 163–171.

[39] HR Wilson, *Spikes, Decisions, and Actions: The Dynamical Foundations of Neuroscience*, Oxford University Press, Oxford, 1999.

5

Leading-Edge Models

To develop an intuitive understanding of a challenging area, it is sometimes useful to bracket the problem, on one hand looking fully at the intricacies and on the other taking the simplest possible perspective. Having considered a rather complete description of a squid axon in Chapter 4, we now turn our attention to simpler models of a nerve fiber that focus attention on the *leading edge* of an impulse.

Although lacking the scope and precision of the Hodgkin–Huxley formulation, these models are easier to grasp and thus useful for appreciating some fundamental aspects of nerve impulse propagation, including stability. Furthermore, we will obtain analytic expressions for impulse velocity and threshold conditions for impulse ignition and show how these features depend on physical parameters of the nerve.

5.1 Leading-Edge Approximation for the H–H Impulse

As we learned in the previous chapter, propagation in a Hodgkin–Huxley squid axon is governed by the nonlinear diffusion equation (or "cable equation") given in Equation (4.10), where j_{ion} is the ionic current flowing out of the fiber per unit of distance in the x-direction. This ionic current, in turn, has three components: sodium, potassium, and leakage.

Because the time for turn-on of the sodium current is about an order of magnitude shorter than the times for sodium turn-off and potassium turn-on, an attractive approximation for representing the leading edge of

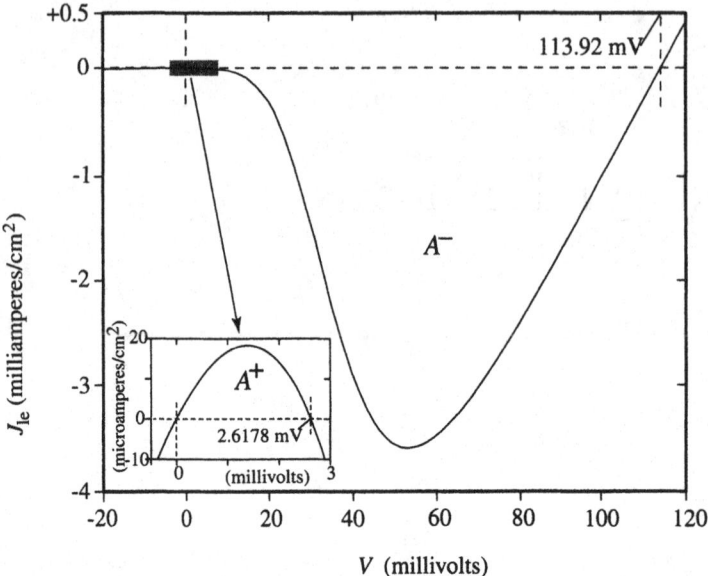

Figure 5.1. An approximation for the leading-edge transmembrane ionic current density J_{le} given by Equation (5.3).

an impulse is to assume that the sodium turn-off and potassium variables remain equal to their resting values, whereas the sodium turn-on variable responds instantly to changes in transmembrane voltage. In other words, we assume the limit

$$
\begin{aligned}
\tau_m &\rightarrow 0, \\
\tau_h &\rightarrow \infty, \\
\tau_n &\rightarrow \infty,
\end{aligned}
\tag{5.1}
$$

which in turn implies that

$$
\begin{aligned}
m &= m_0(V), \\
h &= h_0(0), \\
n &= n_0(0).
\end{aligned}
$$

In this limit, the leading-edge (le) dynamics are described by

$$
\frac{1}{rc}\frac{\partial^2 V}{\partial x^2} - \frac{\partial V}{\partial t} = \frac{j_{le}(V)}{c},
\tag{5.2}
$$

where

$$
j_{le}(V) = 2\pi a J_{le}(V)
$$

is the transmembrane ionic current per unit length of an axon of radius a and, from Equation (4.5),

$$J_{le}(V) = G_{Na}m_0^3(V)h_0(0)(V-V_{Na})+G_Kn_0^4(0)(V-V_K)+G_L(V-V_L) \quad (5.3)$$

is the ionic current per unit area.

From Figure 5.1, we see that J_{le} has the following features: it goes through zero at $V = 0$ with a small positive slope, it becomes sharply negative at $V \sim 30$ mV, indicating the increase of sodium ion permeability above a *threshold* level, and it goes to zero with positive slope again at $V = +113.92$ mV, which is close to the equilibrium (Nernst) potential of $+115$ mV for sodium ions.

Let us normalize Equation (5.2). Near the sodium equilibrium potential, the sodium permeability is fully turned on and

$$j_{le} \doteq g(V - 113.92),$$

where g is the corresponding conductance per unit length of the axon. Thus, it is convenient to rearrange Equation (5.2) by writing

$$\frac{1}{rg}\frac{\partial^2 V}{\partial x^2} - \frac{c}{g}\frac{\partial V}{\partial t} = \frac{j_{le}}{g}$$

and measuring distance in units of $1/\sqrt{rg}$ and time in units of c/g. Because this normalization removes all dependence on the parameters, the final result of any velocity calculation will be proportional to $1/\sqrt{rg}$ divided by c/g. In other words, the leading-edge speed must be

$$v \propto \sqrt{g/rc^2}.$$

What does this result tell us? Recall from Chapter 1 the 1906 lecture and demonstration by Luther [8] pointing out that the velocity of a wave of activity should be given by an expression of the form

$$v \propto \sqrt{D/\tau},$$

where D is a diffusion constant and τ is a time constant for the onset of the active process. In the present example, $1/rc$ is the diffusion constant (in centimeters squared per second) and c/g is the time constant (in seconds).

Finally, suppose that the radius of a smooth axon of circular cross-section is a. Because c/g is independent of the axon radius, $c \propto a$, and $r \propto 1/a^2$, we also see that Luther's factor

$$\sqrt{D/\tau} = \sqrt{g/rc^2} \propto \sqrt{a}.$$

Thus, the impulse speed of a nerve impulse should be approximately proportional to the square root of the fiber radius. In accord with this estimate, my measurements on giant axons of *Loligo vulgaris* indicate that over a temperature range from 15 to 22°C, the impulse velocity is given by

$$v = 20.3\sqrt{\left(\frac{a}{238}\right)}[1 + 0.038(\text{Temp} - 18.5)] \quad \text{m/s} \quad (5.4)$$

to an experimental accuracy of about ±5%, where a is the axon radius in microns and "Temp" is the temperature in degrees Celsius.

5.2 Traveling-Wave Solutions for Leading-Edge Models

Motivated by the dimensional considerations just discussed, let us study the traveling-wave solutions of the *normalized* equation

$$\frac{\partial^2 V}{\partial \tilde{x}^2} - \frac{\partial V}{\partial \tilde{t}} = f(V), \tag{5.5}$$

where

$$\tilde{x} \equiv \sqrt{rg}\, x, \quad \tilde{t} \equiv gt/c,$$

and $f(V)$ is a "cubic-shaped" function having the same qualitative shape as J_{le} plotted in Figure 5.1. More specifically, we will assume that $f(V) = V - V_2$ near $V = V_2$, that $f(V) = 0$ for $V = 0$, and that $f(V)$ has a threshold at $V = V_1$.

Upon determining the traveling-wave speed for the normalized Equation (5.5), we can multiply by Luther's factor of $\sqrt{g/rc^2}$ to obtain the physical impulse velocity. The Hodgkin–Huxley parameters entering into this factor are

Parameter	Value	Units
radius (a)	238	μm
g	.0108	mhos/cm
r	2.0×10^4	ohms/cm
c	1.5×10^{-7}	F/cm

so Luther's velocity factor for the squid axon is

$$\sqrt{g/rc^2} = 49 \text{ m/s}.$$

Returning to Equation (5.5), assume a traveling-wave solution of the form

$$V(\tilde{x}, \tilde{t}) = \tilde{V}(\tilde{x} - \tilde{v}\tilde{t}) = \tilde{V}(\xi),$$

where $\xi \equiv \tilde{x} - \tilde{v}\tilde{t}$ is a *traveling-wave variable*. Just as in the previous chapter for the full Hodgkin–Huxley system, the partial differential equation (PDE) becomes the *ordinary* differential equation (ODE)

$$\frac{d^2\tilde{V}}{d\xi^2} + \tilde{v}\frac{d\tilde{V}}{d\xi} = f(\tilde{V}) \tag{5.6}$$

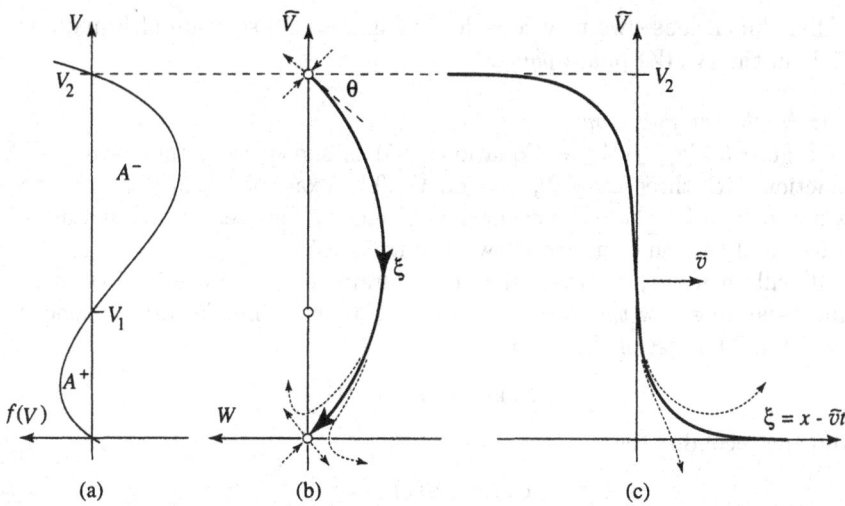

Figure 5.2. (a) A "cubic-shaped" $f(V)$ in Equation (5.5). (b) The two-dimensional *phase plane* for traveling-wave trajectories describing the leading edge of a nerve impulse. (c) A leading-edge traveling wave corresponding to the trajectory in (b).

in which the traveling-wave speed $\tilde{v}$ is again an adjustable parameter.[1] How can we find acceptable solutions for this equation?

5.2.1 Phase-Plane Analysis

Being second order, Equation (5.6) can be written as two autonomous first-order ODEs; thus

$$\frac{d\tilde{V}}{d\xi} \equiv W,$$

$$\frac{dW}{d\xi} = f(\tilde{V}) - \tilde{v}W,$$

(5.7)

where the first equation is a definition and the second is a restatement of Equation (5.6).

Because these two equations correspond to the five first-order ODEs that were developed to describe traveling waves of the full Hodgkin–Huxley system in Equations (4.12), one advantage of the leading-edge approximation becomes evident: it is possible to view solution trajectories in the traveling-wave phase space. Corresponding to Figure 4.4—which could only be schematic because it is not possible to visualize a trajectory in a space

---

[1]I hope that the notation is not confusing. The "tilde" on V indicates that it is a function of the traveling-wave variable ξ, whereas the "tilde" on v implies a velocity in the normalized space and time variables.

of five dimensions—we now consider in Figure 5.2 a solution of Equations (5.7) in the $(\tilde{V}, W)$ phase *plane*.

The Nonlinear Function
In Figure 5.2(a), $f(V)$ in Equation (5.5) is shown as a "cubic-shaped" function with three zeros: 0, V_1, and V_2. The exact shape of this function is not critical for the present analysis; it merely represents the qualitative features of the ionic current shown in Figure 5.1.

It will turn out, however, that two parameter ratios are important in understanding how the shape and speed of a nerve impulse are influenced by $f(V)$. The first of these is the

$$\text{voltage ratio} \equiv V_1/V_2 \,,$$

and the second is the

$$\text{area ratio} \equiv A^-/A^+ \,,$$

where A^+ is the positive-going area under $f(V)$ and A^- is the negative-going area, as shown in Figure 5.2(a).

Indeed, a necessary condition for a wave of increasing voltage (with time), as indicated in Figure 5.2(c), is that[2]

$$A^-/A^+ > 1 \,.$$

A glance at Figure 5.1 confirms that this condition is satisfied for the ionic current of the Hodgkin–Huxley membrane.

The Phase Plane
Central to understanding traveling-wave solutions of Equation (5.5) is an appreciation of solution trajectories of Equations (5.7) on the $(\tilde{V}, W)$ phase plane, which are shown in Figure 5.2(b). Because $f(V) = 0$ for $V = 0$, V_1, and V_2, there are three singular points in this phase plane at

$$(\tilde{V}, W) = (0,0), \ (V_1, 0), \text{ and } (V_2, 0) \,,$$

which satisfy the condition that both $d\tilde{V}/d\xi = 0$ and $dW/d\xi = 0$.

Corresponding to the leading-edge traveling wave shown in Figure 5.2(c), we seek a *heteroclinic* trajectory that starts from $(V_2, 0)$ at $\xi = -\infty$ and approaches $(0,0)$ as $\xi \to +\infty$. In other words, the phase point goes as

---

[2]To see this, note that Equations (5.7) can be written as the differential equation

$$W \, dW = f(\tilde{V}) d\tilde{V} - \tilde{v} W^2 d\xi \,.$$

Integrating from $\xi = -\infty$ to $\xi = +\infty$ and observing that $W^2 = 0$ in these two limits, one finds that

$$A^- - A^+ = \tilde{v} \int_{-\infty}^{\infty} W^2 d\xi \,.$$

Thus, for $\tilde{v}$ to be positive, it is necessary that A^- be greater than A^+.

$(V_2, 0) \to (\tilde{V}, W) \to (0,0)$ while ξ traverses the range $-\infty \to \xi \to +\infty$. This heteroclinic trajectory is shown as a bold line in Figure 5.2(b).

How do we start a solution trajectory near the singular point at $(V_2, 0)$? The *angle* at which the trajectory leaves the singular point can be adjusted merely by varying the assumed value of the traveling wave speed $(\tilde{v})$. In other words, if the angle θ is defined as in Figure 5.2(b), then[3]

$$\theta = \operatorname{arccot}\left[\sqrt{(\tilde{v}/2)^2 + 1} - \tilde{v}/2\right]. \tag{5.8}$$

Using this relationship, we have the opportunity of analytically *aiming* the initial segment of the trajectory in such a direction that it approaches the singular point at the origin as $\xi \to +\infty$. It is this particular value of

---

[3]To see how Equation (5.8) is obtained, note that sufficiently close to this singular point, Equations (5.7) can be approximated by the linear equations

$$d\Delta V/d\xi \;\doteq\; W,$$
$$dW/d\xi \;\doteq\; \Delta V - \tilde{v}W,$$

where $\Delta V \equiv \tilde{V} - V_2$. These equations can be more conveniently written as the matrix system

$$\frac{d}{d\xi}\begin{bmatrix} \Delta V \\ W \end{bmatrix} \doteq \begin{bmatrix} 0 & 1 \\ 1 & -\tilde{v} \end{bmatrix}\begin{bmatrix} \Delta V \\ W \end{bmatrix},$$

in which higher-order (i.e., nonlinear) terms in $\tilde{V}$ and W are neglected. To this linear approximation, the behavior of a trajectory is entirely determined by the properties of the matrix

$$M = \begin{bmatrix} 0 & 1 \\ 1 & -\tilde{v} \end{bmatrix},$$

which in turn depend only on the unknown traveling-wave speed $\tilde{v}$. Because the matrix system is linear, we can assume that the behavior of a solution trajectory near the singular point is as

$$\begin{bmatrix} \Delta V \\ W \end{bmatrix} = \begin{bmatrix} \Delta V_0 \\ W_0 \end{bmatrix} e^{\lambda \xi},$$

where ΔV_0 and W_0 are assumed to be independent of ξ. Thus, the eigenvalues of M are found to be $\lambda_1, \lambda_2 = -\tilde{v}/2 \pm \sqrt{(\tilde{v}/2)^2 + 1}$. These eigenvalues are real and of opposite sign, so the singular point at $(V_2, 0)$ is a *saddle point*. For the positive eigenvalue, in other words, there are two distinct trajectories that leave the singular point as ξ increases from $-\infty$. Along these trajectories,

$$W = \left[\sqrt{(\tilde{v}/2)^2 + 1} - \tilde{v}/2\right]\Delta V,$$

implying Equation (5.8).

traveling-wave speed that corresponds to the propagation of the leading edge shown in Figure 5.2(c).

If $\tilde{v}$ is adjusted to be slightly too small (large), the solution trajectory will pass the singular point $(0,0)$ on the left (right) and then move away in an unbounded manner, which is not physically acceptable. Thus, the term "shooting method" is applied to this procedure in which $\tilde{v}$ is progressively adjusted until a heteroclinic trajectory as shown in Figure 5.2(b) is obtained.

The Leading-Edge Waveform

Figure 5.2(c) shows the desired result of phase-plane analysis: the leading-edge waveform. The voltage of this waveform is obtained directly from the heteroclinic trajectory in Figure 5.2(b), and the horizontal axis indicates the corresponding values of ξ. The leading-edge waveform is closely related to the ionic current function $f(V)$ shown in Figure 5.2(a) because it exhibits a transition between zeros of this function at V_2 and 0.

When an incorrect value for $\tilde{v}$ is assumed in the numerical integration of Equations (5.7), the computed function will diverge to $\pm\infty$. These divergent calculations are indicated by the dashed curves in Figure 5.2(c), which are related to corresponding trajectories in the $(\tilde{V}, W)$ phase plane of Figure 5.2(b).

5.2.2   Analytic Results

Several analytic expressions for leading edge waveforms related to specific nonlinear functions have been recorded in the nonlinear science literature since the 1960s [12, 17, 18]. Figure 5.3 displays two forms of $f(V)$ that can be used for modeling the ionic current of a nerve membrane shown in Figure 5.1. In both examples, $f(V)$ has zeros at 0, V_1, and V_2, and the slope $df/dV = 1$ at $V = V_2$.

1. *Cubic polynomial model:* An analytic solution for the cubic polynomial function

$$f_1(V) = \frac{V(V - V_1)(V - V_2)}{V_2(V_2 - V_1)}, \tag{5.9}$$

shown in Figure 5.3(a), was first obtained in 1938 by Zeldovich and Frank-Kamenetsky in the context of flame-front propagation [21]. Translated into the present notation, they found a speed of

$$v = \sqrt{\frac{g}{rc^2}} \left(\frac{V_2 - 2V_1}{\sqrt{2}V_2} \right) \tag{5.10}$$

for the leading-edge traveling wave (see Figure 5.2(c)) given by

$$\tilde{V}(x - vt) = \frac{V_2}{1 + \exp\left(\sqrt{rg}(x - vt)/\sqrt{2}\right)}, \tag{5.11}$$

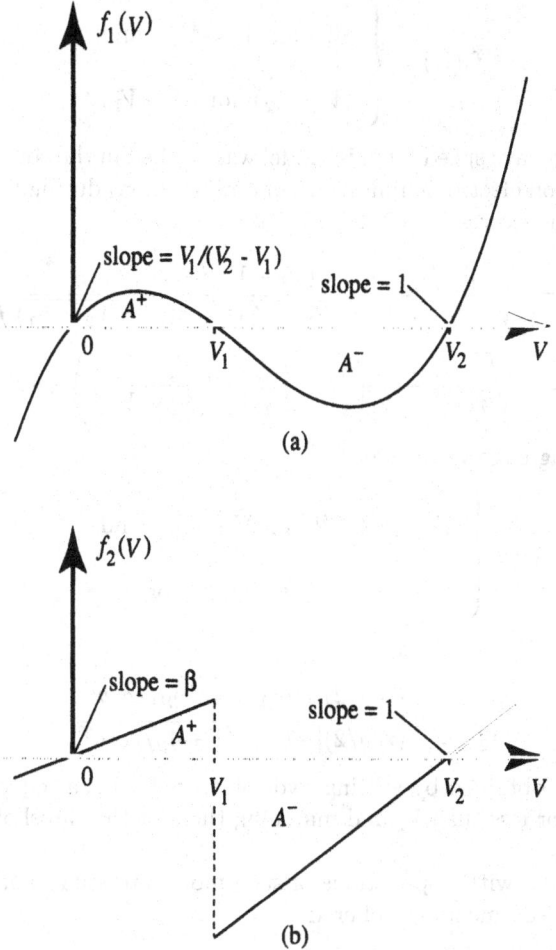

Figure 5.3. Two forms of the function $f(v)$ for which Equation (5.5) has analytic traveling-wave solutions. (a) A cubic function defined in Equation (5.9). (b) A piecewise linear function defined in Equation (5.12).

as can be checked by direct substitution. (In these equations, there is no tilde on v because the velocity is no longer in normalized units.)

2. *Piecewise linear model:* Shortly after the observation by Cole and Curtis that the impedance of a squid membrane decreases by a factor of about 40 during the passage of a nerve impulse [2], Offner, Weinberg, and Young proposed to model a nerve membrane by the "piecewise linear" conductance shown in Figure 5.3(b) and defined by [11]

$$f_2(V) = \begin{cases} \beta V \text{ for } V < V_1, \text{ and} \\ (V - V_2) \text{ for } V > V_1. \end{cases} \quad (5.12)$$

The traveling-wave speed for this model was studied in the course of *neuristor* research (on electronic imitations of a nerve axon) during the 1960s and is given by the expression [7, 14, 15, 19]

$$\begin{aligned} v &= \sqrt{\frac{g}{rc^2}} \left(\frac{(V_2 - V_1)^2 - \beta V_1^2}{\sqrt{V_2 V_1 (V_2 - V_1)^2 + \beta V_2 V_1^2 (V_2 - V_1)}} \right) \quad (5.13) \\ &= \sqrt{\frac{g}{rc^2}} \left(\frac{V_2 - V_1}{\sqrt{V_1 V_2}} - \frac{\beta}{2} \sqrt{\frac{V_1}{V_2}} \left(\frac{V_2 + V_1}{V_2 - V_1} \right) \right) + O(\beta^2) \end{aligned}$$

for the leading-edge wave shape

$$\tilde{V} = \begin{cases} V_1 e^{-\lambda_1 (x - vt)} \text{ for } \tilde{V} < V_1, \text{ and} \\ V_2 - (V_2 - V_1) e^{\lambda_2 (x - vt)} \text{ for } \tilde{V} > V_1, \end{cases} \quad (5.14)$$

where

$$\begin{aligned} \lambda_1 &= (rcv/2)[1 + \sqrt{1 + 4\beta g/rc^2 v^2}], \\ \lambda_2 &= (rcv/2)[-1 + \sqrt{1 + 4g/rc^2 v^2}]. \end{aligned}$$

This result is obtained by writing two solutions for linear equations where $\tilde{V}$ is greater or less than V_1 and matching them at the threshold boundary $(V = V_1)$.

In connection with applications of such models to studies of real nerves, the following comments are offered.

- The two results may be useful in different contexts. The first example has two adjustable parameters: the ratio of threshold to maximum voltage (V_1/V_2) and the Luther factor, $\sqrt{g/rc^2}$. As we will see, this flexibility and simplicity make the assumption of a cubic polynomial conductance convenient for many theoretical estimates.
  There are three adjustable parameters in the second example because, in addition to those of the first example, the ratio of positive membrane conductance near $V = 0$ to its value near $V = V_2$ can be independently adjusted through the parameter β. Although this added flexibility is useful in modeling neuristors, it comes at the cost of added intricacy of Equation (5.13).

- Although it is possible to determine the traveling-wave speed for a piecewise linear model in which $f(V)$ is continuous and has an intermediate region of negative slope, the resulting condition cannot be explicitly solved for v [9].

- If $f(V)$ exhibits hysteresis (is not single-valued), it is necessary to model the system as two coupled nonlinear diffusion equations implying two different wave velocities [4]. As a result of the interaction between these two equations, the combined wave travels at the slower of these two speeds.

- The two examples presented here exhibit different behaviors in the limit $V_1 \to 0$. Thus the normalized velocity for the cubic polynomial model approaches $1/\sqrt{2}$, whereas that of the piecewise linear model diverges as $\sqrt{V_2/V_1}$. This difference should be kept in mind when these models are used to estimate nerve phenomena with small values of the threshold voltage V_1.

- If the functions $f(V)$ in these models are adjusted to match the shape of J_{le} in Figure 5.1, neither gives a particularly good estimate of the impulse-propagation speed of a Hodgkin–Huxley axon. Equation (5.10) implies, for example, that $v = 33$ m/s. Equation (5.13), with $\beta = 0$ and V_2/V_1 rather arbitrarily taken to be 3.25, implies $v = 88$ m/s. Both of these values are significantly larger than the Hodgkin–Huxley result of 18.8 m/s and the experimental values given in Equation (5.4). There are two reasons for these discrepancies, both of which tend to make the calculated speed greater in the leading-edge approximation of Equation (5.2).
  (i) The response times for sodium turn-off and potassium turn-on are not really infinity, as was assumed in Equations (5.1). We will see how such *recovery effects* decrease impulse speed in the following chapter.
  (ii) Because the turn-on delay for sodium ions is not zero as was assumed in Equations (5.1), the qualitative effect of finite sodium turn-on delay is to increase the voltage level at which sodium current begins to flow into the fiber, thereby decreasing the impulse speed. That neglect of sodium turn-on delay is an important factor in determining the velocity of a squid impulse is evident from Equation (5.4), which indicates a speed that increases with increasing temperature. This comes about because near 18.5°C sodium turn-on delay depends on temperature ("Temp") as

$$\tau_m \propto 3^{-(\text{Temp}-18.5)/10} .$$

Thus, the delay time for the inward flow of sodium current to commence decreases with increasing temperature, thereby increasing impulse speed. The effect of finite response time for sodium current was studied analytically in the mid-1970s by Pastushenko, Chizmadzhev, and Markin, who obtained reasonable agreement with the H–H result by assuming that τ_m is independent of the transmembrane voltage [13].
Another way to account for sodium turn-on delay is by taking larger values of (V_1) than are indicated in Figure 5.1. To obtain a speed of 20

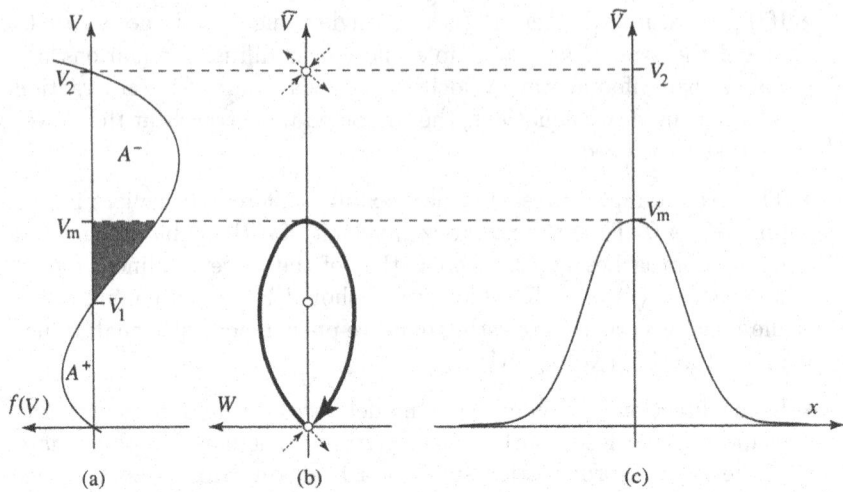

Figure 5.4. (a) A "cubic-shaped" $f(V)$, representing transmembrane ionic current in Equation (5.5). (b) The phase-plane plot for a *homoclinic* traveling-wave trajectory describing a threshold solution of this equation. (c) The "impulse-shaped" stationary voltage wave corresponding to the trajectory in (b).

m/s from Equation (5.10), for example, requires that $V_1/V_2 = 0.21$. The same speed can be obtained from Equation (5.13) with $\beta = 1/40$ and $V_1/V_2 = 0.64$.

In cubic polynomial systems defined by Equation (5.9), it is interesting that $v = 0$ for $V_2 = 2V_1$, which is the condition making the positive-going area under $f_1(V)$ equal to the negative-going area in Figure 5.3(a). Similarly, in piecewise linear systems defined by Equation (5.12), $v = 0$ for $V_2 = 2V_1$ and $\beta = 1$, which again implies

$$A^- = A^+$$

in Figure 5.3(b). This agrees with our previous observation that A^- must exceed A^+ for v to be positive.

What other physically acceptable solutions appear when $v = 0$?

5.3   The Threshold Impulse

In Figure 4.5 of the previous chapter was sketched a low-amplitude, unstable impulse solution of the full Hodgkin–Huxley equations that was obtained by Huxley in 1959 [5]. Referring to Figure 5.4, we can find a corresponding solution of our leading-edge system by setting $\tilde{v} = 0$ in Equations (5.7).

The ionic current function $f(V)$ in Figure 5.4(a) is identical to that in Figure 5.2(a), but Equations (5.7) reduce to

$$\frac{d\tilde{V}}{d\xi} = W,$$

$$\frac{dW}{d\xi} = f(\tilde{V}),$$

determining dynamics of a phase point on the $(\tilde{V}, W)$ phase plane shown in Figure 5.4(b). These equations imply the differential equation

$$W\,dW = f(\tilde{V})d\tilde{V},$$

which can be integrated for *any* form of $f(V)$ to obtain the homoclinic trajectory shown as a bold path in Figure 5.4(b).[4]

The maximum value of $\tilde{V}(\xi)$ (call it V_m) occurs where $W(\tilde{V})$ first returns to zero as indicated in Figure 5.4(b), and it is readily determined as the value at which the shaded portion of the negative-going area in Figure 5.4(a) is equal to A^+.

For the homoclinic trajectory $(0,0) \to (\tilde{V}, W) \to (0,0)$ as $-\infty \to \xi \to +\infty$, in other words, it is necessary that

$$A^- > A^+,$$

allowing $W(\tilde{V})$ to return to zero and fixing V_m by the condition

$$\int_0^{V_m} f(\tilde{V})d\tilde{V} = 0. \tag{5.15}$$

Thus, the zero-velocity solution of Equations (5.7), as shown in Figure 5.4(c), is *impulse-shaped* (or bell shaped), having a maximum value and falling smoothly to zero as $\xi \to \pm\infty$.

In the following section, we will see that the impulse-shaped solution of Figure 5.4(c) is *unstable*, whereas the leading-edge solution of Figure 5.2(c) is *stable*. Why should we be interested in a solution of the PDE equations that is unstable? Because it defines the *threshold condition* for ignition of a nerve impulse.

---

[4]Thus,

$$W(\tilde{V}) = \pm\left(2\int_0^{\tilde{V}} f(\alpha)d\alpha\right)^{1/2},$$

where α is a variable of integration. Having found this trajectory, $\tilde{V}(\xi)$ is determined by the integral

$$\xi = \int^{\tilde{V}} \frac{d\beta}{W(\beta)},$$

which in turn corresponds to the stationary voltage wave that is sketched in Figure 5.4(c).

5.4 Stability of Simple Traveling Waves

In the foregoing discussions, the task of finding solutions for nonlinear partial differential equations (PDEs) has been eased by assuming that all dependent variables (e.g., voltages, currents) are functions of the traveling-wave variable $\xi \equiv x - vt$. This traveling-wave assumption ties time and space derivatives together as

$$-\frac{1}{v}\frac{\partial}{\partial t} = \frac{\partial}{\partial x} = \frac{d}{d\xi},$$

thereby reducing a nonlinear PDE to a more manageable ODE.

Having found a traveling-wave solution, say $\tilde{V}(\xi)$, we have no idea whether it is *stable*. In other words, if $\tilde{V}(\xi)$ is altered by some small change (or deformation) in its shape, will that small change relax to zero, causing no significant alteration of the traveling wave? Or will it grow to such a magnitude that $\tilde{V}(\xi)$ no longer represents the behavior of the system?

To answer such questions, it is necessary to examine the original partial differential equation. As presented in Appendix D, a *linear stability analysis* of a traveling-wave solution proceeds as follows.

- A coordinate (independent variable) transformation is introduced into a frame of reference that moves with the same speed as the traveling wave. This transformation yields a nonlinear PDE describing the dynamics of deviations from the traveling-wave solution.

- Assuming that deformations of the traveling wave are small, the nonlinear PDE is then approximated by a linear PDE, which is obtained by linearizing the nonlinear PDE about the traveling-wave solution.

- Finally, the linear PDE is studied to learn whether small deviations wax or wane with time, thereby determining whether the traveling wave is unstable or stable.

Although the details of this analysis may be intimidating, the results can be simply stated in terms of the following definitions. A traveling-wave solution is said to be *asymptotically stable* if *all* small changes in $\tilde{V}(\xi)$ approach zero with increasing time, *unstable* if *any* deviation from $\tilde{V}(\xi)$ grows without bound in the linear approximation, and *stable* otherwise.

From the discussion in Appendix D, we have the following theorem.

Linear stability theorem: Consider a traveling-wave solution $\tilde{V}(\xi)$ of Equation (5.5). If $d\tilde{V}(\xi)/d\xi$ has no zero crossings, then $\tilde{V}(\xi)$ is stable but not asymptotically stable. If $d\tilde{V}(\xi)/d\xi$ has one or more zero crossings, then $\tilde{V}(\xi)$ is unstable (Zeldovich and Barenblatt, 1959 [20]).

If $\tilde{V}(\xi)$ is a leading-edge waveform as indicated in Figure 5.2(c), then $d\tilde{V}(\xi)/d\xi$ has no zero crossings. Thus, all leading-edge waveforms are stable but not asymptotically stable.

There is a physical reason behind the fact that leading-edge waveforms are not asymptotically stable. Upon experiencing a small change, a leading-edge wave may relax to $\tilde{V}(\xi + \xi_0)$, where ξ_0 is a real constant. Thus, as time increases,

$$\tilde{V}(\xi) + \text{deformation} \rightarrow \tilde{V}(\xi + \xi_0)\,,$$

which is also a traveling wave solution of Equation (5.5).

In other words, altering a leading-edge traveling-wave solution may cause a displacement of that traveling wave in the ξ-direction with respect to where it would have been without the influence of the alteration. Because a displaced traveling wave is still a solution of Equation (5.2), residual effects of the alteration persist in time without growing or decaying.

Consider, on the other hand, the bell-shaped solution shown in Figure 5.4(c), where $\tilde{V}(\xi)$ rises to a maximum value and then returns to zero. In this case, $d\tilde{V}(\xi)/d\xi$ has a zero crossing (at the maximum value of the solution), so $\tilde{V}(\xi)$ is unstable. Small changes in $\tilde{V}(\xi)$ therefore grow with time, eventually distorting it beyond recognition.

To get a physical feeling for this instability, return to Equation (5.15), which determines the amplitude of the impulse-shaped solution. From this equation, V_m is fixed by the requirement that the total integral of $f(V)$ over the waveform is zero, with the positive area just canceled by the negative area. If V_m is slightly decreased, the losses from the positive area exceed energy input from the negative area, and the solution relaxes to zero. If V_m is slightly increased, on the other hand, the increased energy input from the negative-going area overcomes the losses, and the wave grows. It grows in a manner that is qualitatively similar to the Verhulst functions of Figure 1.3, eventually reaching the stable leading-edge solution of Figure 5.2(c).

Thus the impulse-shaped solution of Figure 5.4(c) is recognized as a watershed between those functions from which the fully developed leading-edge solution can emerge and those that relax to zero. Recognition of this fact is a key to understanding the problem of launching (or igniting) a nerve impulse.

5.5 Leading-Edge Charge and Impulse Ignition

How does one initiate an impulse on a nerve axon? From both experimental observations [1, 6, 10] and integration of the full Hodgkin–Huxley system [3], it appears that a certain *threshold* of electric charge (Q_θ) is required. In other words, if a short pulse of current is injected into a nerve fiber, the integral over time of this current pulse must exceed Q_θ for an impulse to emerge. In the standard H–H axon at a temperature of 18.5°C, the threshold charge in coulombs (C) is about

$$Q_\theta = 1.33 \times 10^{-9} \text{ C.}$$

We can understand this theoretically confirmed experimental observation by considering the leading edge of an H–H impulse. From the derivation of the "cable equation" in the previous chapter, Equations (4.8) and (4.9) can be written as the first-order PDEs

$$\frac{\partial V}{\partial x} = -ri, \tag{5.16}$$

$$\frac{\partial i}{\partial x} = -c\frac{\partial V}{\partial t} - j_{ion}. \tag{5.17}$$

So far in this chapter, it has been assumed that the turn-on delay for sodium ion permeability is negligibly small, but from Figure 1.1 we see that this is not so. The action potential has reached almost its full amplitude before the membrane permeability is fully on, and this time lag is due to the sodium turn-on delay, τ_m. In other words,

$$|j_{ion}| \ll \left| c\frac{\partial V}{\partial t} \right|$$

over much of the leading edge of a typical impulse.

If one chooses to neglect ionic current on the leading edge of the propagating impulse, Equation (5.17) becomes the approximate *conservation law*

$$\frac{\partial i}{\partial x} + \frac{\partial(cV)}{\partial t} \doteq 0. \tag{5.18}$$

With reference to the discussion of conservation laws in Appendix A, it is seen that i is the *flow* of the conserved quantity past a fixed point per unit of time, and cV is the corresponding *density* of the conserved quantity per unit of distance along the axon. From the units, the conserved quantity is evidently an electric charge, which is given by the integral expression

$$Q_0 = \int_{le} i\,dt,$$

where the subscript indicates that the integration is across the leading edge of the H–H impulse. Substituting Equation (5.16) and noting that V is a traveling wave of speed v, this expression for the conserved charge becomes

$$Q_0 = -\frac{1}{r}\int_{le} \frac{\partial V}{\partial x}dt = +\frac{1}{rv}\int_{le} \frac{\partial V}{\partial t}dt.$$

This reduces to the simple expression

$$Q_0 = \frac{V_{max}}{rv}, \tag{5.19}$$

where V_{max} is the total amplitude of the nerve impulse.

Upon inserting the values $r = 2 \times 10^4$ ohm/cm and (from Figure 4.3) $v = 1880$ cm/s and $V_{max} = 90.5 \times 10^{-3}$ V gives a leading-edge charge of [16]

$$Q_0 = 2.4 \times 10^{-9} \ \text{C}, \tag{5.20}$$

which is substantially larger than the experimental value for Q_θ recorded previously.

Should we be disappointed that the leading-edge charge that we have just calculated is almost twice the amount of charge needed to launch an impulse? Not at all. The fact that

$$Q_0 = 1.8 Q_\theta$$

tells us that impulse propagation on a squid fiber has a *safety factor* of about 80%. In other words, the leading-edge charge can be reduced by a factor of almost one-half before propagation fails.

Finally, we can use Equation (5.19) and parameters for the small-amplitude *unstable* impulse in Figure 4.3 to calculate Q_θ. Keeping r the same but using a velocity of 566 cm/s and a voltage amplitude of 18×10^{-3} V gives a threshold charge of

$$Q_\theta = 1.6 \times 10^{-9} \ \text{C},$$

which is close to the experimental value quoted at the beginning of this section and implies a safety factor of 50%.

As we will see in Chapter 9, the possibility of impulse failure near branching regions of dendritic and axonal trees opens the doors to more intricate forms of information processing in a single neuron.

5.6 Recapitulation

In a squid giant axon, the turn-on time for sodium conductance is more than an order of magnitude shorter than the turn-off time for sodium conductance and the turn-on time for potassium conductance. Thus, for the leading edge of a nerve impulse, the Hodgkin–Huxley expression for transmembrane ionic current is approximated by an expression that assumes the sodium conductance responds instantly to changes in the membrane potential, whereas the potassium conductance remains at its resting value. In this leading-edge approximation, the ionic current is a direct function of the transmembrane voltage, and the H–H equations reduce to the simple nonlinear diffusion equation of Equation (5.5).

A phase-plane analysis of traveling-wave solutions for this simple nonlinear diffusion equation was used to sketch a "shooting method" of solution showing that only isolated traveling-wave speeds are possible. Simple analytic results were presented that relate the leading-edge velocity to parameters of the nonlinear relationship between ionic current and transmembrane voltage. Zero velocity, low-amplitude, impulse-like solutions of Equation (5.5) are also described.

Stability of traveling-wave solutions is then defined and discussed with the result that leading-edge solutions are found to be stable, whereas the low-amplitude impulse-like solutions are unstable. It is emphasized that the unstable solution is important for understanding the threshold for ignition of an impulse. Starting from this perspective, the quantity of electric charge necessary to launch an H–H impulse is calculated and compared with the charge carried on the leading edge of a fully developed impulse. This comparison indicates that the squid nerve has a safety factor for propagation lying somewhere between 50% and 80%.

References

[1] KS Cole, *Membranes, Ions and Impulses,* University of California Press, Berkeley, 1968.

[2] KS Cole and HJ Curtis, Electrical impedance of a squid giant axon during activity, *J. Gen. Physiol.* 22 (1939) 649–670.

[3] JW Cooley and FA Dodge, Digital computer solutions for excitation and propagation of the nerve impulse, *Biophys. J.* 6 (1966) 583–599.

[4] AV Gurevich, RG Mints, and AA Pukhov, Motion of a kink in a bistable medium with hysteresis, *Physica D* 35 (1989) 382–394.

[5] AF Huxley, Can a nerve propagate a subthreshold disturbance? *J. Physiol. (London)* 148 (1959) 80P–81P.

[6] BI Khodorov, *The Problem of Excitability,* Plenum Press, New York, 1974.

[7] H Kunov, *Nonlinear Transmission Lines Simulating Nerve Axon,* PhD Thesis, Electronics Laboratory, Technical University of Denmark, 1966.

[8] R Luther, Räumliche Fortpflanzung chemischer Reaktionen. *Z. Elektrochem.* 12(32) (1906) 596–600. (English translation in *J. Chem. Ed.* 64 (1987) 740–742.)

[9] OA Mornev and AC Scott, unpublished calculations.

[10] D Noble and RB Stein, The threshold conditions for initiation of action potentials by excitable cells, *J. Physiol. (London)* 187 (1966) 129–142.

[11] F Offner, A Weinberg, and C Young, Nerve conduction theory: Some mathematical consequences of Bernstein's model, *Bull. Math. Biophys.* 2 (1940) 89–103.

[12] VF Pastushenko, YA Chizmadzhev, and VS Markin, Speed of excitation in the reduced Hodgkin–Huxley model: I. Rapid relaxation of sodium current, *Biophysics* 20 (1975) 685–692.

[13] VF Pastushenko, YA Chizmadzhev, and VS Markin, Speed of excitation in the reduced Hodgkin–Huxley model: II. Slow relaxation of sodium current, *Biophysics* 20 (1975) 894–901.

[14] AC Scott, Analysis of nonlinear distributed systems, *Trans. IRE* CT–9 (1962) 192–195.

[15] AC Scott, Neuristor propagation on a tunnel diode loaded transmission line, *Proc. IEEE* 51 (1963) 240.

[16] AC Scott, Strength duration curves for threshold excitation of nerves, *Math. Biosci.* 18 (1973) 137–152.

[17] AC Scott, The electrophysics of a nerve fiber, *Rev. Mod. Phys.* 11 (1975) 487–533.

[18] AC Scott, *Nonlinear Science: Emergence and Dynamics of Coherent Structures,* Oxford University Press, Oxford, 1999.

[19] YL Vorontsov, MI Kozhevonikova, and IV Polyakov, Wave processes in active RC-lines, *Radio Eng. Electron. Phys. (USSR)* 11 (1967) 1449–1456.

[20] YB Zeldovich and GI Barenblatt, Theory of flame propagation, *Combust. and Flame* 3 (1959) 61–74.

[21] YB Zeldovich and DA Frank-Kamenetsky, K teorii ravnomernogo raspros-tranenia plameni, *Dokl. Akad. Nauk SSSR* 19 (1938) 693–697.

6
Recovery Models

Propagation of a nerve impulse is often compared with the burning of a candle, of which the leading-edge models considered in the previous chapter provide examples. This is a flawed metaphor, however, because a candle burns only once, spending (like H.C. Andersen's little match girl) its entire store of chemical energy to keep the flame bright and hot, with no possibility of transmitting a second flame. As we have seen both from Cole's classic oscillogram of Figure 1.1 and the more detailed data of Figure 4.8, a nerve impulse exhibits *recovery* over an interval of a few milliseconds, allowing subsequent impulses to be transmitted by the nerve. Without this feature, our nervous systems would be useless for processing information, and the animal kingdom could not have developed.

In this chapter, we explore some simple models for the recovery phenomenon that are useful not only for broadening our physical and mathematical understanding of nerve impulse propagation but also for making better estimates of nerve behavior.

6.1 The Markin–Chizmadzhev (M–C) Model

One of the simplest means for representing recovery of a propagating nerve impulse was introduced by Kompaneyets and Gurovich in the mid-1960s [23] and developed in detail by Markin and Chizmadzhev in 1967 [24]. This M–C model assumes the diffusion equation (or "cable equation") with which we began the previous chapter; thus the transmembrane voltage V

is governed by the PDE[1]

$$\frac{1}{rc}\frac{\partial^2 V}{\partial x^2} - \frac{\partial V}{\partial t} = \frac{j_{mc}(x,t)}{c}. \tag{6.1}$$

In this model, however, the ionic membrane current is not represented as a voltage-dependent variable, as in Equation (5.3), but by one of the following prescribed functions of time.

(1) If V does not reach the threshold value of V_θ, then

$$j_{mc}(x,t) = 0.$$

(2) If, on the other hand, V does reach the threshold value of V_θ at some instant (which is defined as $t = 0$), then at $x = 0$

$$
\begin{aligned}
j_{mc}(0,t) &= 0 \quad \text{for} \quad t < 0, \\
&= -j_1 \quad \text{for} \quad 0 < t < \tau_1, \\
&= +j_2 \quad \text{for} \quad \tau_1 < t < \tau_1 + \tau_2, \text{ and} \\
&= 0 \quad \text{for} \quad t > \tau_1 + \tau_2.
\end{aligned}
\tag{6.2}
$$

Whereas Equation (5.2) is a nonlinear diffusion equation, Equation (6.1) is a *piecewise linear inhomogeneous diffusion equation*, which is easier to solve. Thus, this is evidently a helpful assumption to make, but how do we choose the parameters $(j_1, j_2, \tau_1, \tau_2)$ that define $j_{mc}(0,t)$?

Recalling that the positive direction for ionic current is outward, the early current $-j_1$ represents the inward flow of sodium ions, whereas the later component $+j_2$ describes outward flow of potassium ions. Thus, τ_1 and τ_2 can be obtained from the waveform of the squid impulse in Figure 1.1, and it is possible to estimate j_1 from the leading-edge charge Q_0, which we obtained in Equation (5.20).

Noting that j_1 has the units of current per unit of distance along the axon (amperes per centimeter), it follows that the spatial width over which inward current flows is $v\tau_1$, where v is the impulse speed. Assuming further that the flow of j_1 across the membrane supplies the leading-edge charge—defined in Equation (5.19)—implies $j_1 v\tau_1 = Q_0/\tau_1$, or

$$j_1 = \frac{Q_0}{v\tau_1^2} = \frac{Q_0^2 r}{V_{max}\tau_1^2} \quad \text{A/cm}. \tag{6.3}$$

Finally, the condition

$$j_1\tau_1 = j_2\tau_2$$

---

[1] An even simpler version of the M–C concept is the "integrate and fire" model of a neuron, in which the entire cell is approximated as a single switch in parallel with a capacitor [1, 21, 22]. The capacitor integrates incoming charge until a threshold voltage is reached, whereupon the switch closes briefly, discharging the capacitor and restarting the process. A more realistic version is Gerstner's "spike response model" [12, 13], which is convenient for approximate numerical studies of large neural networks.

requires that the total ionic charge crossing the membrane during an impulse be zero. This ensures that the impulse voltage returns to zero at the end of the impulse and implies that

$$j_2 = \frac{Q_0}{v\tau_1\tau_2} = \frac{Q_0^2 r}{V_{\max}\tau_1\tau_2} \quad \text{A/cm}. \tag{6.4}$$

With these values and the additional parameters (r and c) given in Table 4.2, we can proceed to a quantitative analysis of Equation (6.1) for the standard squid axon at 18.5°C.

Assuming that the solution is a traveling wave of the form[2]

$$V(x,t) = V(x - vt) = V(\xi),$$

where $\xi \equiv x - vt$, Equation (6.1) reduces to

$$\frac{1}{rc}\frac{\partial^2 V}{\partial \xi^2} + v\frac{\partial V}{\partial \xi} = \frac{j_{mc}(\xi)}{c}, \tag{6.5}$$

a piecewise linear ODE. Further assuming that the threshold voltage V_θ is reached at $\xi = 0$, the ionic current in Equation (6.5) is defined as a function of ξ as

$$
\begin{aligned}
j_{mc}(\xi) &= \quad 0 \quad \text{for} \quad \xi > 0, \\
&= \quad -j_1 \quad \text{for} \quad -v\tau_1 < \xi < 0, \\
&= \quad +j_2 \quad \text{for} \quad -v(\tau_1 + \tau_2) < \xi < -v\tau_1, \text{ and} \\
&= \quad 0 \quad \text{for} \quad \xi < -v(\tau_1 + \tau_2),
\end{aligned}
$$

which is displayed in Figure 6.1(a).

For this ODE, the propagation problem can be solved as follows [24].

- With reference to Figure 6.1(b), consider the ξ-axis to be divided into four regions: region #1, where $\xi > 0$; region #2, where $-v\tau_1 < \xi < 0$; region #3, where $-v(\tau_1 + \tau_2) < \xi < -v\tau_1$; and region #4, where $\xi < -v(\tau_1 + \tau_2)$.

- For each region, write a solution of Equation (6.5) of the form

$$V_i(\xi) = A_i + B_i\xi + C_i \exp(-vrc\xi),$$

where $i = 1, 2, 3,$ or 4 indicates one of the four regions.

- Require the expressions for both $V_i(\xi)$ and $dV_i/d\xi$ to be continuous at the boundaries between regions. (It is necessary that $dV_i/d\xi$ be continuous at the boundaries to avoid divergence of the first term in Equation (6.5).)

---

[2]For typographical convenience, we dispense in this chapter and subsequently with the use of a tilde to distinguish between a general function of both x and t and a traveling-wave function of $x - vt$.

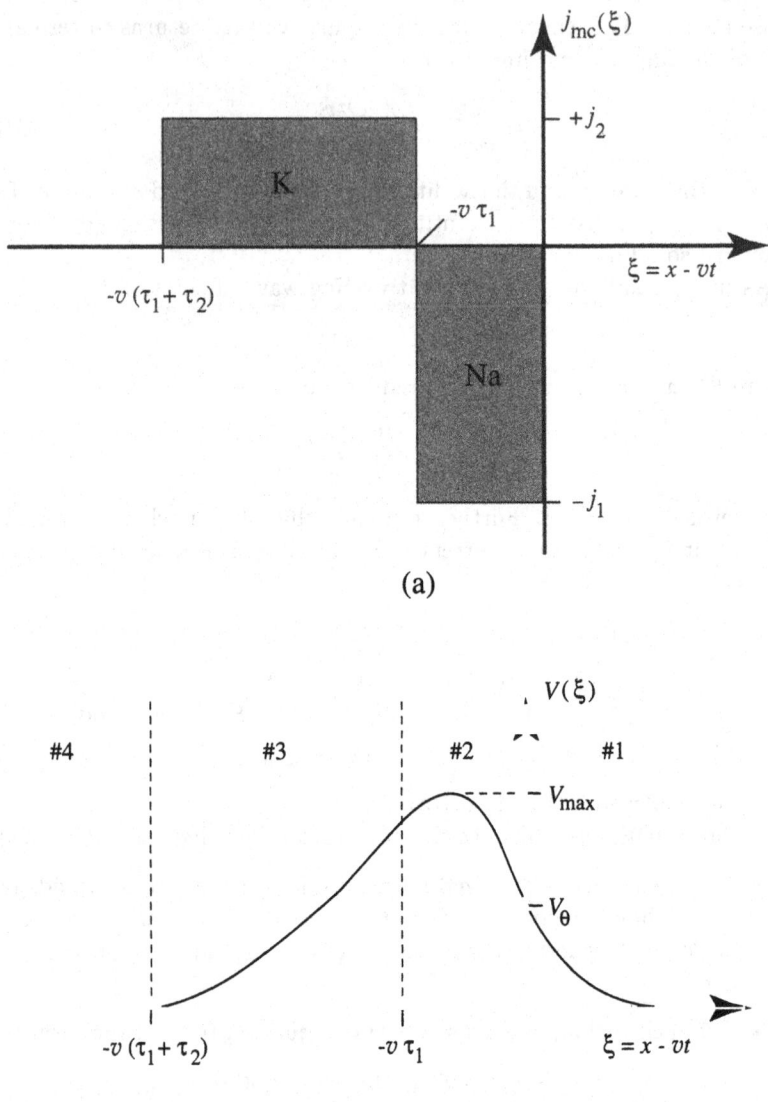

Figure 6.1. (a) Ionic current in the M–C model as a function of the traveling-wave variable (ξ). (b) Structure of the associated nerve impulse.

- Because $V_4(\xi) = 0$ and $V_1(\xi) = C_1 \exp(-vrc\xi)$, there are a total of seven constants to determine: C_1, A_2, B_2, C_2, A_3, B_3, and C_3. (The impulse speed v appears as a parameter in Equation (6.5), so these "constants" are actually functions of the traveling-wave speed.) The boundary conditions between regions #1 and #2, #2 and #3, and #3 and #4 comprise six constraints in addition to the above-mentioned condition $j_1 \tau_1 = j_2 \tau_2$. Thus it is straightforward, if tedious, to solve for the seven constants that define $V_i(\xi)$.

After going through the details of this analysis, one finds

$$V_1(\xi) = C_1(v) \exp(-vrc\xi) \tag{6.6}$$

in region #1, where $j_{\mathrm{mc}}(\xi) = 0$, and

$$C_1(v) = \frac{1}{v^2 rc^2} \left[j_1 + j_2 e^{-v^2 rc(\tau_1 + \tau_2)} - (j_1 + j_2) e^{-v^2 rc\tau_1} \right]. \tag{6.7}$$

Reference to Figure 1.1 confirms that such behavior occurs for a squid nerve. In other words, an exponentially decreasing *precursor* (or "skirt") precedes the voltage waveform of a real nerve impulse into the region where the membrane permeability has not yet begun to change. This is an important observation and merits additional comment.

From the perspectives of linear dynamics, the presence of an exponential skirt extending into a region of undisturbed membrane seems problematic. How does a patch of membrane at (say) $x = 0$ know that its transverse voltage should begin rising with time as

$$\exp(-vrc\xi) \propto f(x) \exp(v^2 rct)$$

before there is any change in the permeability of the nerve membrane? Yet Figure 1.1 provides clear empirical evidence that this is indeed what occurs, so we must explain this *nonlocal* phenomenon.

From the broader perspectives of *nonlinear* science, it turns out, the nerve impulse is a *coherent structure* that emerges from the underlying dynamics. All aspects of the impulse are dynamically interrelated, each making appropriate contributions to its global nature [34]. In more physical terms, the exponential precursor is being pushed into region #1 by the dynamics of the membrane in regions #2 and #3.

The significance of Equation (6.6) was not lost on Hodgkin and Huxley in their classic paper of 1952 [18]. Differentiating this equation with respect to ξ and setting $\xi = 0$, they noted that

$$v = -\frac{1}{rcV_\theta} \left[\frac{dV_1(\xi)}{d\xi} \right]_{\xi=0},$$

where $V_\theta = V_1(0)$ is the threshold voltage for the onset of sodium ion current. In other words, the conduction velocity of a nerve impulse can be calculated from a knowledge of the threshold voltage for sodium ion conductance and the spatial derivative of the impulse voltage at threshold. This is the motivating idea of M–C analysis.

A key item of information for applying this idea is the impulse voltage at threshold, which must equal the function $C_1(v)$ defined in Equation (6.6) and given in Equation (6.7). Using the estimates for j_1 and j_2 from

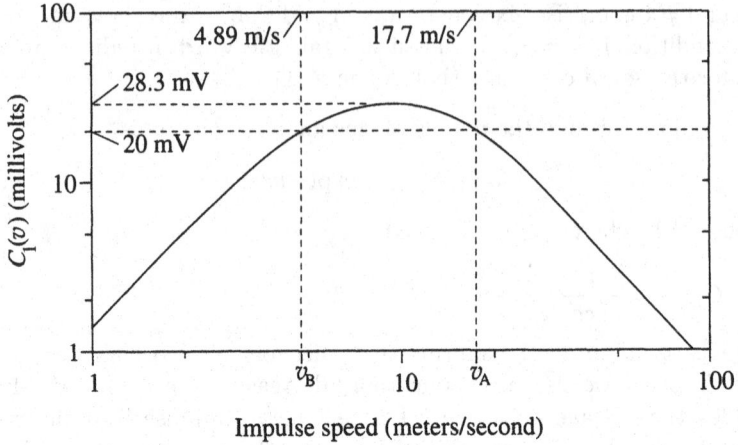

Figure 6.2. The function C_1 appearing in Equation (6.6) is plotted from Equation (6.8) as a function of v for parameter values corresponding to a standard squid axon that are given in Table 6.1.

Table 6.1. Markin–Chizmadzhev parameters for a Hodgkin–Huxley axon at 18.5°C.

Parameter	Value	Units
Q_0	2.4×10^{-9}	C
$V_{\max}$	90.5×10^{-3}	V
V_θ	20×10^{-3}	V
τ_1	2×10^{-4}	s
τ_2	2×10^{-3}	s
r	2×10^4	ohms/cm
c	1.5×10^{-7}	F/cm

Equations (6.3) and (6.4), Equation (6.7) takes the form

$$C_1(v) = \frac{Q_0^2}{v^2 c^2 \tau_1 V_{\max}} \left[\frac{1}{\tau_1} + \frac{1}{\tau_2} e^{-v^2 rc(\tau_1+\tau_2)} - \left(\frac{1}{\tau_1} + \frac{1}{\tau_2} \right) e^{-v^2 rc\tau_1} \right]. \tag{6.8}$$

Parameters of this equation for the standard Hodgkin–Huxley axon are posted in Table 6.1 and used to construct the plot of C_1 vs. v in Figure 6.2.

To find allowed values of the impulse speed, recall that at $\xi = 0$ the threshold voltage for the onset of sodium conductance is given by the condition

$$V_\theta = C_1(v).\tag{6.9}$$

Thus as indicated in Figure 6.2, the propagation velocity of a self-consistent impulse can be found by seeking the intersections between $C_1(v)$ and a horizontal line drawn at $V_1(0) = V_\theta$. For $V_\theta < 28.3$ mV, there are two such intersections, an upper one at $v = v_A$ and a lower one at $v = v_B$.

From the following argument, we see that the higher value of conduction velocity is stable. On the upper range (right-hand side) of Figure 6.2, $C_1(v)$ approaches the asymptote

$$C_1(v) \sim \left(\frac{Q_0^2}{V_{\max}c^2\tau_1^2} \right) \frac{1}{v^2}.$$

If v is raised above v_A, $C_1(v)$ becomes less than V_θ because the precursor is less able to penetrate into the undisturbed region of the axon at higher speeds. This makes the impulse slow down. If, on the other hand, v is smaller than v_A, $C_1(v)$ becomes greater than V_θ (because the precursor is better able to penetrate into region #1), causing the impulse to accelerate.

Similarly, the lower value of traveling-wave speed is unstable. Near and below $v = v_B$, $C_1(v)$ has the asymptote

$$C_1(v) \sim \left(\frac{Q_0^2 r^2(\tau_1 + \tau_2)}{2V_{\max}\tau_1} \right) v^2.$$

Thus, if v falls below v_B, the system is *less* able to satisfy the threshold condition, and the impulse dies out. For $v > v_B$, however, $C_1(v)$ becomes greater than the threshold voltage, and the impulse speeds up, eventually reaching the stable value, v_A.

In this manner, we see how well the M–C analysis manages to predict the occurrences of both the stable and unstable impulses shown in Figure 4.5. For the parameter values indicated in Table 6.1 (with the threshold voltage somewhat arbitrarily chosen to be $V_\theta = 20$ mV), these speeds (V_A and V_B) are found from the intersections of Equations (6.8) and (6.9) in Figure 6.2. As indicated in the following table, the values thus obtained are in reasonable agreement with the corresponding Hodgkin–Huxley velocities.

Impulse	M–C	H–H	Units
Stable	17.7	18.8	m/s
Unstable	4.89	5.66	m/s

M–C analysis also suggests a *safety factor* for impulse propagation on a squid nerve. To see this, note from Figure 6.2 that the maximum threshold voltage above which there is no traveling wave is 28.3 mV. Because this maximum allowable threshold is about 42% above the actual value, the M–C model has a safety factor that is in approximate accord with the estimates obtained from leading-edge analysis in Section 5.5.

In summary, the central point of the Markin–Chizmadzhev analysis is this:

> The exponentially decreasing precursor into region #1 *guides* the impulse (thereby determining its speed), and (retroactively) the main body of the impulse *generates* the precursor.

From the perspectives of Chapter 1, this is yet another example of a *closed causal loop*, which can be represented by the following diagram

Impulse generates precursor
↓   ↑
Precursor guides impulse

Evidently, the M–C formulation is able to capture qualitative aspects of nerve impulse propagation that are in approximate agreement with those obtained from the H–H equations in Chapter 4. We will return to this description when we consider ephaptic coupling in Chapter 8 and impulse propagation through regions of changing fiber diameter in Chapter 9.

6.2 FitzHugh–Nagumo (F–N) Models

In the previous section, the recovery effect in a nerve fiber was represented by assuming that the transmembrane ionic current follows a prescribed function of time after the transmembrane voltage reaches a threshold value. Although this M–C model gives useful results, it has a certain *ad hoc* quality, lacking the dynamic character of a nonlinear partial differential equation (PDE) such as the Hodgkin–Huxley system. In the subsequent sections of this chapter, we consider a simple PDE system that represents recovery dynamics on a nerve fiber.

Because the present formulation is aimed at obtaining insight into analytic tools and methods rather than generating relationships that predict the quantitative behavior of real axons, we assume that the basic nonlinear diffusion equation is normalized as in Equation (5.5). In other words, the space variable is measured in units of $1/\sqrt{rg}$ (where r is the series resistance per unit length of the axon and g is a transmembrane ionic conductance per unit length), and time is measured in units of a membrane time constant c/g (where c is the transmembrane capacitance per unit length). For typographical convenience, tildes are no longer used to indicate the normalized independent variables, x and t.

Following a formulation suggested by FitzHugh [10] and explicitly described by Nagumo, Arimoto, and Yoshizawa in 1962 [27], a *recovery variable* (R) is added to the right-hand side of Equation (5.5), which is then allowed to follow its own dynamics. Thus, the system we now consider is the FitzHugh–Nagumo (F–N) equation[3]

$$\frac{\partial^2 V}{\partial x^2} - \frac{\partial V}{\partial t} = f(V) + R, \qquad (6.10)$$

in which the recovery variable is governed by

$$\frac{\partial R}{\partial t} = \varepsilon(V + c - bR). \qquad (6.11)$$

In particular applications of this system, it may be convenient to represent $f(V)$ by the cubic polynomial of Equation (5.9) or the piecewise linear model of Equation (5.12).

Upon differentiating Equation (6.10) with respect to time and inserting Equation (6.11), one obtains for V the equivalent third-order nonlinear PDE

$$\frac{\partial^3 V}{\partial x^2 \partial t} + \varepsilon b \frac{\partial^2 V}{\partial x^2} - \frac{\partial^2 V}{\partial t^2} - [\varepsilon b + f'(V)] \frac{\partial V}{\partial t} - \varepsilon [V + c + bf(V)] = 0, \quad (6.12)$$

where $f'(V) \equiv df/dV$.

With $\varepsilon = 0$, the recovery variable remains constant, and Equation (6.10) reduces to the simple nonlinear diffusion equation that was used in the previous chapter to model leading-edge dynamics. With $\varepsilon > 0$, on the other hand, R can vary with time and space.

The fundamental idea of the F–N model is that a positive voltage causes R to increase with time through the action of Equation (6.11) and that an increase of R in Equation (6.10) eventually forces V back to its resting value.

Although analytic simplicity was the main motivation in constructing the right-hand sides of Equations (6.10) and (6.11), there are some crude physical interpretations, which can be seen by relating

$$f(V) + R \sim j_{\text{ion}},$$

where j_{ion} is the transmembrane ionic current defined in Equation (4.5). From this perspective,

$$f(V) \quad \sim \quad \text{sodium ion current, and}$$
$$R \quad \sim \quad \text{potassium ion current},$$

---

[3]This system was independently introduced by the present author to describe an electronic nerve fiber or *neuristor* [32, 33].

with the symbol "$\sim$" implying "roughly equivalent to" or "approximately representing." More particularly,

$$R \quad \sim \quad n\,,$$
$$\varepsilon b \quad \sim \quad \kappa/\tau_n\,,$$
$$\varepsilon V \quad \sim \quad \kappa n_0(V)/\tau_n\,,$$

with τ_n being the time constant for turn-on of potassium ion current at 6.3°C, n and n_0, respectively, the time-dependent and resting values of the potassium turn-on variable, and

$$\kappa = 3^{(\text{Temp}-6.3)}\,.$$

Thus, $\varepsilon \sim \kappa$ is sometimes referred to as a "temperature parameter."

During the 1970s, the F–N system became recognized by the applied mathematics community as a model that is both sufficiently flexible to model the recovery dynamics of a real nerve and simple enough for detailed analyses [4, 14, 15, 16, 17, 26, 28, 29, 30]. Let us take a look at some of their results.

6.3   Phase-Space Analysis of an F–N Model

To effect a traveling-wave analysis of the F–N equations, we transform the independent variables (x and t) to a moving coordinate system, as indicated in Appendix D, and assume temporal independence in the moving system. The dependent variables (V and R) then become functions of the traveling-wave coordinate $\xi \equiv x - vt$ with partial derivatives transforming as

$$\frac{\partial}{\partial x} \longrightarrow \frac{d}{d\xi} \quad \text{and} \quad \frac{\partial}{\partial t} \longrightarrow -v\frac{d}{d\xi}\,.$$

In this formulation, v is the speed of a traveling-wave solution to be determined in the course of the analysis. Thus the third-order F–N system can be written as the three first-order ODEs

$$\frac{dV}{d\xi} \;\equiv\; W\,,$$
$$\frac{dW}{d\xi} \;=\; f(V) + R - vW\,, \qquad\qquad (6.13)$$
$$\frac{dR}{d\xi} \;=\; \frac{\varepsilon}{v}(bR - V - c)\,,$$

defining a solution trajectory in the three-dimensional phase space (V, W, R).

Singular points (where all derivatives with respect to ξ are zero) occur at $W = 0$ with V and R solutions of

$$R + f(V) \;=\; 0\,,$$

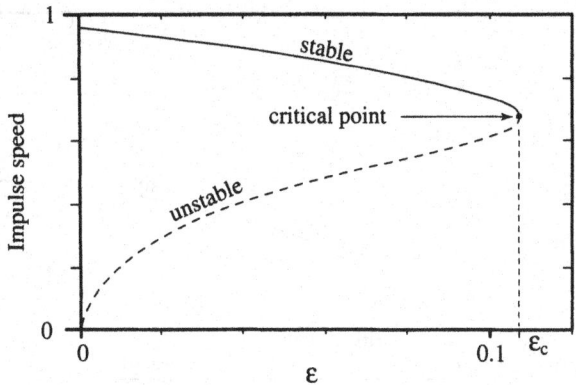

Figure 6.3. Propagation speeds for traveling-wave impulse solutions of a FitzHugh–Nagumo system plotted against the "temperature parameter" ε. (Redrawn from [11] with $f(V) = V^3/3 - V$, $b = 0.8$, and $c = 0.7$.)

$$ bR - V \;=\; c. $$

For $f(V)$ a "cubic-shaped" function and b sufficiently small, there is only one singular point, which can be rescaled to the origin.

Just as in the Hodgkin–Huxley analysis of Chapter 4, impulse-like solutions of the original PDE system correspond to *homoclinic trajectories* of the ODE system. In other words, it is of central interest to find solution trajectories of Equations (6.13) that issue from the SP at $\xi = -\infty$ and return to it again as $\xi \to +\infty$.

In the 1960s, FitzHugh did numerical calculations on the PDE system of Equations (6.10) and (6.11) and the ODE system of Equations (6.13) with a variety of parameters. This work demonstrated the existence of two homoclinic trajectories for distinctly different values of the impulse velocity, the faster being stable and the slower unstable. From this work, we sketch in Figure 6.3 a typical dependence of these two traveling-wave speeds on the "temperature parameter" ε, which controls the rate of change of the recovery variable R. In using this term, however, it should be remembered that the F–N system does not include the effects of a decrease in sodium turn-on time (τ_m) with increasing temperature that were discussed in the previous chapter. Thus, Figure 6.3 misses the experimentally observed increase in impulse speed with temperature given in Equation (5.4).

In Figure 6.3, the upper curve corresponds to the stable higher-amplitude H–H impulse of Figure 4.3 and the lower curve to the unstable lower-amplitude impulse. Figure 6.3 suggests that traveling-wave solutions are no longer possible at a sufficiently high temperature, in qualitative agreement with experimental observations on real nerves.

Interestingly, homoclinic solution trajectories of Equations (6.13) are not the only possibility for obtaining bounded traveling-wave solutions. Fam-

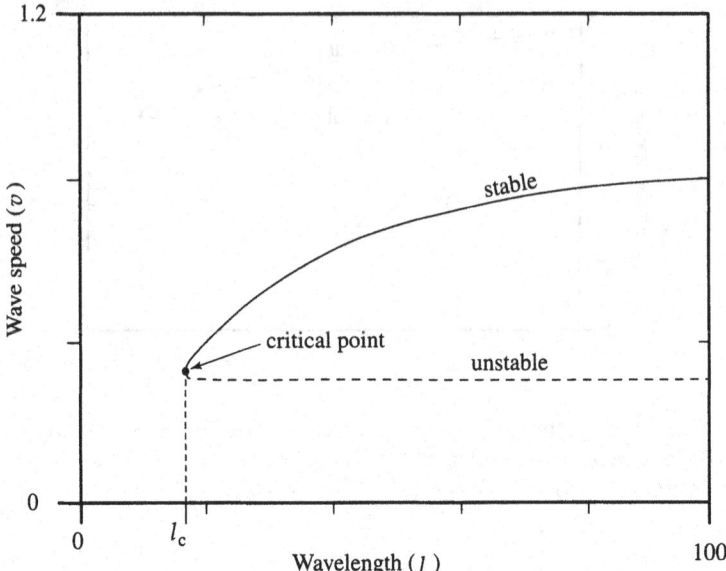

Figure 6.4. Propagation speeds for periodic traveling-wave solutions of a FitzHugh–Nagumo system plotted against the wavelength l. (Redrawn from [30] with $b = 0$, $c = 0$, $\varepsilon = 0.05$, and $f(V)$ the piecewise linear function of Equation (5.12) with $V_1 = 0.3$, $V_2 = 1$, and $\beta = 1$.)

ilies of closed trajectories (or *cycles*) in the phase space are also possible, and these correspond to *periodic* traveling-wave solutions of the original PDE system. For such periodic solutions, an important parameter is the wavelength l, defined by the condition

$$V(\xi) = V(\xi + l).$$

Periodic solutions of the F–N system were studied numerically by Rinzel and Keller, and a typical result is sketched in Figure 6.4 [30].

The numerical calculations by Rinzel and Keller show four salient features: (1) a higher-speed curve that is numerically stable, (2) a lower-speed curve that is numerically unstable, (3) a critical value of wavelength (l_c) below which periodic traveling waves are not found, and (4) as $l \to \infty$, the two traveling-wave speeds approach those of isolated impulses.

How can we understand the physical significance of the numerical results plotted in Figures 6.3 and 6.4? Return to the discussion of Section 4.6 in which a variety of mechanisms for degradation of impulse propagation on a squid nerve were sketched. Note in particular the calculations in Figure 4.6 showing the effects of "narcotizing" an H–H model by reducing the maximum allowed values of sodium and potassium conductances. Just as in Figures 6.3 and 6.4, a critical value of some experimental parameter is observed at which the fast and slow traveling-wave solutions merge and beyond which no traveling-wave solutions are found.

From the F–N perspective, the critical value of ε in Figure 6.3 is qualitatively equivalent to the maximum temperature at which a real squid axon can support a nerve impulse. At higher temperatures, the time constant for turn-on of potassium current becomes so short that sustained propagation of an isolated impulse is no longer possible.

Similarly, the critical value of wavelength (l_c) in Figure 6.4 indicates a minimum spacing between impulses. For squid axons, such an effect is seen in the double impulse data of Figure 4.8 showing how the ratio

$$\frac{\text{speed of second impulse}}{\text{speed of first impulse}}$$

of traveling-wave speeds of a pair of impulses depends on their temporal spacing (T). In the context of Figure 6.4, $v = l/T$; thus, if

$$T < l_c/v,$$

the preceding speed ratio is zero. In other words, if the second impulse follows too closely behind the first, it is unable to propagate in the *absolute refractory zone* of the first impulse.

6.4 Power Balance for Traveling Waves

In Chapter 4, it was noted that a traveling-wave impulse on a smooth nerve fiber establishes a condition of *power balance* between the rate at which energy is being dissipated and the rate at which it is released. In physical terms, energy is generated by the "ionic batteries" (V_{Na}, V_{K}, and V_{L}) in Equation (4.5) and dissipated by circulating ionic currents. An advantage of the F–N formulation is that the algebra becomes simple enough to follow the details.

To see how this goes, note that Equation (6.12) can be written in the form

$$\frac{\partial^2 V}{\partial t^2} + \varepsilon V = \frac{\partial^3 V}{\partial x^2 \partial t} - F'(V)\frac{\partial V}{\partial t},$$

where it is assumed that parameters b and c equal 0, and conservative terms are collected on the left-hand side. (A corresponding result, with four extra terms, is obtained for b and c not equal to zero. Because these terms add little physical insight, the simpler form is presented here, but the reader may wish to include them as an exercise.)

Conservative terms are recognized because they can be derived by substituting the Lagrangian density

$$\mathcal{L} = \frac{1}{2}\int_{-\infty}^{\infty}\left(\varepsilon V^2 - \left(\frac{\partial V}{\partial t}\right)^2\right)\,dx$$

into the Euler equations (see Appendix A), which implies that the energy transported by the traveling wave is

$$\mathcal{E} = \frac{1}{2} \int_{-\infty}^{\infty} \left(\varepsilon V^2 + \left(\frac{\partial V}{\partial t} \right)^2 \right) dx. \tag{6.14}$$

The first term in this expression represents electrostatic energy that is stored in the capacitance of the cell membrane. The second energy term stems from a phenomenological inductance of the membrane, which has been studied by Mauro and his colleagues [25] and was mentioned at the close of Chapter 4.

Differentiating Equation (6.14) with time, using the FitzHugh–Nagumo equation, and integrating by parts, one finds that

$$\frac{d\mathcal{E}}{dt} = - \int_{-\infty}^{\infty} \left(\left(\frac{\partial^2 V}{\partial x \partial t} \right)^2 + \frac{dF(V)}{dV} \left(\frac{\partial V}{\partial t} \right)^2 \right) dx. \tag{6.15}$$

Assuming a traveling-wave solution with

$$V(x,t) = V(x - vt) \equiv V(\xi),$$

the right-hand side of Equation (6.15) must be zero. Because the first term on the right-hand side of Equation (6.15) is always negative, a necessary condition for a traveling wave is a range of voltage over which

$$\frac{dF(V)}{dV} < 0.$$

The right-hand side of Equation (6.15) being zero thus expresses the condition of *power balance* for a traveling wave.

The concept of power balance is so important for understanding the nature of nerve impulse dynamics that more discussion seems appropriate. First, note that the power balance constraint depends on two assumptions: that the nerve fiber is a uniform structure in the direction of impulse propagation (x-direction) and that the nerve impulse propagates without change of speed or shape. (If power balance were not established under these two assumptions, then the energy of the wave would change, thereby causing changes in the speed or shape of the impulse.) Interestingly, *any* uniform PDE can be analyzed in this manner because each term is either energy-conserving or dissipative.

At this point, the reader may ask: Isn't conservation of energy a general law? How can a nerve impulse break a fundamental law of physics? The answer, of course, is that conservation of energy is not violated by the dynamics of a nerve impulse. The energy released by the passage of an impulse is dissipated through ohmic losses of the circulating ionic currents, thereby slightly heating the external medium, but this general energy conservation is of no help in understanding the behavior of a nerve impulse. The dynamics of an impulse are governed by a condition such as Equation

(6.15) for which a more general conservation of energy is not wrong but irrelevant.

It was noted in Chapter 2 that a system for which the dynamics conserve energy has the property that time is bidirectional. In describing the motions of the planets, recall that one can reverse the sign of time and still have solutions of Newton's laws of motion. Thus, it is possible for astronomers not only to predict where the planets will be some hundreds of years in the future but also to retrodict where they were hundreds of years in the past.

For diffusion problems in general and nerve problems in particular, energy is not conserved and time is unidirectional, so retrodiction is not feasible. All of us eventually come to recognize this "arrow of time" as an unavoidable feature of biological dynamics.

An example of time's arrow in nerve dynamics is provided by the considerations of impulse stability to be discussed in Section 6.5.2. Note that an impulse on the upper curve of the traveling-wave locus in Figure 6.3 is stable, whereas an impulse on the lower curve is unstable. In the context of Equation (6.15), this observation can be qualitatively described as follows.

- *Upper curve.* Suppose that the impulse amplitude increased slightly above the value at which power balance is established. Then the dissipative term

$$-\int_{-\infty}^{\infty} \left(\frac{\partial^2 V}{\partial x \partial t}\right)^2 dx$$

becomes greater in magnitude than the energy-producing term

$$-\int_{-\infty}^{\infty} \frac{dF(V)}{dV}\left(\frac{\partial V}{\partial t}\right)^2 dx,$$

causing $\mathcal{E}$ to decrease. The impulse amplitude then relaxes back to its power balance value.

- *Lower curve.* A corresponding amplitude increase on the lower curve will cause the dissipative term to become less than the energy-producing term. Thus the impulse amplitude will continue to grow until it reaches the stable value of the upper curve. If, on the other hand, the amplitude is decreased below the power balance level, energy production fails to match the rate of dissipation, and the amplitude falls to zero.

If the direction of time were reversed, interestingly, the upper curve would become unstable and the lower curve stable.

6.5   Structure of an F–N Impulse

While providing a reasonable description of recovery in a nerve, the FitzHugh–Nagumo model is also a useful context for understanding the dynamics of impulse propagation. In this section, we consider the various components of a nerve impulse and study how their relationships contribute to impulse stability.

6.5.1   Rapid and Relaxing Regimes

If the temperature parameter (ε) in Equations (6.13) is set to zero, R remains constant. In this case, as we have seen in the previous chapter, the only traveling-wave solution with nonzero velocity is a level change from one of the outer zeros of $F(V) + R$ to the other; a moving impulse-like solution does not exist.

With $0 < \varepsilon \ll 1$, however, there is always an impulse solution. Be ε ever so small, this impulse continues to exist. Let us take advantage of this fact by viewing the very slow changes in the recovery variable as a small perturbation to the leading-edge solution. To simplify the description, set b and c to zero in Equations (6.13) and assume that

$$f(V) = V(V - a)(V - 1),$$

with $a < 1/2$.

As $\varepsilon \to 0$, the solitary-wave solution appears as indicated in Figure 6.5, with four distinctly different regimes. Each of these regimes can be related to the corresponding homoclinic trajectory in the (V, W, R) phase space shown in Figure 6.6.

To see how this goes, let us walk around the homoclinic trajectory, starting at the singular point

$$(V, W, R) = (0, 0, 0)$$

and proceeding in the $-\xi$ direction. Along each regime of this trajectory, the reader is invited to identify the corresponding features in Figures 6.5 and 6.6.

Regime #1. This corresponds to the leading edge of the impulse, where $R \approx 0$ and V makes a rapid transition from 0 to 1. As $\varepsilon \to 0$, the traveling wave approaches

$$V(\xi) \to V_0(\xi) = \frac{1}{1 + \exp(\xi/\sqrt{2})}$$

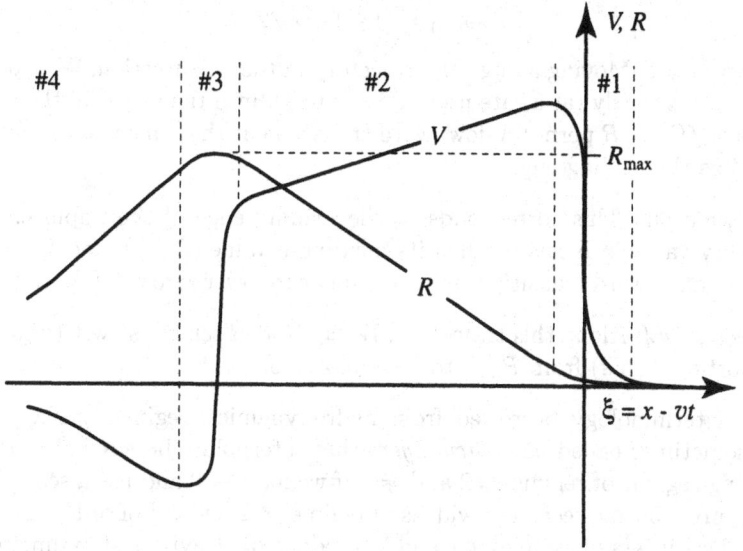

Figure 6.5. V and R as functions of the traveling-wave variable ξ as $\varepsilon \to 0$.

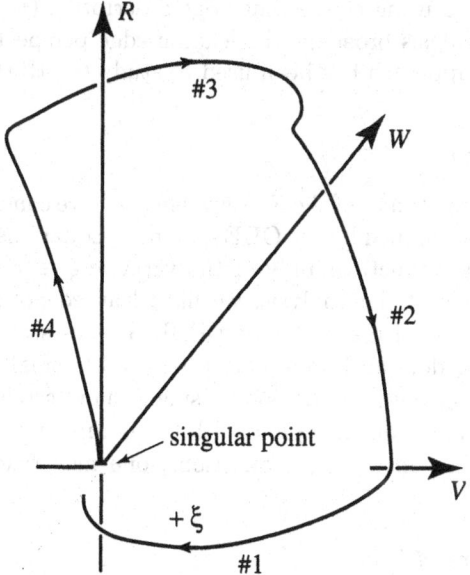

Figure 6.6. The homoclinic solution trajectory of Equations (6.13) corresponding to the traveling-wave impulse shown in Figure 6.5.

with velocity approaching[4]

$$v \to v_0 = (1 - 2a)/\sqrt{2}\,.$$

Regime #2. Moving along this trajectory in the $-\xi$ direction, $W \approx 0$ and R increases slowly (at a rate proportional to ε) until the values of the outer zeros of $f(V) + R$ permit a downward transition at the same traveling-wave speed as the leading edge.

Regime #3. This corresponds to the trailing edge of the impulse. The recovery variable R has reached its maximum value (R_{max}), and V makes a rapid downward transition from the upper to lower zeros of $f(V) + R_{max}$.

Regime #4. Along this trajectory, $W \approx 0$ and R relaxes slowly (at a rate proportional to ε) from R_{max} to 0 as $\xi \to -\infty$.

In a terminology borrowed from hydrodynamics, regimes #1 and #3 are sometimes called *boundary layers* that interpolate between the slowly varying regions of regimes #2 and #4. If we let $\varepsilon \to 0$ and use a scale for ξ in Figure 6.5 that keeps the widths of regimes #2 and #4 of order one, the boundary layers of regimes #1 and #3 reduce to Heaviside step functions. If we use a scale that keeps regime #1 of order one as $\varepsilon \to 0$, then regime #2 becomes a plateau of unbounded length.

In Appendix E, it is shown how *perturbation theory* can be used to extend the present ideas to allow an estimate of the first-order dependence of F–N impulse speed on ε using the leading-edge description (v_0 and $V_0(\xi)$) as a basis [3]. This analysis broadens the leading-edge perspectives of Chapter 5, and a related approach has been used to study the H–H system [2].

6.5.2   Stability

To this point in our studies of the F–N system, we have considered traveling-wave solutions determined by the ODE system of Equations (6.13). Because traveling waves are functions only of the variable $\xi \equiv x - vt$, where v is fixed in the course of the analysis, we have learned nothing about their *stability* as solutions of the underlying PDE. In this section, we ask such questions as: How does an F–N equation respond to small deviations from the exact traveling-wave solutions? Do small changes relax back to zero, implying stability? Or do they grow with time, implying instability?

In attempting to answer such questions, one may take three different approaches.

Numerical Studies of Stability
One way to assess the stability of a traveling-wave solution is to use it

---

[4]This follows from Equations (5.10) and (5.11), where $a = V_1/V_2$ and Luther's factor is removed by the normalization.

as the initial condition for numerical computations that are based on the original PDE system given in Equation (6.12). Assuming that the numerical procedure is reliable, divergence of the numerical solution from traveling-wave initial conditions implies instability; otherwise, the impulse is stable. Such computations have now been carried out many times for the F–N system for a wide range of parameters, and all results have been found to be consistent with the following statements: traveling wave impulses corresponding to the upper curve in Figure 6.3 are stable; and impulses corresponding to the lower curve are unstable [11, 28, 29, 30].

Because there may be some unexplored combinations of parameters giving contradictory results, such numerical studies do not provide mathematical proofs but offer circumstantial evidence.

Qualitative Stability Estimates
The above-mentioned numerical observations are in accord with qualitative (or common sense) considerations that help one gain a better appreciation for the physical nature of nerve impulse stability.

First, as we have seen in Section 5.4 and Appendix D, mathematical studies of leading-edge dynamics show the upper curve to be stable and the lower curve unstable in the limit $\varepsilon \to 0$ of Figure 6.3.

Second, for the F–N impulse shown in Figure 6.5, we can ask what happens if an impulse gets a little too long or too short. Along the upper branch of Figure 6.3, the answer to this question is that an excessively long impulse will become shorter, and an impulse that is too short will become longer, with both converging on the traveling wave shown in Figure 6.5. To see this, consider the following argument.

From the discussion in Section 6.5.1, the correct value of R_{max} is determined by the condition that the speed of the trailing edge (regime #3) must equal that of the leading edge (regime #1). Moving in the $-\xi$-direction, the leading edge is an upward transition between the two outer zeros of $f(V)$, whereas the trailing edge is a downward transition between the two outer zeros of

$$\tilde{f}(V) = f(V) + R_{\mathrm{max}} \,.$$

Now, consider the following two cases.

(1) R_{max} is larger than the correct value, which indicates that the impulse is too long. Then the positive-going area (A^{+}) under $\tilde{f}(V)$ is increased, making it more dissipative. Because the trailing edge is a downward transition, requiring the dissipation of energy, this change in the shape of $\tilde{f}(V)$ accelerates the trailing edge, which shortens the impulse and returns R_{max} to its correct value.

(2) R_{max} is smaller than the correct value, which indicates that the impulse is too short. Then the positive-going area under $\tilde{f}(V)$ is decreased, which slows down the trailing edge, thereby lengthening the impulse and again returning R_{max} to its correct value.

Thus do the nonlinearities in the system allow an impulse to automatically adjust itself for the correct length, as shown in Figure 6.5. These qualitative considerations suggest that the upper curve of Figure 6.3 is stable for some range $0 < \varepsilon \ll \varepsilon_c$.

Mathematical Analyses

Because the initial growth of an instability is governed by a PDE obtained by linearizing the nonlinear PDE about the nerve impulse, a general stability study can be based on a mode analysis of that linearized PDE. This is the approach presented in Appendix D.

The key idea in this analysis is to look at individual modes (also called *eigenfunctions*) of the linearized PDE and ask whether they grow or decay with time. (Familiar examples of such modes are the various ways that a guitar string or a drum head can vibrate, each mode having a characteristic shape and frequency of vibration.)

Because the PDE is linear, the temporal behavior of several modes will be determined by a sum of their individual behaviors. Thus, a sufficient condition for instability is that at least one mode of the linearized PDE grows with time. A necessary condition for asymptotic stability, on the other hand, is that all modes decay with time.

The first step therefore is to express the original nonlinear PDE in a *moving coordinate system,* for which the independent variables become

$$x \;\;\rightarrow\;\; \xi = x - vt\,,$$
$$t \;\;\rightarrow\;\; \tau = t\,,$$

where v is the speed of the undisturbed traveling wave. Thus, ξ and τ are, respectively, the distance and time in the moving system, where τ is measured by a clock in the stationary system. It is in the moving coordinate system that the original nonlinear PDE is linearized about the traveling wave.

Because the linearized PDE does not depend explicitly on time in the moving system, the method of separation of variables (sketched in Appendix D) implies that the temporal behavior of a mode will be as

$$e^{\lambda \tau}\,,$$

where λ is an *eigenvalue* of the corresponding eigenfunction (or mode). (In a different language, λ is an element of the *spectrum* of the linearized PDE and its boundary conditions.)

Eigenvalues may be of two types: *discrete* eigenvalues, occurring at isolated points in the complex λ-plane, corresponding to localized dynamics (internal oscillations, for example) of an impulse, or *continuous* eigenvalues, which are dense on lines in the λ-plane and correspond to modes of radiation from the traveling wave.

If the real part of λ is positive for any mode, that mode will grow with time, implying instability. If, on the other hand, the real parts of the eigenvalues of *all* modes are less than zero, then the corresponding traveling wave will be asymptotically stable.

As noted in Appendix D, there is always one mode for which $\lambda = 0$ is a discrete eigenvalue: the derivative of the traveling wave with respect to the traveling-wave variable ξ. Adding this mode to a traveling wave merely translates that traveling wave in the ξ direction; thus, it is called a *translation mode*. Because this translation mode neither grows nor decays with time, a nerve impulse is at most stable but not asymptotically stable.[5]

The mathematical basis for stability analysis of nerve axon models was established during the 1970s in an important series of papers by Evans. In the first of these, he formulated a general set of nonlinear PDEs for which both the H–H and F–N systems are special cases, showing that the impulse is stable relative to the nonlinear PDE if and only if it is stable in the linearized PDE [5]. The second paper studied stability of the resting state, from which it is seen that the resting states of both the H–H and F–N systems are stable [6]. In the third paper of the series, Evans showed that disturbed impulses relax exponentially back into (possibly displaced) traveling waves if and only if [7]:

- There are no eigenvalues for which $\mathrm{Re}[\lambda] > 0$.

- $\mathrm{Re}[\lambda] = 0$ only if $\lambda = 0$.

- The eigenvalue at $\lambda = 0$ is simple. (In other words, the translation mode is the only mode for which $\lambda = 0$.)

Finally, in the fourth paper, Evans showed how to construct an analytic function that allows the preceding stability conditions to be deduced [8]. Thus, zeros of this "Evans function" are eigenvalues, and multiple zeros indicate that the corresponding eigenvalues are not simple.

The work of Evans gave criteria for the stability of impulses but left open whether any traveling wave representing a nerve impulse in, say, the H–H or F–N systems, is actually stable. The first full result of impulse stability is due to Jones [19, 20] for the fast F–N pulse. He verified stability in the limit $\varepsilon \ll 1$, for which the leading edge is separated from the trailing edge by a long intermediate phase. By well-known results in reaction–diffusion equations (see Fife and McLeod [9]), the front and back are individually stable for their relevant reduced systems, with the recovery variable fixed. The proof of stability then involves showing that the front and back lock to each other under the dynamics of the PDE rather than drift apart.

---

[5]Some choose to define stability with respect to a measure that permits arbitrary translations in the ξ-direction, thereby allowing asymptotic stability. In the present work, this artifice is avoided because an impulse at a different location is considered to be a different impulse.

Discerning which of these effects actually occurs amounts to showing that the eigenvalue associated with relative motion of the front and back lies in the left half-plane, where $\text{Re}[\lambda] < 0$. Jones used the Evans function to show that the corresponding zero lies in the left half of the complex λ-plane. Yanagida [35] took a similar approach to give a somewhat more qualitative proof of fast impulse stability.

There is no known stability proof away from the singular limit because the decomposition of the impulse into front and back is then lost. A continuation argument can be used as ε is increased to show that no bifurcations of other impulses take place up to the parameter values at which the fast and slow pulses coalesce. Internal oscillations, however, may destabilize the impulse, and it is an analytic challenge to find techniques that rule this out, as is suggested by numerical evidence.

6.6 Recapitulation

This chapter began with the simplest formulation of nerve impulse dynamics that includes the phenomenon of recovery. Developed by Markin and Chizmadzhev in the 1960s and 1970s for a variety of neural interactions, the M–C model assumes that upon reaching threshold the membrane goes through a prescribed time course of inward (sodium) current and outward (potassium) currents, after which the net ionic charge crossing the membrane is zero. Rather good estimates for the traveling-wave speeds of both the larger (stable) and the smaller (unstable) impulses on the standard Hodgkin–Huxley axon are obtained by this method. In addition, a reasonable safety factor for impulse propagation is obtained.

At a somewhat higher level of mathematical sophistication, recovery is represented by a single dynamic variable in a formulation introduced by Nagumo and FitzHugh in the early 1960s. This F–N model can be viewed as a reduced version of the Hodgkin–Huxley description, with one variable representing the dynamics of both sodium turn-off and potassium turn-on. The F–N model can be conveniently analyzed in a phase space of three dimensions and helps one grasp the qualitative behavior of the more complex H–H model. The F–N system also provides a context in which to discuss the power balance principle on which nerve impulse dynamics are based.

Next, the structure of an F–N impulse was considered in some detail, paying particular attention to the relationships among the fast phases of the leading and trailing edges, leading to a qualitative understanding of impulse stability. Finally, several approaches to the study of impulse stability were sketched, emphasizing an intuitive understanding of the underlying dynamic behavior. Although it is not as accurate as the H–H equations, familiarity with the F–N model helps one to better understand the related concepts of threshold for impulse ignition, all-or-nothing propagation, and nerve impulse stability.

References

[1] PC Bressloff, A dynamical theory of spike train transitions in networks of integrate-and-fire oscillators, *SIAM J. Appl. Math.* 60 (2000) 820–841.

[2] GA Carpenter, A geometric approach to singular perturbation problems with applications to nerve impulse equations, *J. Differential Eq.* 23 (1977) 152–173.

[3] RG Casten, H Cohen, and PA Lagerstrom, Perturbation analysis of an approximation to Hodgkin–Huxley theory, *Q. Appl. Math.* 32 (1975) 365–402.

[4] H Cohen, Nonlinear diffusion problems. In *Studies in Applied Mathematics*, AH Taub (ed), Prentice-Hall, Englewood Cliffs, NJ, 1971, pp 27–64.

[5] JW Evans, Nerve axon equations: I. Linear approximations, *Indiana Univ. Math. J.* 21 (1972) 877–885.

[6] JW Evans, Nerve axon equations: II. Stability at rest, *Indiana Univ. Math. J.* 22 (1972) 75–90.

[7] JW Evans, Nerve axon equations: III. Stability of the nerve impulse, *Indiana Univ. Math. J.* 22 (1972) 577–593.

[8] JW Evans, Nerve axon equations: IV. The stable and unstable impulse, *Indiana Univ. Math. J.* 24 (1975) 1169–1190.

[9] P Fife and JB McLeod, The approach of solutions of nonlinear diffusion equations to travelling front solutions, *Arch. Rational Mech. Anal.* 65 (1977) 335–361.

[10] R FitzHugh, Impulses and physiological states in theoretical models of nerve membrane, *Biophys. J.* 1 (1961) 445–466.

[11] R FitzHugh, Mathematical models of excitation and propagation in nerve. In *Biological Engineering,* HP Schwan (ed), McGraw–Hill, New York, 1969, pp 1–85.

[12] W Gerstner and JL van Hemmen, Associative memory in a network of 'spiking' neurons, *Network* 3 (1992) 139–164.

[13] W Gerstner, R Ritz, and JL van Hemmen, A biologically motivated and analytically soluble model of collective oscillations in the cortex: I. Theory of weak locking, *Biol. Cybern.* 68 (1993) 363–374.

[14] JM Greenberg, A note on the Nagumo equation, *Q. J. Math.* 24 (1973) 307–314.

[15] SP Hastings, On a third order differential equation from biology, *Q. J. Math.* 23 (1972) 435–448.

[16] SP Hastings, The existence of periodic solutions to Nagumo's equation, *Q. J. Math.* 25 (1974) 369–378.

[17] SP Hastings, The existence of homoclinic and periodic orbits for the FitzHugh–Nagumo equations, *Q. J. Math.* 27 (1976) 123–134.

[18] AL Hodgkin and AF Huxley, A quantitative description of membrane current and its application to conduction and excitation in nerve, *J. Physiol. (London)* 117 (1952) 500–544.

[19] CKRT Jones, Some ideas in the proof that the FitzHugh–Nagumo pulse is stable. In *Nonlinear Partial Differential Equations,* J Smoller (ed), Contemporary Mathematics 17, American Mathematical Society, Providence, 1984, pp 287–292.

[20] CKRT Jones, Stability of the travelling wave solution of the FitzHugh–Nagumo system, *Trans. Am. Math. Soc.* 286 (1984) 431–469.

[21] BW Knight, Dynamics of encoding in a population of neurons, *J. Gen. Physiol.* 59 (1972) 734–766.

[22] BW Knight, The relationship between the firing rate of a single neuron and the level of activity in a population of neurons, *J. Gen. Physiol.* 59 (1972) 767–778.

[23] AS Kompaneyets and VT Gurovich, Propagation of an impulse in a nerve fiber, *Biophysics* 11 (1966) 1049–1052.

[24] VS Markin and YA Chizmadzhev, On the propagation of an excitation for one model of a nerve fiber, *Biophysics* 12 (1967) 1032–1040.

[25] A Mauro, F Conti, F Dodge, and R Schor, Subthreshold behavior and phenomenological impedance of the squid giant axon, *J. Gen. Physiol.* 55 (1970) 497–523.

[26] HP McKean, Jr, Nagumo's equation, *Adv. Math.* 4 (1970) 209–223.

[27] J Nagumo, S Arimoto, and S Yoshizawa, An active impulse transmission line simulating nerve axon, *Proc. IRE* 50 (1962) 2061–2070.

[28] J Rinzel, Spatial stability of traveling wave solutions of a nerve conduction equation, *Biophys. J.* 15 (1975) 975–988.

[29] J Rinzel, Neutrally stable traveling wave solutions of nerve conduction equations, *J. Math. Biol.* 2 (1975) 205–217.

[30] J Rinzel and JB Keller, Traveling wave solutions of a nerve conduction equation, *Biophys. J.* 13 (1973) 1313–1337.

[31] DH Sattinger, On the stability of waves of nonlinear parabolic systems, *Adv. Math.* 22 (1976) 312–355.

[32] AC Scott, Analysis of nonlinear distributed systems, *Trans. IRE* CT–9 (1962) 192–195.

[33] AC Scott, Neuristor propagation on a tunnel diode loaded transmission line, *Proc. IEEE* 51 (1963) 240.

[34] AC Scott, *Nonlinear Science: Emergence and Dynamics of Coherence Structures,* Oxford University Press, Oxford, 1999.

[35] E Yanagida, Stability of fast travelling pulse solutions of the FitzHugh–Nagumo equations, *J. Math. Biol.* 22 (1985) 81–104.

7
Myelinated Nerves

Following the Hodgkin–Huxley formulation of nerve impulse dynamics for the giant axon of the squid [31], most mathematical studies have focused on smooth nerve fibers, as in the previous three chapters. Although this picture is appropriate for the squid axon, many vertebrate nerves—including the frog motor nerve studied by Galvani and axons of mammalian brains—are bundles of discrete, periodic structures, comprising active nodes (also called "nodes of Ranvier") separated by relatively long fiber segments that are insulated by a fatty material called myelin. In such *myelinated* nerves, the wave of activity jumps from one node to the next, and should be modeled by nonlinear difference-differential equations rather than by PDEs.

Impulse propagation on myelinated nerves (called *saltatory* conduction by the electrophysiologists) is qualitatively similar to a row of falling dominos or to the signal fires of coastal warning systems during the Middle Ages. In an evolutionary context, myelinated nerve structures are useful because they allow an increase in the speed of a nerve impulse while decreasing the diameter of the nerve fiber. Thus, the motor nerves of vertebrates may comprise several hundred individual saltatory fibers, each serving as an independent signaling channel [76]. The rabbit sciatic nerve shown in Figure 1.2, for example, can transmit information about three orders of magnitude faster than a squid axon of the same diameter while expending much less energy in transmitting an individual impulse than does a smooth fiber.

Over the past century, studies of impulse propagation on myelinated nerves have been carried on in three different professional areas, among which there has been less than ideal communication. Electrophysiology, of course, is the foremost of these groups [7, 11, 33, 35, 61, 62, 68, 69], and since the 1960s these researchers have assembled a trove of data and corresponding-ing analytical interpretations, demonstrating the physiological importance of myelination [10, 16, 22, 23, 24, 29, 32, 34, 48, 52, 53, 60, 70]. In the 1960s, the engineering community began to study the phenomenon of saltatory conduction, motivated by the possibility of using recently invented semi-conductor devices to construct electronic nerve models. Whimsically called *neuristors*, such structures were anticipated to provide a novel basis for computer design [40, 41, 42, 49, 58, 59, 65]. Finally in the 1980s, saltatory conduction became of interest to applied mathematicians, who have added their special insights to the collective understanding [3, 5, 6, 20, 21, 38, 39]. It is hoped that the present discussion may help to bring these scattered centers of activity together, encouraging future researchers to draw on efforts that have gone before.

From these previous studies, it is seen that the phenomenon of saltatory conduction on myelinated nerves introduces two qualitatively important features. On the up side is the above-mentioned increase in speed of con-duction with reduced fiber diameter and energy dissipation. On the down side, however, is the possibility of *failure* when the distance—or electrical resistance—between successive active nodes becomes too large.

The aim in this chapter is to present a simple yet physically reasonable model for impulse propagation on myelinated nerve fibers and demon-strate the ability of this model to describe both theoretical results and experimental observations.

7.1   An Electric Circuit Model

In Figure 7.1(a) is sketched a single myelinated nerve fiber showing active nodes separated by regions of the fiber that are insulated by myelin. In 1978, a rather thorough numerical calculation of conduction velocity on this structure was published by Moore et al. [48] in which the ionic cur-rents crossing the active nodes were given by Hodgkin–Huxley equations and each myelinated region was represented by ten repeated segments of series resistance and shunt capacitance. While giving a good value for the conduction velocity (22.65 m/s compared with a measured value of about 23 m/s), the model of Moore et al. showed this velocity to be relatively independent of a seemingly key parameter: the internode spacing (s). One aim of this chapter is to present a simpler numerical model from which this rather surprising parameter insensitivity can be intuitively understood.

To this end, consider the difference-differential equations (DDEs)

$$V_n - V_{n+1} = (R_i + R_o)I_n \tag{7.1}$$

and

$$I_{n-1} - I_n = C\frac{dV_n}{dt} + I_{\text{ion},n}, \tag{7.2}$$

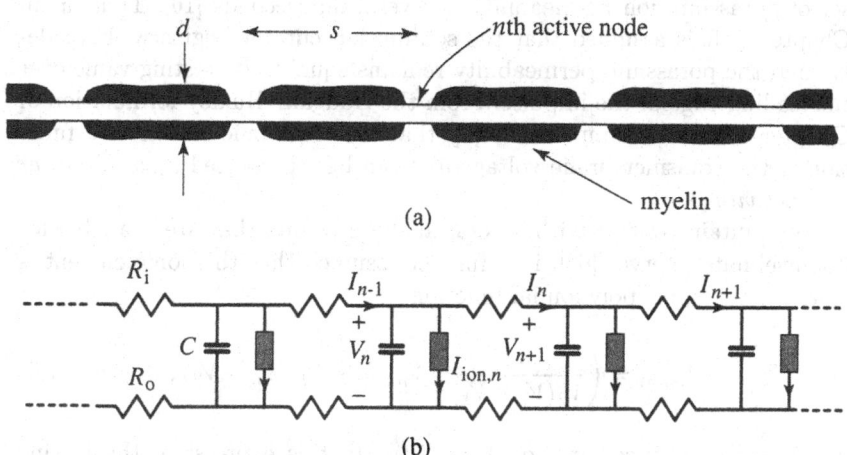

Figure 7.1. (a) A single myelinated nerve fiber (not to scale). (b) The corresponding electric circuit diagram.

for which a corresponding electric circuit diagram is shown in Figure 7.1(b).[1]

In these equations, the index n indicates successive active nodes, each characterized by a transverse voltage across the membrane (V_n). A second dynamic variable is the current (I_n) flowing longitudinally through the fiber from node n to node $n + 1$. Thus Equation (7.1) is merely Ohm's law, which relates the voltage difference between two adjacent nodes of the current flowing between them times the sum of the inside and outside resistances, R_i and R_o.

Equation (7.2) says that the current flowing into the nth node from the $(n - 1)$th node (I_{n-1}) minus the current flowing out of it to the $(n + 1)$th node (I_n) is equal to the following two components of transverse (inside to outside) current leaving the node: capacitive current, $C \, dV_n/dt$, and ionic current, $I_{\text{ion},n}$, comprising mainly a sodium component [31].

The time delay for the onset of sodium ion permeability is rather short (in the frog nerve it is about 0.1 ms), whereas the time delay for the on-

---

[1]More correctly, the passive fiber joining two active nodes should be represented by a linear diffusion equation (see Section 9.1.1), as was approximately done by Moore et al. [48]. In Equations (7.1) and (7.2), however, the passive internode fiber is modeled by a single series resistance (R_i) and a single shunt capacitor (equal to the capacitance of the myelin sheath), which is simply added to the node capacitance to obtain the total capacitance C. Although this approximation neglects shunt conductance of the myelin sheath, Moore et al. have shown that it has a negligible effect on conduction velocity. Such a "Π-network approximation" for the internode fiber greatly eases computational problems while reducing the number of parameters to be considered, thereby facilitating interpretations of numerical results.

set of potassium ion permeability is several milliseconds [16]. Thus, as in Chapter 5, it is assumed that the sodium ion current begins without delay and the potassium permeability remains equal to its resting value over the leading edge of the impulse. From the Hodgkin–Huxley formulation of Chapter 4, these assumptions imply that the total ionic current is a function of the transmembrane voltage and can be represented by a *nonlinear conductivity* [31].

To maintain contact with several analytic results that are available for nonmyelinated nerves [66], it is further assumed that this ionic current is given by the cubic polynomial function

$$I_{\text{ion},n} = \left(\frac{G}{V_2(V_2 - V_1)}\right) V_n(V_n - V_1)(V_n - V_2),\qquad(7.3)$$

which was introduced in Equation (5.9). In this expression, the resting potential of the active membrane is zero, and the parameters are defined as follows. The threshold voltage at which sodium current begins to flow *into* an active node is V_1. The Nernst (diffusion) potential at which total (primarily sodium) ion current returns to zero is V_2. The total (primarily sodium) ionic conductance near V_2 is G.

Consider next how the parameters of this model are obtained from the following detailed measurements on a single frog motor axon [16, 32, 70]. We will refer to this structure as the standard frog axon.

Standard frog axon

Distance between nodes $(s) = 2$ mm.

Outside fiber diameter $(d) = 14$ μm.

Internal resistance/length $(r_i) = 140$–145 megohm/cm.

External resistance/length $(r_o) \ll r_i$.

Capacity of myelin/length $(c_m) = 10$–16 pF/cm.

Capacity of active node $(C_n) = 0.6$–1.5 pF.

Experimental impulse speed $v_e = 23$ m/s.

From these data,

$$C = C_n + sc_m = 3.7 \pm 1 \text{ pF},$$

where the distributed capacitance of the internode myelin sheath has been lumped together with nodal membrane capacitance.

For an isolated nerve fiber in an experimental chamber, the cross section for external current flow is much greater than that for internal current;

Table 7.1. Standard membrane permeability parameters for an active node of a frog's sciatic nerve [16].

Parameter	Value	Units
G_{Na}	0.57	μmhos
G_K	0.104	μmhos
G_L	0.025	μmhos
V_K	0	mV
V_{Na}	+122	mV
V_L	0	mV

thus, the internode resistance is taken as[2]

$$R = R_i + R_o \approx R_i = sr_i = 28 \pm 1 \text{ megohm.}$$

As indicated in Table 7.1, Cole has reported the maximum sodium conductance of a frog node to be

$$G = 0.57 \ \mu\text{mhos,}$$

and the Nernst potential for sodium ions as [16]

$$V_2 = 122 \text{ mV.}$$

Also he gives the threshold potential for a typical frog membrane as about

$$V_1 \approx 25 \text{ mV.}$$

Potassium ion current carries positive charge out of an active node; thus, it is a *recovery variable*. As we saw in Chapter 4, a detailed expression for this current was presented by Hodgkin and Huxley [31], and the simple representation proposed by FitzHugh was studied in the previous chapter [22]. Here we take advantage of the fact that the time delay for the onset of potassium current is about 3 ms [16]. Because the length of the impulse is about equal to this time delay times its speed (about 2.3 cm/ms), the trailing edge of an impulse is expected to lag behind its leading edge by about 6 cm, or 30 nodes. Thus it is reasonable to neglect effects of the trail-

---

[2]Larger values of R_o are to be expected in nerve bundles, where many individual fibers are situated together [75]. In Chapter 8, we consider impulse couplings in adjacent fibers caused by currents flowing through this external resistance.

ing edge of the impulse on its leading-edge dynamics.[3] We will employ this approximation throughout this chapter by assuming that the potassium ion permeability remains equal to its resting value.

It is convenient to measure voltages in units of the Nernst potential (V_2) to obtain the dimensionless voltage variables

$$\mathcal{V}_n \equiv V_n/V_2 . \qquad (7.4)$$

Then, Equations (7.1) and (7.2) become the *discrete reaction diffusion* system

$$RC\frac{d\mathcal{V}_n}{dt} = (\mathcal{V}_{n+1} - 2\mathcal{V}_n + \mathcal{V}_{n-1}) - \left(\frac{RG}{1-a}\right)\mathcal{V}_n(\mathcal{V}_n - a)(\mathcal{V}_n - 1), \quad (7.5)$$

where $a \equiv V_1/V_2$ and $R = R_i + R_o$.

At this point, the model is normalized in a manner that allows the internode spacing to be an independent parameter and maintains contact with notations in recent studies of discrete nonlinear diffusion by applied mathematicians. To these ends, s is taken to be a variable internode distance, and a *discreteness parameter* is defined as

$$D \equiv \left(\frac{2\ \text{mm}}{s}\right) = \frac{R_f}{R} .$$

Under this definition, it is intended that

$$R_f = 28\ \text{megohms} ,$$

which is the internode resistance of the standard frog axon. In other words, $1/D$ is the spacing between nodes in units of 2 mm, so $D = 1$ implies the discreteness of the standard frog axon.

In this formulation, the dynamic equation becomes

$$D(\mathcal{V}_{n+1} - 2\mathcal{V}_n + \mathcal{V}_{n-1}) = R_f C\frac{d\mathcal{V}_n}{dt} + \left(\frac{R_f G}{1-a}\right)\mathcal{V}_n(\mathcal{V}_n - a)(\mathcal{V}_n - 1). \quad (7.6)$$

Although the experimental values of the parameters upon which Equation (7.6) is based are known only approximately, they provide reasonable estimates for numerical studies of myelinated motor nerves of a frog.

7.2 Impulse Speed and Failure

Equation (7.6) is a discrete reaction diffusion system modeling a myelinated nerve. This section presents some numerical calculations of impulse speeds

---

[3]Interestingly, potassium current seems to be lacking in dynamics of some rabbit nodes, with recovery generated by leakage currents [15]. The time course for recovery, however, is about the same as for the frog.

obtained in [4] for comparison with corresponding theoretical results [3, 20, 42, 58, 59].

Broadly speaking, the nature of the wave propagation on a discrete nerve model can be characterized by looking at the relative change in voltage between two adjacent nodes. If this relative change everywhere satisfies the inequality

$$\left| \frac{(\mathcal{V}_{n+1} - \mathcal{V}_n)}{\mathcal{V}_n} \right| \ll 1 \,,$$

then the voltages and currents are relatively smooth functions of distance and the system can be described by partial differential equations—the corresponding continuum system. If, on the other hand, the maximum value

$$\left| \frac{(\mathcal{V}_{n+1} - \mathcal{V}_n)}{\mathcal{V}_n} \right| \gg 1 \,,$$

then the conduction process is saltatory, jumping from one active node to the next in a discontinuous manner. We refer to these two cases as the *continuum limit* and *saltatory limit*, respectively.

7.2.1 Continuum Limit

If the internode spacing s is small enough so that the continuum limit is reached, Equation (7.6) can be written as the partial differential equation

$$s^2 D \frac{\partial^2 \mathcal{V}}{\partial x^2} - R_f C \frac{\partial \mathcal{V}}{\partial t} = \left(\frac{R_f G}{1-a} \right) \mathcal{V}(\mathcal{V} - a)(\mathcal{V} - 1) \,, \qquad (7.7)$$

where we have assumed $D \gg 1$ and let

$$ns \to x \,.$$

This PDE was discussed in Section 5.2.2. Thus, if time is measured in units of $(1-a)C/G$ and distance in units of $s\sqrt{(1-a)/RG}$, then a traveling wave front (the leading edge of the impulse) has shape [77]

$$\mathcal{V}(x, t) = \frac{1}{1 + e^{(x-vt)/\sqrt{2}}} \qquad (7.8)$$

and speed

$$v = \frac{(1 - 2a)}{\sqrt{2}} \,.$$

Although the condition $D \gg 1$ is not satisfied for the standard frog nerve, it is convenient to have an explicit expression for the wave speed in the continuum limit as a benchmark for numerical calculations. From Equation (5.10), the wave speed $v \to v_c$, where

$$v_c = \sqrt{\frac{G}{RC^2}} \left(\frac{1 - 2a}{\sqrt{2(1-a)}} \right) \quad \text{nodes/s} \,. \qquad (7.9)$$

To get the corresponding impulse speed in (say) meters/second, as is plotted in Figure 7.2, merely multiply this expression by s, the number of
meters between adjacent nodes.

7.2.2 Saltatory Limit

For D equal to or less than unity, the wave of excitation jumps from node
to node in a discontinuous manner, allowing the speed of conduction to be
greatly increased without a corresponding increase in fiber diameter. An
additional feature of the saltatory limit is the possibility that the switching
of one node may be unable to ignite the adjacent node. In this situation,
called *failure*, the impulse ceases to propagate [3, 5, 6, 20, 21, 38, 39].
Because failure of impulse conduction is an undesired property of a real
nerve, we expect the node spacing for frog axons to lie comfortably beyond
this range.

   If the internode spacing s is increased so that D is reduced to a critical
value D^*, failure of impulse propagation occurs because the fully developed
voltage at one node is unable to bring the next node above threshold. Basically, this occurs when the internode distance, and therefore R, becomes
too large with respect to the node resistance $1/G$.

   With the cubic form of the sodium ion current in Equation (7.3), Erneux
and Nicolis [20] have shown that the critical value of the discreteness
parameter is given to lowest order in a ($\equiv V_1/V_2$) by

$$D^* \approx \frac{R_f G a^2}{4(1-a)} = 0.21 \,. \tag{7.10}$$

For D slightly larger than D^*, these same authors show that the impulse
velocity $v \to v_s$, where

$$v_s = \frac{1}{\pi C} \sqrt{\frac{G(D-D^*)}{R_f(1-a)}} \quad \text{nodes/s} \,. \tag{7.11}$$

Again, the corresponding impulse speed in meters/second, as plotted in
Figure 7.2, is obtained from multiplying this expression by s, the number
of meters/node.

   In the saltatory limit, each segment of the myelinated nerve is modeled
much like the "integrate and fire" switch, which was mentioned at the
beginning of the previous chapter. Thus node n fires, whereupon its voltage
forces current through the internode resistance and into the capacitance of
node $n+1$. When the voltage across the node $n+1$ reaches threshold, it
fires, and the process is repeated.

   This perspective on mechanism of failure goes back to the seminal
work of Kunov and Richer in the mid-1960s, where the piecewise linear
model of Equation (5.12) leads to a similar value for D^* [40, 58]. Taking
$G_L = 0.025\,\mu$mho from Table 7.1 to be the conductance below threshold,

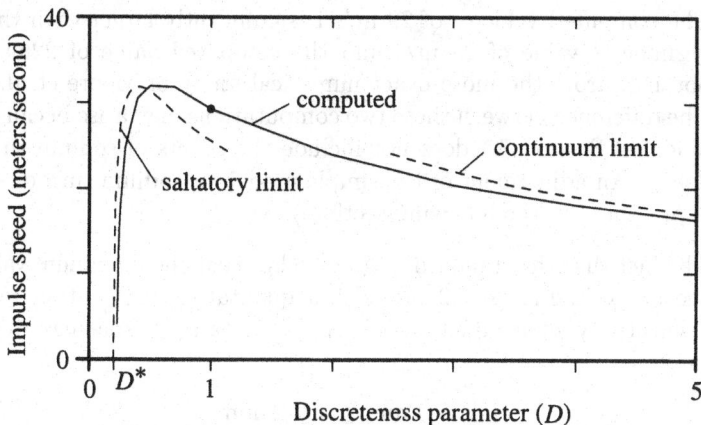

Figure 7.2. Leading-edge impulse velocity on a myelinated axon as a function of the discreteness parameter $D = 2\,\mathrm{mm}/s$. The black dot at $D = 1$ indicates a standard frog axon of outside fiber diameter equal to 14 μm. (Redrawn from [4].)

a simple voltage-divider calculation shows that failure is expected to occur for $V_2/(RG_L + 1) < V_1$ or $RG_L > (1 - a)/a$ [67]. Thus to lowest order in a,

$$D^* \approx \frac{R_f G_L a}{(1 - a)} = 0.18\,.$$

Using the piecewise linear approximation of Equation (5.12) with $\beta = 1$, Keener has obtained a related result holding for $0 < a < 1/2$ [38, 39].

7.2.3 Numerical Results

Equation (7.6) has been used to compute the wave-front velocity for a nerve impulse, which is plotted against the discreteness parameter ($D = 2\,\mathrm{mm}/s$) in Figure 7.2 [4]. From these numerical data, one can make the following observations.

- At larger values of the discreteness parameter ($D > 5$ or $s < 0.4$ mm), the continuum approximation holds and impulse velocity is given by Equation (7.9).

- Failure in the model is accurately predicted by Equation (7.10) to occur at $D^* = 0.21$ or $s = 9.5$ cm, in accord with the value of "almost 10 cm" observed by Moore et al. [48]. Near the failure point, impulse velocity is well-represented by Equation (7.11).

- At $D = 1$ (corresponding to the parameters of a real frog nerve), neither the continuum approximation formula nor Equation (7.11) give a satisfactory estimate of the impulse velocity.

- The computed velocity of 29 m/s is significantly larger than the experimental value of 23 m/s and the computed value of 22.65 m/s obtained from the more exact numerical study of Moore et al. [48]. The difference between these two computations may arise because the model of Section 7.1 does not include the effects of sodium turn-on delay. (An adjustment of the simple model for sodium turn-on delay is presented in the following section.)

- The fact that the standard frog nerve lies near the maximum value of the curve in Figure 7.2 provides a qualitative explanation for the insensitivity of conduction velocity to internode spacing over the range

$$1\,\mathrm{mm} \leq s \leq 2\,\mathrm{mm}$$

observed by Moore et al. [48]. Because the effects of the saltatory and continuum limits are in balance at this maximum, Figure 7.2 shows that s can change by a factor of more than 2, whereas the conduction velocity varies by less than 10%.

7.3   Biological Considerations

In this last section, we consider some applications of the foregoing formulation to motor nerves of several different vertebrate species, with emphasis on means that evolutionary pressures may have used to optimize myelinated axon structures. The discussion concludes with a testable prediction of the conduction velocities to be found in arctic vertebrates. Interestingly, a century of effort has not exhausted the possibilities for fruitful research on the dynamics of myelinated nerves.

7.3.1   Frog Motor Nerves

Before drawing biological conclusions from our model, it is necessary to check whether Equation (7.6) gives a value for impulse speed that agrees with experimental observations. In this connection, Tasaki and his colleagues have reported measurements of the conduction velocities of 49 individual frog axons with outside diameters ranging from 3 to 17 μm at 24°C [70, 71]. These data are displayed in Figure 7.3 and indicate that the experimental values of impulse velocity (v_e) are related to the outside diameter of the axon (d) by the linear relationship [70]

$$v_e = 2d \tag{7.12}$$

to an accuracy of about ±25%, where the velocity is in meters per second and the diameter is in microns. Because the standard frog axon has an

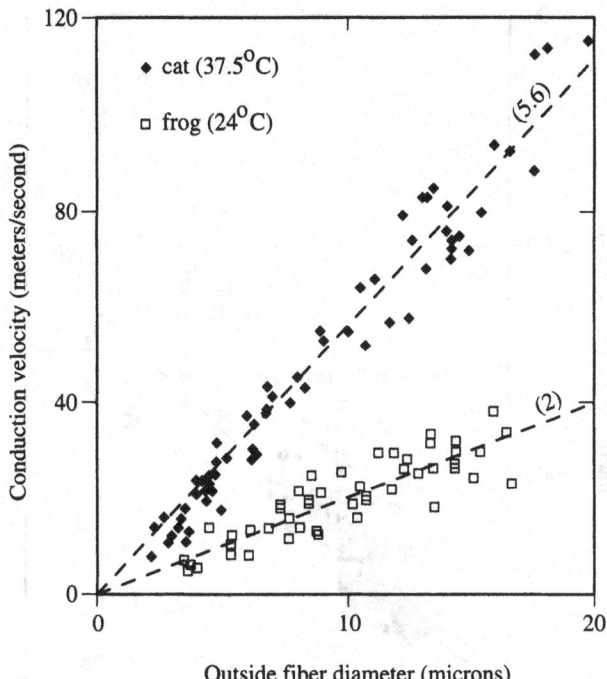

Figure 7.3. Empirical conduction velocities (v_e) vs. outside fiber diameters (d) for myelinated axons of two different vertebrate species: the frog at a temperature of 24°C (from data in [71]), and the cat at a temperature of 37.5°C (from data in [33]).

outer fiber diameter of 14 μm, the calculated conduction velocity of 29 m/s is in accord with the data of Figure 7.3.

Assured that the simple model of Section 7.1 is not unreasonable, we are led to two observations of biological significance. First, failure of an impulse on the standard frog axon is expected to occur at an internode spacing of 9.5 mm (corresponding to $D^* = 0.21$), whereas the normal spacing is 2 mm. The evolutionary design of this axon thus provides a comfortable margin of safety against failure. Second, Figure 7.2 shows that at $D = 1$ the impulse velocity of a normal frog nerve is close to the maximum possible value, again suggesting that an optimal design has evolved.

Although the preceding results for varying D (or internode spacing $s = 2\,\text{mm}/D$) have been obtained under the assumption that other properties of a nerve fiber remain fixed, this is a mathematical fiction. In real nerves, some sort of design optimization has occurred over the course of biological evolution that simultaneously adjusts all parameters in appropriate ways.

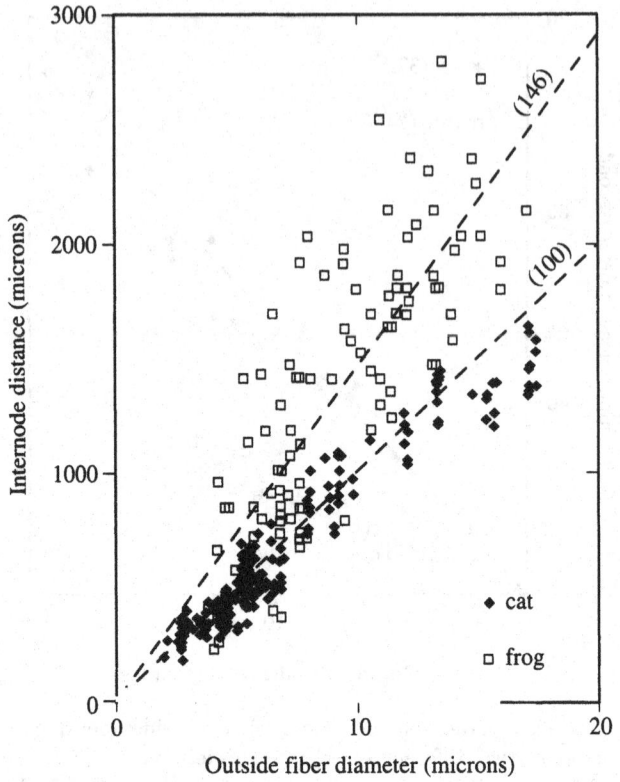

Figure 7.4. Average internode distance (s) vs. outside fiber diameter (d) for myelinated axons of two different vertebrate species: the frog (from data in [71]), and the cat (from data in [33]).

To see whether the optimal design features displayed for the standard frog axon in Figure 7.2 should hold for fibers of different diameters, consider the following line of logic.

(1) As indicated in Figure 7.4, internode spacings and fiber diameters of frog axons obey the approximately linear relation [33, 71]

$$s = 146d.\qquad(7.13)$$

(2) From the simple perspective introduced in Equation (1.2), the impulse speed of a myelinated axon is

$$v_e = s/T.\qquad(7.14)$$

As Tasaki et al. have observed [71], Equations (7.12) and (7.13) imply that

$$T = \frac{146}{2} \times 10^{-6}\ \text{s} = 0.073\ \text{ms}$$

is the *internode conduction time*, or the time required for an impulse to jump from one node to the next. (This value compares favorably with the value of 0.069 ms for the standard frog model.) From Equations (7.12) and (7.13), T is approximately independent of fiber diameter.[4]

(3) In the model of Section 7.1, $RC = 0.1$ ms, whereas $C/G = 0.006$ ms; thus, the internode conduction time

$$T \propto RC,$$

implying that RC is independent of fiber diameter.

(4) The internode resistance $R \propto s/d^2 \propto 1/d$, and for nodes of constant length the node capacitance $G \propto d$, implying that RG is also insensitive to changes of fiber diameter.

Thus, Equation (7.5) appears to be invariant to differences in the fiber diameter, suggesting that the optimal properties displayed in Figure 7.2 may hold for the many axons of Figure 7.3. In other words, frog motor axons of widely different diameters are expected to have conduction velocities near their maximum values. What about axons from other species?

7.3.2 Other Vertebrates

For several vertebrate species, it has been observed that impulse speed (v_e) and internode spacing (s) are proportional to fiber diameter (d). For myelinated motor axons of the cat, for example, it is seen from Figure 7.3 that [33]

$$v_e = 5.6d, \tag{7.15}$$

where d is measured in microns and v_e in meters per second. Also, from Figure 7.4 [33],

$$s = 100d, \tag{7.16}$$

where s is also in microns.

Let us write these relations as

$$v_e = \alpha d \text{ and } s = \beta d,$$

implying that

$$v_e = s/T \tag{7.17}$$

with

$$T = \beta/\alpha.$$

---

[4]There are some typographical errors in Tasaki's references. Equation (5.5) of [70] incorrectly gives $v_e = 2.5d$ instead of $v_e = 2d$, and in [71], the coefficients in the equations for v_e and s as linear functions of d are interchanged.

Table 7.2. Myelinated nerve parameters for several vertebrate species. (Computed values of $T = \beta/\alpha$ are underlined.)

Species	Temp	α	β	T	References
frog[a]	24	2	146	0.073–0.082	[7, 35, 70, 71]
toad[b]	24	2	100–150	0.05–0.075	[25, 34, 51, 72]
cat[c]	37.5	4.6–5.6	65–100	0.012–0.022	[8, 33, 47, 64]
mouse[d]	38–39	6	74–84	0.012–0.014	[7, 9, 63, 64]
rat[e]	30–39	6.1	94–146	0.02	[26, 57, 63, 64]
rabbit[f]	38	4.5	59–105	0.013–0.023	[19, 46, 51, 74]
guinea pig[g]	38–39	6	69–103	0.012–0.017	[27, 63, 64]
sheep[h]	37	4–5	60	0.012–0.015	[28, 44]
human[i]	37	5.9	84–100	0.014–0.017	[37, 43, 51, 73]
Units:	°C	(m/s)/μm	μm/μm	ms	

[a] α is from Figure 7.3 [70, 71], and β is from Figure 7.4 [71]. The lower value of T is computed, whereas the upper value is a direct measurement [35]. (For the standard frog axon of Section 7.1, $T = 0.069$ ms.)
[b] Considered similar to the frog [70], the toad value of $\alpha = 2$ at 24°C is confirmed in [34], where $\beta = 100$ (probably too low) was used for numerical studies. T is computed.
[c] α is from Figure 7.3 [33] and [8], β from Figure 7.4 [33], [47], and [64]. T is computed.
[d] α is from [63] and β from [7] and [64]. T is computed.
[e] T is directly measured [56, 57] at 30°C. α is from [63] at 38–39°C. β is from [26] and [64].
[f] α is the average of five values given in [19]. Measured values for β were given in [19] and [74], from which T is computed.
[g] α is measured in [63], and β is measured in [27] and [64]. T is computed.
[h] β averages unpublished measurements of 367 internodes of 20 nerves from 13 sheep [28], α is estimated from velocity and diameter distributions in [44], and T is computed.
[i] α is from [37], β from [43] and [73], and T is computed.

In Table 7.2 are collected some of the values that have been measured for these parameters on nerves of various vertebrate species and from which the following observations can be made. Warm-blooded creatures have significantly greater conduction velocities for a given fiber diameter (α) than the frog and toad, which are cold-blooded.[5] Warm-blooded creatures have shorter internode spacings (β) for a given fiber diameter. Internode conduction times (T) are about four times larger for cold-blooded animals than for warm-blooded ones.

This difference in T might be thought to arise from the fact that nerves of warm-blooded animals operate at higher temperatures, but there are two reasons for finding this explanation incomplete. First, measurements by Hutchinson et al. [34] on the toad show that impulse velocity varies with fiber diameter and temperature as

$$v_e = 0.6(1 + 0.1\,\mathrm{Temp})d. \tag{7.18}$$

Extrapolating from 24°C to 37.5°C implies $\alpha = 2.9$, which is about half the value observed for the cat. Second, measurements by Paintel [50] on cat nerves show that conduction velocity is decreased by 58% at 24°C, implying $\alpha = 3.2$, which is significantly greater than the frog value of $\alpha = 2$.

In other words, it appears that only a fraction of the decrease in internode conduction time for warm-blooded over cold-blooded animals can be directly attributed to temperature increase. How might the remaining difference be explained?

7.3.3 An Evolutionary Perspective

In an early formulation for the evolutionary optimization of myelinated nerves, Rushton supposed that structural dimensions could be uniformly scaled, with all electric potentials remaining invariant under the scaling factor [61]. Although his theory implies that both conduction velocity and internode spacing should be nearly proportional to fiber diameter, it does not allow predictions of internode conduction times or of their dependence upon temperature.

Here a somewhat less restrictive approach to the optimization problem is taken based on an evolutionary interplay between the RC time constant, defined in Equations (7.1) and (7.2), and τ_m, the delay for rise of the sodium ion turn-on variable (m) at an active node. Sodium turn-on delay is an intrinsic property of sodium channels, remaining essentially unchanged over a wide range of biological organisms [15, 30].[6] The circuit time constant (RC), on the other hand, depends strongly on a number of structural parameters, including the internode distance (s), the outside fiber diameter

---

[5]This generalization is supported by a measurement of $\alpha = 2.4$ for trout [18].

[6]Although recent genetic investigations reveal several different sodium channels [14, 54], a particular channel, from a gene variously called $Na_v1.6$, Scn8a, PN4, and cer3, is concentrated at nodes of Ranvier [12], with little phylogenic variation [1, 45].

(d), the area of the active node membrane, and the ratio

$$g \equiv \frac{\text{axon diameter}}{\text{outside fiber diameter}} \qquad (7.19)$$

which seems to have an optimum value around 0.6 [32, 48, 61, 64]. These parameters present many options for adjusting the conduction velocity of a myelinated nerve [2].

Among the effects of structural changes are the following. (1) R is proportional to s and inversely proportional to $(gd)^2$. (2) Node capacitance C_n is equal to the node area times the capacitance per unit area of the membrane (about 1 μF/cm^2). (3) Node area, in turn, is proportional to gd and also to the length of the node. (4) Internode (myelin) capacitance C_m is proportional to both s and d, and it decreases with decreasing g because this implies a thicker myelin sheath. (5) Under the Π-network approximation of Equations (7.1) and (7.2), $C = C_n + C_m$. With all of these avenues for possible change, how is an optimum value of RC established, and why should it be independent of fiber diameter?

For the standard frog nerve model of Section 7.1, sodium turn-on time is neglected, but it was noted in Section 5.2.2 that this delay has a significant effect on the speed of a squid nerve impulse. To bring sodium turn-on delay into the formulation, assume in Equations (7.14) and (7.17) that

$$T = 2.2a\sqrt{(RC)^2 + \tau_m^2} + T_{\text{sw}}, \qquad (7.20)$$

where the first term is the time required for the membrane to go from zero to the threshold voltage (V_1) and T_{sw} is the time to switch from V_1 to V_2. Also $a \equiv V_1/V_2$ and the factor $\sqrt{(RC)^2 + \tau_m^2}$ interpolates between a circuit response time of RC (which dominates the dynamics at $RC \gg \tau_m$) and a sodium ion response time of τ_m (which dominates at $RC \ll \tau_m$). For the standard frog model, it is expected that T_{sw} will be small compared with T because it is governed by the time constant $C/G \ll RC$.[7]

---

[7] Equation (7.20) is a rather crude expression for T introduced to reduce the number of parameters to two, RC and τ_m. The derivation assumes that below threshold the sinusoidal steady-state response function is proportional to

$$\frac{1}{(1 + i\omega RC)(1 + i\omega \tau_m)},$$

where the first factor represents the frequency response stemming from circuit behavior and the second from response of the sodium turn-on variable m. This response function has magnitude

$$\frac{1}{\sqrt{[1 + (\omega RC)^2][1 + (\omega \tau_m)^2]}} = \frac{1}{\sqrt{1 + \omega^2[(RC)^2 + \tau_m^2]}} + O[\omega^4(RC\tau_m)^2].$$

Thus, the half-power frequency is approximately $1/\sqrt{(RC)^2 + \tau_m^2}$, with a corresponding 10% to 90% rise time of about $2.2\sqrt{(RC)^2 + \tau_m^2}$. From an "integrate and fire" perspective, the time for the next node to reach threshold is a times the 10% to 90% rise time.

To maximize impulse speed, it is necessary to reduce T. Because τ_m is fixed by properties of the sodium channel proteins, RC will plausibly be pushed down by evolutionary pressures on the structure until

$$RC = K\tau_m\,,$$

where K is a factor of the order of unity. At this point of diminishing returns,

$$T - T_{sw} = 2.2a\tau_m\sqrt{K^2+1} = 2.2aRC\sqrt{1+1/K^2}\,. \qquad (7.21)$$

The temperature at which this equality is established—called the *evolutionary design temperature* (EDT)—is expected to be about twenty degrees Celsius higher for warm-blooded animals than for cold-blooded ones.

Four items of empirical evidence can be advanced in support of this optimization principle.

1. Assuming that evolutionary pressures adjust the structure of a myelinated fiber until RC is of the order of the sodium ion response time (a fixed, albeit temperature-dependent, property of the sodium channel proteins) provides a qualitative explanation for observations that internode conduction times are independent of fiber diameters.

2. For the standard frog nerve, Equation (7.21) (with $K = 1$ and $T_{sw} = 0.01$ ms [15]) implies $T = 0.076$ ms in agreement with the values of 0.068–0.084 ms in Table 7.2.

3. Equation (7.21) also suggests that the optimum internode conduction times should be smaller for warm-blooded animals than for a cold-blooded frog. Because the body temperatures of mammals are about 37°C, whereas a frog might prefer to swim about in water of (say) 24°C, mammalian EDTs are some 13 degrees higher than for a frog. From Equation (B.3) of Appendix B, the Hodgkin–Huxley temperature dependence

$$\tau_m \propto 3^{-\text{Temp}/10} \qquad (7.22)$$

in turn implies that the sodium turn-on time for a warm-blooded nerve should be about 4.2 times smaller than for a frog nerve. Equation (7.21) then gives $T = 0.076/4.2 = 0.018$ ms at 37°C, in accord with the empirical values for mammals given in Table 7.2.

4. Experimental observations of variations of conduction velocity with temperature include the following. Measurements between 15°C and 30°C on six selected toad axons by Hutchinson et al. suggest the linear dependence of Equation (7.18), which is plotted in Figure 7.5 [34]. Studies on toads between 5°C and 20°C by Tasaki and Fujita consistently showing

$$v_e = k1.8^{\text{Temp}/10} \qquad (7.23)$$

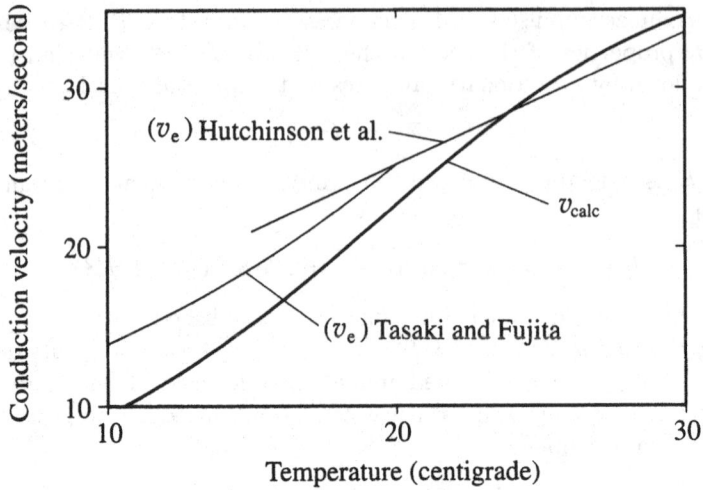

Figure 7.5. A plot of calculated conduction velocity (v_{calc}) vs. temperature from Equation (7.24). This plot is compared with experimental data (v_e) from Hutchinson et al. and from Tasaki and Fujita in Equations (7.18) and (7.23), respectively. (In Equation (7.23), k is chosen to equal the velocity from Equation (7.18) at $20°$C.)

over many different preparations [72]. These data are also plotted on Figure 7.5, where k has been chosen so Equations (7.18) and (7.23) give the same value of conduction velocity at $20°$C. Evidently, these two sets of data diverge between $15°$C and $20°$C.

Together with the Hodgkin–Huxley temperature dependence of Equation (7.22), Equations (7.20) and (7.21) allow calculation of conduction velocity as a function of temperature. Thus with $K = 1$ at an EDT of $24°$C,

$$v_{calc} = \frac{s}{T} = \frac{\sqrt{2}s}{(T_0 - T_{sw})\sqrt{1 + 3^{-(\text{Temp}-24)/5}} + \sqrt{2}T_{sw}}. \qquad (7.24)$$

To match the experimental conduction velocities at $24°$C, $s = 2$ mm, $T_0 = 0.07$ ms, and $T_{sw} = 0.01$ ms, allowing Equation (7.24) to be plotted in Figure 7.5. Thus, the simple analysis embodied in Equation (7.24) is in fair agreement with empirical observations between $20°$C and $30°$C.

Having presented arguments in support of the proposed optimization principle, it is appropriate to consider and respond to some questions that might be raised.

(1) In the derivation of Equation (7.20), it was assumed that one active node would "integrate and fire" before the next node goes through the same cycle. Although this assumption may hold for a frog nerve (with $T_{sw} \approx 0.01$

ms and $T \approx 0.07$ ms), it appears unlikely for the mammalian nerve, with $T \approx 0.02$ ms or less. Thus, Equation (7.20) should not be used for detailed predictions concerning mammalian nerves.

(2) It is not clear from the present formulation why the parameter β (see Table 7.2) should be smaller for warm-blooded animals. Careful measurements along cat fibers, however, show that β decreases greatly as one moves out on axonal trees, perhaps because lower β brings a needed increase in the safety factor [55]. Could a high safety factor be more important for warm-blooded animals?

(3) Assuming $T_0 = 0.07$ ms, $T_{sw} = 0.01$ ms, and an EDT of 24°C, Equation (7.24) implies that v_{calc} changes by a factor of 4 as temperature increases from 5°C to 20°C, whereas Tasaki and Fujita observed a factor of 2.4. Thus, Equation (7.20) does not provide a good model for the dynamics below 20°C, where $\tau_m \gg RC$. Because Equation (7.24) was derived without reference to the details of sodium ion dynamics, such disagreement is not unexpected. On the other hand, the disagreement noted earlier between the data of Hutchinson et al. [34] and that of Tasaki and Fujita [72] between 15°C and 20°C suggests that temperature dependence of conduction velocity needs to be more carefully measured between 5°C and 30°C before firm conclusions are drawn.

(4) Returning to the measurements plotted in Figure 7.4, the assumption that internode distance is proportional to fiber diameter ($s = \beta d$) is seen to be not quite true for the cat. Fibers of larger diameter appear to have relatively shorter internode distances (smaller β) than those of smaller diameters, suggesting a relationship of the form $s = s_0 + \tilde{\beta}d$, as has also been observed for the guinea pig [27] and the rabbit [74]. Together with the linear relations ($v_e = \alpha d$) in Figure 7.3, this observation implies that the internode conduction times are shorter for larger fibers, as has been confirmed for cat fibers by Coppin and Jack [17, 36].

By directly measuring internode delay as a function of conduction velocity on cat axons, Coppin and Jack found $T \approx 0.025$ ms for smaller fibers having speeds of 30 m/s, diminishing to $T \approx 0.01$ ms for larger fibers of 120 m/s. Going further, Boyd and Kalu have shown that α increases slightly for larger cat fibers [8].

From the present perspective, such data might be accounted for by supposing that RC becomes somewhat smaller (relative to τ_m) on the larger fibers than is indicated by Equation (7.21) in order to compensate for the relatively smaller values of β. In other words, K is smaller for larger fibers.

(5) It is to be noted that equations in the preceding discussion indicate statistical propensities rather than fixed functional laws. To appreciate this feature, turn back to Figure 7.3 and compare these data with the expressions $v_e = 5.6d$ and $v_e = 2d$. All of the preceding equations are similarly intended as averages over such variations of about ±25%. Although some of the randomness stems from measurement errors, much arises from differences in the degree of optimization actually achieved by individual fibers

[51]. As was mentioned in connection with the original Hodgkin–Huxley data presented in Chapter 4, such variations are not unusual in the realms of biology, reflecting different responses of individual fibers to the happenstance of growth. From this perspective, Equation (7.20) represents an evolutionary pressure, acting in an average manner over the members of a species.

Although it must be admitted that this formulation for the influence of evolution on axonal structures is speculative, leaving many aspects to be improved through better measurements and more thorough numerical and theoretical studies, it provides a context for organizing structural and dynamic data. Furthermore, the model makes a testable prediction.

7.3.4   The Evolution of Arctic Fish

Consider the internode conduction times of arctic fish that have evolved in near-freezing water. If only the direct effects of temperature are considered, Equation (7.23) suggests that T at 4°C should increase over that of a frog at 24°C by a factor of 3.2 from the range 0.073–0.082 ms (see Table 7.2) to

$$T = 0.23 - 0.26 \text{ ms}.$$

If evolutionary pressures associated with Equation (7.20) come into play, on the other hand, RC should also be reduced as changes in T respond to the H–H temperature dependence of τ_m. From Equation (7.22), the internode conduction times of arctic fish measured at 4°C should be nine times larger than for a frog at 24°C, implying an increase of T to

$$T = 0.6 - 0.74 \text{ ms},$$

a factor of about 40 longer than the internode delays of typical mammals.

Using a moving probe to monitor external voltage waveforms along a single fiber, Huxley and Stämpfli have shown that internode conduction time can be directly measured at a frog node [35]. As we have seen, this technique has been successfully employed by Coppin and Jack on cats [17] and also by Raminsky and Sears on rats [56, 57]; thus, it seems feasible to do the same with single motor axons of arctic fish.

One candidate for such a study is the Greenland shark [13]. Typically described as "sluggish," *Somniosus microcephalus* lives deep in the Atlantic polar regions, where the water temperature is 2–7°C, surfacing only during the winter. Does this sleepy pea-brain offer little resistance to fishermen because its nerves are so slow?

7.4   Recapitulation

The nature of saltatory conduction was formulated for axons of the frog sciatic nerve, which was studied by Galvani and Helmholtz. Using experimental data published by Hodgkin, Cole, and Tasaki, a simple quantitative model of a single axon was developed, depending on only four experimentally determined parameters and demonstrating the relationship of real myelinated nerves to impulse failure and to the continuum models considered in Chapter 5. Numerical studies of this model confirm theoretical calculations of impulse speeds in the continuum and saltatory limits, which show that conduction velocity on the myelinated axon of a frog nerve lies between these two limits, close to its maximum value.

In accord with a qualitative theory proposed a half-century ago by Rushton, this frog axon model provides an example of evolutionary optimization in which an engineering balance is established between the twin design requirements of increasing impulse speed and avoiding failure.

To enlarge the biological perspectives, some results for motor nerves of warm-blooded vertebrates were presented and compared with frog and toad data. Based on an interplay between axon circuit constants (resistance and capacitance) and the turn-on delay for sodium ion current, an evolutionary mechanism for optimizing conduction velocity was proposed that is consistent with the following empirical observations: the fact that internode conduction time is approximately independent of fiber diameter for myelinated nerves of different vertebrate species, internode conduction times of frog axons, faster conduction velocities for warm-blooded animals than for the frog and toad, and the increase in toad conduction velocity with temperature. Finally, the evolutionary theory implies that internode conduction times for arctic fish should be about nine times longer than for the frog—a prediction that can be tested on single axons.

References

[1] PAV Anderson and RM Greenberg, Phylogony of ion channels: clues to structure and function, *Comp. Biochem. and Physiol.* B 129 (2001) 17–28.

[2] ER Arbuthnott, IA Boyd, and KU Kalu, Ultrastructure dimensions of myelinated peripheral nerve fibres in the cat and their relation to conduction velocity, *J. Physiol. (London)* 308 (1980) 125–157.

[3] ARA Anderson and BD Sleeman, Wave front propagation and failure in coupled systems of discrete bistable cells modeled by FitzHugh–Nagumo dynamics, *Int. J. Bifurcation Chaos* 5 (1995) 63–74.

[4] S Binczak, JC Eilbeck, and AC Scott, Ephaptic coupling of myelinated nerve fibers, *Physica D* 148 (2001) 159–174.

[5] V Booth and T Erneux, Mechanisms for propagation failure in discrete reaction-diffusion systems, *Physica A* 188 (1992) 206–209.

[6] V Booth and T Erneux, Understanding propagation failure as a slow capture near a limit point, *SIAM J. Appl. Math.* 55 (1995) 1372–1389.

[7] AE Boycott, On the number of nodes of Ranvier in different stages of the growth of nerve fibres in the frog, *J. Physiol. (London)* 30 (1903) 370–380.

[8] IA Boyd and KU Kalu, Scaling factor relating conduction velocity and diameter for myelinated afferent nerve fibres in the cat hind limb, *J. Physiol. (London)* 289 (1979) 277–297.

[9] MET Boyle, EO Berglund, KK Murai, L Weber, E Peles, and B Ranscht, Contactin orchestrates assembly of the septate-like junctions at the paranode in myelinated peripheral nerve, *Neuron* 30 (2001) 385–397.

[10] PC Bressloff and G Rowlands, Exact travelling wave solutions of an 'integrable' discrete reaction-diffusion lattice, *Physica D* 106 (1997) 255–269.

[11] M Brill, SG Waxman, JW Moore, and RW Joyner, Conduction velocity and spike configuration in myelinated fibers, *J. Neurol. Neurosurg. Psychiatry* 40 (1977) 769–774.

[12] JH Caldwell, KL Schaller, RS Lasher, E Peles, and SR Levinson, Sodium channel $Na_v 1.6$ is localized at nodes of Ranvier, dendrites and sysapses, *Proc. Natl. Acad. Sci. USA* 97 (2000) 5616–5620.

[13] JA Castro, *The Sharks of North American Waters*, Texas A&M University Press, College Station, 1983.

[14] WA Catterall, From ionic currents to molecular mechanisms: The structure and function of voltage gated sodium channels, *Neuron* 26 (2000) 13–25.

[15] SY Chiu, JM Ritchie, RB Rogart, and D Stagg, A quantitative description of membrane currents in rabbit myelinated nerve, *J. Physiol. (London)* 292 (1979) 149–166.

[16] KS Cole, *Membranes, Ions and Impulses*, University of California Press, Berkeley, 1968.

[17] CML Coppin and JJB Jack, Internodal length and conduction velocity of cat muscle afferent nerve fibers, *J. Physiol. (London)* 222 (1972) 91P–93P.

[18] BG Cragg and PK Thomas, The relationships between conduction velocity and the diameter of internodal length of peripheral nerve fibres, *J. Physiol. (London)* 136 (1957) 606–614.

[19] BG Cragg and PK Thomas, The conduction velocity of regenerated peripheral fibres, *J. Physiol. (London)* 171 (1964) 164–175.

[20] T Erneux and G Nicolis, Propagating waves in discrete bistable reaction-diffusion systems, *Physica D* 67 (1993) 237–244.

[21] G Fáth, Propagation failure of traveling waves in a discrete bistable medium, *Physica D* 116 (1998) 176–180.

[22] R FitzHugh, Impulses and physiological states in theoretical models of nerve membrane, *Biophys. J.* 1 (1961) 445–466.

[23] R FitzHugh, Computation of impulse initiation and saltatory conduction in a myelinated nerve fiber, *Biophys. J.* 2 (1962) 11–21.

[24] B Frankenhaeuser and AF Huxley, The action potential in the myelinated nerve fiber of *Xenopus laevis* as computed on the basis of voltage clamp data, *J. Physiol. (London)* 171 (1964) 302–325.

[25] B Frankenhaeuser and B Waltman, Membrane resistance and conduction velocity of large myelinated nerve fibres from *Xenopus laevis*, *J. Physiol. (London)* 148 (1959) 677–682.

[26] PM Fullerton and JM Barnes, Peripheral neuropathology in rats produced by acrylamide, *Br. J. Ind. Med.* 23 (1966) 210–221.

[27] PM Fullerton, RW Gilliatt, RG Lascelles, and JA Morgan-Hughes, The relation between fibre diameter and internodal length in chronic neuropathology, *J. Physiol. (London)* 178 (1965) 26P–28P.

[28] Michael Glasby, Private communication, July, 2001.

[29] L Goldman and JS Albus, Computation of impulse conduction in myelinated fibers, *Biophys. J.* 8 (1968) 596–607.

[30] B Hille, *Ion Channels of Excitable Membranes*, third edition, Sinauer Associates, Sunderland, MA, 2001.

[31] AL Hodgkin and AF Huxley, A quantitative description of membrane current and its application to conduction and excitation in nerve, *J. Physiol. (London)* 117 (1952) 500–544.

[32] AL Hodgkin, *The Conduction of the Nervous Impulse*, Liverpool University Press, Liverpool, 1964.

[33] JB Hursh, Conduction velocity and diameter of nerve fibres, *Am. J. Physiol.* 127 (1939) 131–139.

[34] NA Hutchinson, ZJ Koles, and RS Smith, Conduction velocity in myelinated nerve fibers of *Xenopus laevis*, *J. Physiol. (London)* 208 (1970) 279–289.

[35] AF Huxley and R Stämpfli, Evidence for saltatory conduction in peripheral myelinated nerve fibers, *J. Physiol. (London)* 108 (1949) 315–339.

[36] JJB Jack, D Noble, and RW Tsien, *Electric Current Flow in Excitable Cells*, Oxford University Press, Oxford, 1975.

[37] ER Kandel, JH Schwartz, and TM Jessell, *Principles of Neural Science*, Fourth Edition, Appleton & Lange, Norwalk, CT 2000.

[38] JP Keener, Propagation and its failure in coupled systems of discrete excitable cells, *SIAM J. Appl. Math.* 47 (1987) 556–572.

[39] JP Keener and J Sneyd, *Mathematical Physiology*, Springer-Verlag, New York, 1998.

[40] H Kunov, Controllable piecewise-linear lumped neuristor realization, *Electron. Lett.* 1 (1965) 134.

[41] H Kunov, Nonlinear transmission lines simulating nerve axon, PhD Thesis, Electronics Laboratory, Technical University of Denmark, Lyngby, 1966.

[42] H Kunov, On recovery in a certain class of neuristors, *Proc. IEEE* 55 (1967) 427–428.

[43] RG Lascelles and PK Thomas, Changes due to age in internodal length in the sural nerve in man, *J. Neurol. Neurosurg. Psychiatry* 29 (1966) 40–44.

[44] GM Lawson and MA Glasby, Peripheral nerve reconstruction using freeze-thawed muscle grafts: A comparison with fascicular nerve grafts in a large animal model, *J. R. Coll. Surg. Edinburgh* 43 (1998) 295–302.

[45] GF Lopreato, Y Lu, A Southwell, NS Atkinson, DM Hills, and TP Wilcox, Evolution and divergence of sodium channel genes in vertebrates, *Proc. Natl. Acad. Sci. USA* 98 (2001) 7588–7592.

[46] L Lubińska, Demyelination and remyelination in the proximal parts of regenerating nerve fibers, *J. Comp. Neurol.* 117 (1961) 275–289.

[47] WI McDonald and GD Ohrlich, Quantitative anatomical measurements on single isolated fibers from the cat spinal cord, *J. Anat.* 110 (1971) 191–202.

[48] JW Moore, RW Joyner, MH Brill, SG Waxman, and M Najar-Joa, Simulations of conduction in uniform myelinated fibers: Relative sensitivity to changes in nodal and internodal parameters, *Biophys. J.* 21 (1978) 147–160.

[49] J Nagumo, S Arimoto, and S Yoshizawa, An active impulse transmission line simulating nerve axon, *Proc. IRE* 50 (1962) 2061–2070.

[50] AS Paintal, Effects of temperature on conduction in single vagal and saphenous myelinated nerve fibres of the cat, *J. Physiol. (London)* 180 (1965) 20–49.

[51] AS Paintal, Conduction properties of normal peripheral mammalian axons. In *Physiology and Pathobiology of Axons,* SG Waxman (ed), Raven Press, New York, 1978.

[52] WF Pickard, On the propagation of the nervous impulse down medulated and unmedulated fibers, *J. theor. Biol.* 11 (1966) 30–45.

[53] WF Pickard, Estimating the velocity of propagation along myelinated and unmyelinated fibers, *Math. Biosci.* 5 (1969) 305–319.

[54] NW Plummer and MH Meisler, Evolution and diversity of mammaliam sodium channel genes, *Genomics* 57 (1999) 323–331.

[55] DC Quick, WR Kennedy, and L Donaldson, Dimensions of myelinated nerve fibers near the motor and sensory terminals in cat tenuissimus muscles, *Neuroscience* 4 (1979) 1089–1096.

[56] M Rasminsky and TA Sears, Internodal conduction in normal and demyelinated mammalian single nerve fibres, *J. Physiol. (London)* 217 (1971) 66P–67P.

[57] M Rasminsky and TA Sears, Internodal conduction in undissected demyelinated nerve fibers, *J. Physiol. (London)* 227 (1972) 323–350.

[58] I Richer, Pulse transmission along certain lumped nonlinear transmission lines, *Electron. Lett.* 1 (1965) 135–136.

[59] I Richer, The switch-line: A simple lumped transmission line that can support unattenuated propagation, *IEEE Trans. Circuit Theory* CT-13 (1966) 388–392.

[60] JM Ritchie, On the relation between fiber diameter and conduction velocity in myelinated nerve fibres, *Proc. R. Soc. London* B217 (1982) 29–35.

[61] WAH Rushton, A theory of the effects of fibre size in medullated nerve, *J. Physiol. (London)* 115 (1951) 101–122.

[62] FK Sanders and D Whitteridge, Conduction velocity and myelin thickness in regenerating nerve fibers, *J. Physiol. (London)* 105 (1946) 152–174.

[63] G Schnepp, P Schnepp, and G Spaan, Faseranalytische Untersuchungen an peripheren Nerven bei Tieren verschiedener Grösse. I. Fasergesamtzahl, Faserkaliber und Nervenleitungsgeschwindigkeit, *Z. Zellforsch.* 119 (1971) 77–98.

[64] P Schnepp and G Schnepp, Faseranalytische Untersuchungen an peripheren Nerven bei Tieren verschiedener Grösse. II. Verhältnis Axondurchmesser/gesamtdurchmesser und Internodallänge, *Z. Zellforsch.* 119 (1971) 99–114.

[65] AC Scott, Neuristor propagation on a tunnel diode loaded transmission line, *Proc. IEEE* 51 (1963) 240.

[66] AC Scott, The electrophysics of a nerve fiber, *Rev. Mod. Phys.* 47 (1975) 487–533.

[67] AC Scott, *Nonlinear Science: Emergence and Dynamics of Coherent Structures,* Oxford University Press, Oxford, 1999.

[68] I Tasaki, *Nervous Transmission,* Thomas, Springfield, 1953.

[69] I Tasaki, Conduction of the nerve impulse. In *Handbook of Physiology,* Section I. Neurophysiology, American Physiological Society, Washington, DC, 1959.

[70] I Tasaki, *Physiology and Electrochemistry of Nerve Fibers,* Academic Press, New York, 1982.

[71] I Tasaki, K Ishii, and H Ito, On the relation between the conduction-rate, the fibre-diameter and the internodal distance of the medullated nerve fibre, *Jpn. J. Med. Sci.* 9 (1944) 189–199

[72] I Tasaki and M Fujita, Action currents of single nerve fibers as modified by temperature changes, *J. Neurophysiol.* 11 (1948) 311–315.

[73] AD Vizoso, The relationship between internodal length and growth in human nerves, *J. Anat.* 84 (1950) 342–353.

[74] AD Vizoso and JZ Young, Internode length and fibre diameter in developing and regenerating nerves, *J. Anat.* 82 (1948) 110–134.

[75] SG Waxman, JD Kocsis, and PK Stys, *The Axon: Structure, Function and Pathophysiology,* Oxford University Press, New York, 1995.

[76] JZ Young, *Doubt and Certainty in Science,* Oxford Clarendon Press, Oxford, 1951.

[77] YaB Zeldovich and DA Frank-Kamenetsky, K teorii ravnomernogo rasprostranenia plameni, *Dokl. Akad. Nauk SSSR,* 19 (1938) 693–697.

8

Ephaptic Interactions Among Axons

In Chapter 2, were mentioned some of the ways by which neurons can communicate with each other, including chemical transmission via synapses and direct electrical connections through gap junctions, but there are other possibilities. Here, we consider coupling of individual impulses through their external current loops. This is called *ephaptic* coupling (as opposed to synaptic), from a Greek verb meaning "to touch."

A chapter is devoted to this phenomenon for two reasons. First, the possibility of nonsynaptic modifications of neuronal activity in the brain has received less than its fair share of attention [15], and second, this is an area in which mathematical perspectives are particularly helpful [3, 5, 6, 7, 8, 9, 20, 21, 22, 23, 24, 30, 31, 32]. But what is the experimental evidence on which theories of ephaptic interactions are based?

8.1 Empirical Evidence

Since the work of Ewald Hering in 1882 [13],[1] it has been known that nerve impulses on adjacent fibers can influence one another, and the cross section of a sciatic nerve shown in Figure 1.2 suggests that real nerves offer opportunities for such interactions. Shortly after the cathode-ray oscilloscope became available to electrophysiologists in the mid-1930s, the

---

[1]Anticipating some of the ideas discussed at the close of the preceding chapter, Hering slowed the dynamics of his experiments by using frogs that had been kept near 0°C for several months.

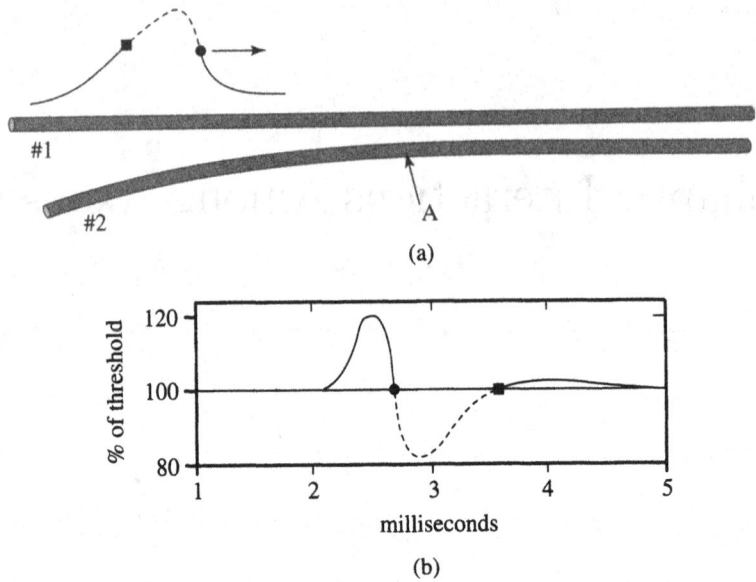

Figure 8.1. (a) Experiment of Katz and Schmitt to measure impulse interactions on parallel fibers. (b) Change in threshold on fiber #2 (at point A) caused by the presence of an impulse on fiber #1. (Redrawn from [17].)

pioneering work of Arvanitaki [1] inspired several observations of ephaptic interactions [4, 10, 12, 14, 17, 18, 19, 25, 26, 28, 29]. More recent references include both theoretical and experimental studies [2, 7, 8, 11, 27] and the important review by Jefferys [15].

An early investigation by Katz and Schmitt provides particularly clear evidence for nonsynaptic interactions [17, 18, 19]. From a variety of experiments on a pair of naturally adjacent, unmyelinated fibers from the limb nerve of a crab, these authors presented the following results.

- Using the experiment sketched in Figure 8.1(a), a reference impulse was launched on fiber #1 from the left, traveling toward the right, and at various later times the relative threshold on fiber #2 was measured at point A.

  Their observations are sketched in Figure 8.1(b), from which it is seen the threshold on fiber #2 changes in a manner that is related to the second derivative of the impulse voltage on fiber #1. (To emphasize this relationship, the impulse voltage in Figure 8.1(a) is dashed where its second derivative is negative, and the corresponding range of reduced threshold in Figure 8.1(b) is also dashed.)

- If impulses are launched at about the same time on two parallel fibers with independent impulse speeds that do not differ by more than about 10%, these impulses become "locked together," or *synchronized*. In other words, they are observed to move at exactly the same speed.

- Both of these interaction effects are strengthened by increasing the ionic resistivity of the external medium.

The aim here is to employ some of the tools that we have learned from previous chapters to find formulations for these observations.

8.2   M–C Analysis of Ephaptic Coupling

In 1970, Markin used the M–C model (see Section 6.1) to describe impulse coupling on parallel fibers [22, 23]. To this end, he considered two fibers to be represented as in Equation (4.10), allowing different values for their parameters. In addition, he assumed that the two fibers share an *external* series resistance per unit length (r_3) that is proportional to the ionic resistivity of the external medium [7], thereby obtaining the equivalent circuit shown in Figure 8.2. Then, using Kirchhoff's circuit laws, just as in Section 4.4, he derived the coupled nonlinear diffusion equations

$$\left(\frac{r_2 + r_3}{\gamma}\right)\frac{\partial^2 V_1}{\partial x^2} - c_1 \frac{\partial V_1}{\partial t} = j_{\text{ion}1} + \frac{r_3}{\gamma}\frac{\partial^2 V_2}{\partial x^2},$$

$$(8.1)$$

$$\left(\frac{r_1 + r_3}{\gamma}\right)\frac{\partial^2 V_2}{\partial x^2} - c_2 \frac{\partial V_2}{\partial t} = j_{\text{ion}2} + \frac{r_3}{\gamma}\frac{\partial^2 V_1}{\partial x^2},$$

where $\gamma \equiv r_1 r_2 + r_1 r_3 + r_2 r_3$. For $r_3 = 0$, these equations evidently reduce to the form of Equation (4.10). For r_3 not zero, on the other hand, an impulse on fiber #1 can influence the dynamics on fiber #2 and vice versa.

Interestingly, these equations readily explain the observations in Figure 8.1. Assuming that V_1 is a fully developed impulse and V_2 is sufficiently small, they become

$$\left(\frac{r_2 + r_3}{\gamma}\right)\frac{\partial^2 V_1}{\partial x^2} - c_1 \frac{\partial V_1}{\partial t} \doteq j_{\text{ion}1},$$

$$(8.2)$$

$$\frac{\partial V_2}{\partial t} \doteq -\frac{r_3}{c_2 \gamma}\frac{\partial^2 V_1}{\partial x^2}.$$

Thus, the region of V_1 that has a negative second derivative (dashed on Figure 8.1(a)) causes V_2 to rise (or depolarize), thereby reducing the threshold of fiber #2, as indicated by the dashed region on Figure 8.1(b).

Inspired in part by Markin's analysis, Ramón and Moore revisited the experimental study of ephaptic interactions between squid giant axons in

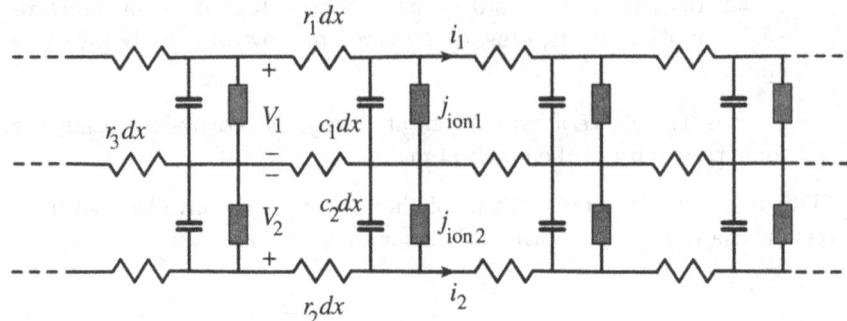

Figure 8.2. Markin's equivalent circuit for two ephaptically coupled nerve fibers [22, 23].

the late 1970s using internal and external voltage recordings to obtain data confirming the second of Equations (8.2). Although ephaptic interactions are unlikely to permit direct transmission of an impulse from one nerve fiber to another, they concluded, impulse coupling is feasible under normal physiological conditions.

Key to the M–C description of nerve impulse propagation is the assumption that

$$j_{\text{ion}} = j_{\text{mc}}(\xi),$$

where $j_{\text{mc}}(\xi)$ follows the piecewise constant function shown in Figure 6.1 whenever V reaches the threshold voltage. Thus, any influence that reduces (increases) the time for an impulse to reach threshold will increase (decrease) its speed.

To apply this concept, let us assume that an impulse on fiber #2 is *leading* an impulse on fiber #1 by a distance δ. In other words, the impulse on fiber #1 goes through threshold at $\xi_1 = 0$, where

$$\xi_1 = x - v_1 t,$$

and the impulse on fiber #2 goes through threshold at $\xi_2 = 0$, where

$$\xi_2 = x - v_2 t - \delta.$$

Now note two additional facts that are evident from the general shape of a nerve impulse: (i) ahead of the point where an impulse goes through threshold, its second space derivative is positive; and (ii) behind this point the second space derivative is negative.

Consider the first of Equations (8.1), and assume that $\xi_1 \approx 0$. Because V_2 has already gone through its threshold, $\partial^2 V_2 / \partial x^2$ is negative. Thus, the influence of V_2 on impulse #1 is to increase $\partial V_1 / \partial t$, thereby raising V_1 above what it would be without the interaction. This has the effect of speeding up impulse #1 (increasing v_1), which causes δ to decrease.

Next, consider the second of Equations (8.1), and assume that $\xi_2 \approx 0$. Because V_1 has not yet gone through its threshold, $\partial^2 V_1/\partial x^2$ is positive. Then, the influence of V_1 on impulse #2 is to decrease V_2 below what it would be without the interaction, which has the effect of slowing down impulse #2 (decreasing v_2), again causing δ to decrease.

Thus, the M–C model predicts a stable coupling (synchronization) of the two impulses with $\delta = 0$. For those who are uncomfortable with qualitative arguments, a dynamic study of impulse synchronization is sketched in the following section.

8.3   Leading-Edge Analysis of Ephaptic Coupling

As an introduction to the application of perturbation to the study of ephaptic interactions, the results of such a study of the simple leading model of Chapter 5 is sketched. This is followed by a qualitative description of the impulse interactions along the lines of the previous section, which—it is hoped—will make the perturbation results seem reasonable.

8.3.1   Sketch of the Perturbation Theory

In Appendix F, perturbation theory is used to study the synchronization of nerve impulses on parallel fibers. In this section, it is anticipated that a leading-edge analysis should be sufficient to obtain the experimental results of Katz and Schmitt because the qualitative argument presented in the previous section depends primarily on impulse behavior near threshold.

To simplify the algebra, it is convenient to assume that the two fibers are identical and described by normalized leading-edge equations as in Equation (5.5). Defining a *coupling parameter*

$$\alpha \equiv \frac{r_3}{r_1 + r_3} = \frac{r_3}{r_2 + r_3}$$

as the ratio of outside (shared) series resistance to total series resistance per unit length of each axon, Equations (8.1) reduce to

$$(1 - \alpha)\frac{\partial^2 V_1}{\partial x^2} - \alpha\frac{\partial^2 V_2}{\partial x^2} - \frac{\partial V_1}{\partial t} \doteq f(V_1),$$

$$(1 - \alpha)\frac{\partial^2 V_2}{\partial x^2} - \alpha\frac{\partial^2 V_1}{\partial x^2} - \frac{\partial V_2}{\partial t} \doteq f(V_2),$$

(8.3)

where the "$\doteq$" signs indicate that terms of order α^2 have been neglected.[2]

---

[2]Some authors have used the system

$$\frac{\partial^2 V_1}{\partial x^2} - \frac{\partial V_1}{\partial t} \doteq f(V_1) + \eta(V_1 - V_2),$$

In the limiting case of $\alpha = 0$, Equations (8.3) evidently reduce to a pair of identical uncoupled nonlinear diffusion equations, which were considered in Chapter 5. In Appendix F, α is taken to be a small parameter in a perturbation analysis; thus, all dependent variables are expressed as power series in α.

To appreciate the fruits of this analysis, consider the cubic representation of the sodium ion current

$$f(V) = V(V - a)(V - 1),$$

for which the leading-edge waveform is

$$V_0(\xi) = \frac{1}{1 + e^{\xi/\sqrt{2}}}, \tag{8.4}$$

and the corresponding traveling-wave speed is

$$v_0 = (1 - 2a)/\sqrt{2}.$$

Defining δ as the distance by which an impulse on fiber #2 leads (is ahead of) an impulse on fiber #1, it is expected that the impulse speeds on the two fibers will be some function of δ. Denoting by $v^{(1)}(\delta)$ and $v^{(2)}(\delta)$ the impulse speed on fiber #1 and #2, respectively, perturbation analysis shows that to first order in α (i.e., neglecting effects of order α^2)

$$\left[v^{(1)}(\delta) - v^{(2)}(\delta)\right] = \frac{3\alpha}{\sqrt{2}} \left(\frac{\Delta}{a(1 - a)(\Delta - 1)^3} \right) \times$$

$$\left(\frac{2\left(\Delta^{1-2a} - \Delta^{1+2a}\right)}{(\Delta - 1)} - \Delta - 1 \right.$$

$$\left. + \frac{\left(\Delta^{1-2a} - \Delta^{-1+2a}\right)\left[2\left(\Delta - 1\right)^2 a^2 + \Delta\left(4a + 1\right)\right]}{(\Delta - 1)(1 - 2a)} \right), \tag{8.5}$$

where $\Delta \equiv \exp(\delta/\sqrt{2})$.

The velocity difference $[v^{(1)}(\delta) - v^{(2)}(\delta)]/\alpha$ is plotted as a function of δ in Figure 8.3 for several values of the threshold parameter a. To comprehend this figure, note three facts. First, the function is odd; that is

$$\left[v^{(1)}(\delta) - v^{(2)}(\delta)\right] = -\left[v^{(1)}(-\delta) - v^{(2)}(-\delta)\right].$$

$$\frac{\partial^2 V_2}{\partial x^2} - \frac{\partial V_2}{\partial t} \doteq f(V_2) + \eta(V_2 - V_1)$$

(where η is a multiplicative constant) as a model for interacting nerve fibers [5, 6, 16, 20]. Unfortunately, this model is biologically unrealistic because it implies that gap junctions (see Section 2.3.2) are uniformly distributed between the two fibers. Such "η-coupling" was also studied by Steve Luzader using the perturbation techniques of Appendix F and comparing it to the "α-coupling" of Equations (8.3) [21].

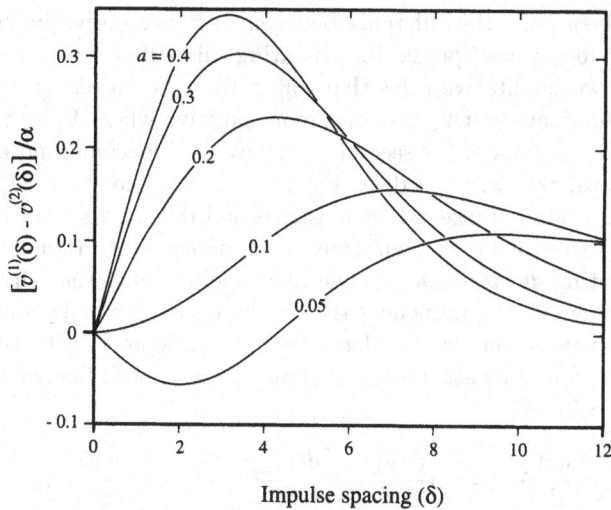

Figure 8.3. A plot of $[v^{(1)} - v^{(2)}]/\alpha$ against δ from Equation (8.5) for several values of the threshold parameter a.

Second, the change in impulse spacing is necessarily related to the velocity difference by

$$\frac{d\delta}{dt} = - \left[v^{(1)}(\delta) - v^{(2)}(\delta) \right]. \tag{8.6}$$

Finally, a detailed analysis of Equation (8.5) shows that for

$$a > a_c \equiv 1/(6 + 2\sqrt{6}) = 0.0918\ldots,$$

the slope of Equation (8.5) at the origin is positive, implying that δ decays to zero and indicating a stable locking of pairs of impulses at $\delta = 0$, just as in the M–C analysis of the previous section. For $a < a_c$, on the other hand, the slope at the origin is negative, indicating that a pair of impulses with $\delta = 0$ is unstable [3].

Why does the M–C representation miss this effect? Referring back to Figure 6.1, the key idea is that the M–C model assumes that the point of maximum slope on the leading edge of an impulse equals the threshold voltage. For the analysis of the present section, however, this is not so. It is seen from Equation (8.4) that the maximum slope is always at $V_0 = 1/2$, whereas the threshold can lie anywhere in the range $0 < a < 1/2$. How might one use this fact and qualitative arguments from the previous section to obtain an intuitive understanding of Figure 8.3?

8.3.2 A Qualitative Analysis

As we have seen in Figure 8.1, it is the second derivative of the impulse voltage on one nerve that speeds up or slows down an impulse on the other (and vice versa), and the difference between these two derivatives generates a difference in impulse speeds. In estimating this difference, it is helpful to remember two qualitative rules that follow from Equations (8.3). First, if the second derivative of V_2 becomes more positive where $V_1 = a$, this will slow down V_1. Second, if the second derivative of V_1 becomes more negative where $V_2 = a$, this will speed up V_2. Thus, if δ is the distance by which impulse #2 leads impulse #1, it is the second derivative of V_2 minus the second derivative of V_1 that will cause δ to increase with time.

In computing $d\delta/dt$, however, one must include only the difference between the way that V_1 influences the second derivative of V_2 and the way that V_2 influences the second derivative of V_1. Near $\delta = 0$, these cross terms can be found by assuming traveling waves on both fibers and writing Equations (8.3) as

$$v\frac{d^2V_2}{d\xi^2} = \alpha\frac{\partial^3V_1}{\partial\xi^3} + \left[\frac{df(V_2)}{d\xi} - (1-\alpha)\frac{\partial^3V_2}{\partial\xi^3}\right],$$

$$v\frac{d^2V_1}{d\xi^2} = \alpha\frac{\partial^3V_2}{\partial\xi^3} + \left[\frac{df(V_1)}{d\xi} - (1-\alpha)\frac{\partial^3V_1}{\partial\xi^3}\right].$$

Because the bracketed terms are equal and do not contribute to the δ-dependent difference of second derivatives (they are not cross terms), it follows that

$$\frac{d\delta}{dt} \propto \frac{d^3V_1}{d\xi^3} - \frac{d^3V_2}{d\xi^3}, \tag{8.7}$$

where V_1 is the amplitude of impulse #1 at the threshold of impulse #2 and V_2 is the amplitude of impulse #1 at the threshold of impulse #1.

At the threshold of impulse $V_2(\xi) = V_0(\xi)$,

$$\frac{d^3V_1(\xi)}{d\xi^3} = \frac{d^3V_0(\xi+\delta)}{d\xi^3} = \frac{d^3V_0(\xi)}{d\xi^3} + \delta\frac{d^4V_0(\xi)}{d\xi^4} + O(\delta^2),$$

and at the threshold of impulse $V_1(\xi) = V_0(\xi)$,

$$\frac{d^3V_2(\xi)}{d\xi^3} = \frac{d^3V_0(\xi-\delta)}{d\xi^3} = \frac{d^3V_0(\xi)}{d\xi^3} - \delta\frac{d^4V_0(\xi)}{d\xi^4} + O(\delta^2).$$

Substituting the indicated difference between these equations into Equation (8.7) leads to the main result of this section: for $\delta \approx 0$,

$$\frac{d\delta}{dt} \propto +\delta\frac{d^4V_0(\xi)}{d\xi^4} + O(\delta^2). \tag{8.8}$$

In using this expression, the fourth derivative is to be computed at threshold voltage $V_0(\xi) = a$, which lies in the range $0 < a \leq 1/2$.

With these ideas in mind, consider four cases.

Case (i): $1/2 > a > a_c$ and $\delta \approx 0$

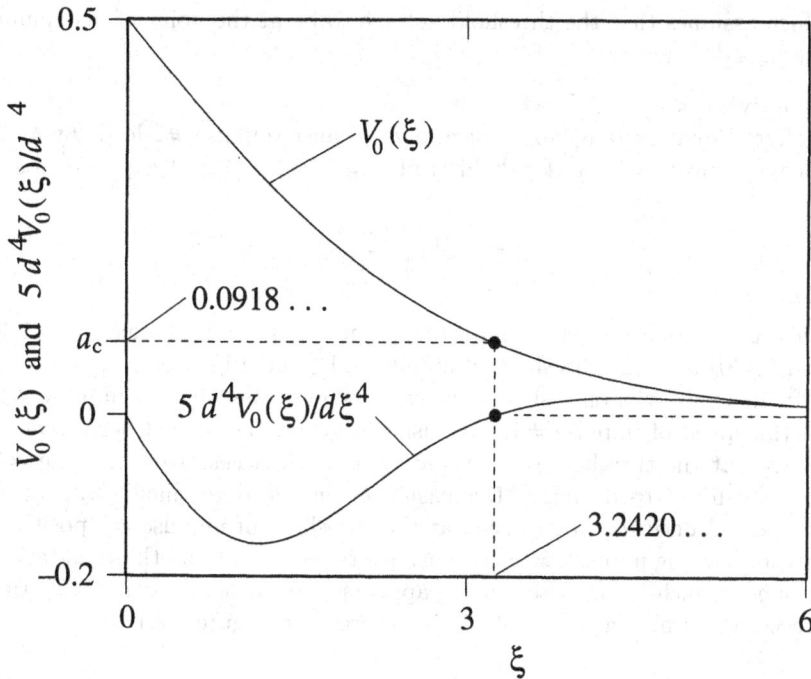

Figure 8.4. Plots of $V_0(\xi) = 1/[1+\exp(\xi/\sqrt{2})]$ and $5\,d^4V_0/d\xi^4$ for the leading-edge waveform of Equation (8.4). At $V_0 = a_c$, $\xi = \sqrt{2}\ln(5 + 2\sqrt{6}) = 3.2420\ldots$. (The factor of 5 in the fourth derivative is introduced for visual clarity.)

As indicated in Figure 8.4, $d^4V_0/d\xi^4$ is negative over the corresponding values of leading-edge voltage, $1/2 > V_0 > a_c$. Thus, Equation (8.8) shows that a small increase in δ relaxes back to zero, indicating that $\delta = 0$ is *asymptotically stable*. This is the situation treated by M–C analysis in Section 8.2.

Case (ii): $a = a_c$ and $\delta \approx 0$

If V_0 is equal to the critical threshold voltage, $d^4V_0/d\xi^4 = 0$.[3] From Equation (8.8), increasing or decreasing δ by a small amount leaves $d\delta/dt$ unchanged to first order, implying *neutral stability* (neither stable nor unstable).

Case (iii): $a_c > a > 0$ and $\delta \approx 0$

Over this range of V_0, $d^4V_0/d\xi^4$ is positive, so Equation (8.8) indicates that δ grows away from zero, implying $\delta = 0$ is *unstable*, in accord with Figure 8.3. This phenomenon is missed by the M–C analysis of Section 8.2,

---

[3]More precisely, the fourth derivative of V_0 goes to zero at $\exp(\xi/\sqrt{2}) = 5 + 2\sqrt{6} = (1 - a_c)/a_c$, where $a_c = 1/(6 + 2\sqrt{6}) = 0.0918\ldots$.

which assumes that the threshold voltage to be at the point of maximum leading-edge slope.

Case (iv): $a < a_c$ and $\delta \gg 0$

Here, Equation (8.8) does not apply. Because impulse #2 leads by δ, it is convenient to let the threshold of fiber #1 be $a = 1/[1 + \exp(\delta/\sqrt{2})]$ so

$$\delta = \sqrt{2}\ln\left(\frac{1-a}{a}\right).$$

This ensures that the central inflection point of impulse #2 (where $V_2 = 1/2$ and $\xi = 0$) is at the threshold of impulse #1 (where $V_1 = a$).

In this case, the only significant cross term is the effect of impulse #2 on the speed of impulse #1. Increasing δ makes the second derivative of V_2 seen at the threshold of V_1 negative, thereby accelerating impulse #1 and causing δ to decrease. Decreasing δ, on the other hand, will make the second derivative of V_2 seen at the threshold of impulse #1 positive, thereby slowing impulse #1 and causing δ to increase. From this qualitative argument, such an impulse spacing appears to be *stable*. (For $a = 0.05$, the preceding formula gives $\delta = 4.164$ in accord with Figure 8.3.)

8.4   Ephaptic Coupling in an F–N Model

Perturbation calculations corresponding to those of the previous section can be carried through for coupled FitzHugh–Nagumo fibers, as was done by Steve Luzader in the late 1970s [21, 32]. The details of this analysis are also included in Appendix F, where it is seen that difficulties arise because certain linear operators that are second order in the leading-edge analysis of Equations (8.3) become third order. Thus, it is not possible to find an analytic expression for the velocity difference corresponding to Equation (8.5). Nonetheless, a sketch of the numerical results is as follows.

Starting with the coupled F–N system

$$(1-\alpha)\frac{\partial^2 V_1}{\partial x^2} - \alpha\frac{\partial^2 V_2}{\partial x^2} - \frac{\partial V_1}{\partial t} \doteq f(V_1) + R_1,$$

$$\frac{\partial R_1}{\partial t} = \varepsilon V_1,$$

$$(1-\alpha)\frac{\partial^2 V_2}{\partial x^2} - \alpha\frac{\partial^2 V_1}{\partial x^2} - \frac{\partial V_2}{\partial t} \doteq f(V_2) + R_2,$$

$$\frac{\partial R_2}{\partial t} = \varepsilon V_2,$$

(8.9)

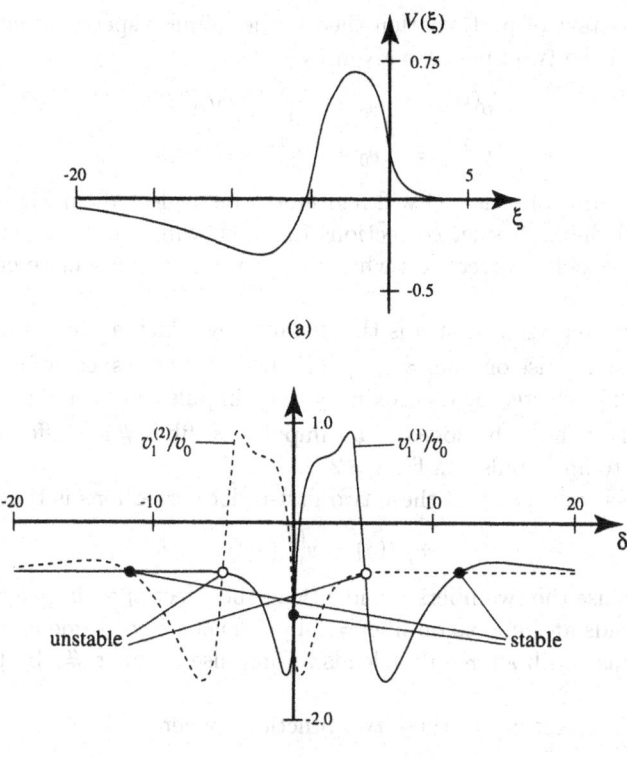

Figure 8.5. (a) A single impulse solution of Equations (8.9) for $\varepsilon = 0.1$, $\alpha = 0.1$, and $a = 0.3$. (b) First-order velocity corrections $v_1^{(1)}$ (full line) and $v_1^{(2)}$ (dashed line) as functions of the impulse separation δ. Note that $v_1^{(1)}(\delta) = v_1^{(2)}(-\delta)$. (Redrawn from [21].)

the nonlinear function was assumed to be the piecewise linear function

$$f(V) = \begin{cases} V \text{ for } V < a, \text{ and} \\ \\ (V-1) \text{ for } V > a. \end{cases}$$

With $a = 0.3$ in $f(V)$ and with $\alpha = 0.1$ and $\varepsilon = 0.1$ in Equations (8.9), a fully developed impulse on one of the fibers appears as in Figure 8.5(a) as a function of $\xi = x - vt$ with velocity

$$v = \frac{v_0}{\sqrt{1+\alpha}} = 0.8323\ldots, \tag{8.10}$$

where $v_0 = 0.8729\ldots$. Because this F–N impulse has two regions where the second derivative $(d^2V/d\xi^2)$ is positive and two where it is negative, the M–C analysis of Section 8.2 suggests that the ephaptic interaction will be more involved than for leading edges.

In the context of perturbation theory, the impulse speeds of interacting impulses on the two fibers are given by

$$v^{(1)} = v_0 + \alpha v_1^{(1)} + O(\alpha^2),$$
$$v^{(2)} = v_0 + \alpha v_1^{(2)} + O(\alpha^2),$$

where the terms of order α^2 will henceforth be neglected. In Figure 8.5(b) are plotted the first-order corrections to v_0, showing what happens to the first-order velocity corrections when the two impulses are close enough to interact.

Assuming yet again that δ is the distance by which an impulse on fiber #2 leads an impulse on fiber #1, $v_1^{(2)}(\delta)$ shows how the speed of an impulse on fiber #2 is affected by its proximity to an impulse on fiber #1. Similarly, $v_1^{(1)}(\delta)$ tells us how the speed of an impulse on fiber #1 is affected by its proximity to an impulse on fiber #2.

A necessary property of these two first-order corrections is that

$$v_1^{(1)}(\delta) = v_1^{(2)}(-\delta).$$

Why? Because the two fibers are identical, the effect of an impulse on fiber #2 that leads an impulse on fiber #1 by δ must be the same as the effect of an impulse on fiber #1 that leads an impulse on fiber #2 by the same distance.

It is at intersections of these two functions, where

$$v_1^{(1)}(\delta) = v_1^{(2)}(\delta),$$

that the two impulses can travel together at the same speed, but only three of these intersections are stable (at $\delta = 0$ and $\delta = \pm 12$, indicated by black dots), whereas two are unstable (at $\delta = \pm 5.8$, indicated by white dots).

To appreciate the nature of this stability, consider first the black dot at $\delta = 0$, indicating that the two impulses are traveling at the same speed with no spatial separation. Suppose that δ happens to be increased slightly from zero so the impulse on fiber #2 is slightly ahead of the impulse on fiber #1. Figure 8.5(b) then shows that

$$v_1^{(1)}(\delta) > v_1^{(2)}(\delta),$$

implying that impulse #1 will go faster than impulse #2, thereby closing the gap. Corresponding arguments show that all of the black dots indicate stable interactions. Similarly, the white dots indicate unstable impulse separations because a slight increase of δ makes

$$v_1^{(2)}(\delta) > v_1^{(1)}(\delta),$$

thereby further increasing δ.

From a general perspective, this situation is analogous to the ignition of an impulse, which we encountered in previous chapters. On a single fiber, the fully developed impulse and the zero state are stable solutions of the PDE system (H–H, F–N, M–C, leading edge, or whatever), and the unstable threshold solution is a separatrix dividing the two. Similarly, the white dot at $\delta = 5.8$ in Figure 8.5(b) indicates a critical value above which δ will grow to the stable value of 12 and below which it will relax to the stable value of zero.

Importantly, the results of the perturbation calculation in [21] have been checked by comparing solutions of the equation

$$\frac{d\delta}{dt} = -\left[v_1^{(1)}(\delta) - v_1^{(2)}(\delta)\right]$$

with computations of $\delta(t)$ from numerical integrations of the full PDE system of coupled F–N equations. Good agreement between the two approaches was observed [9].

The black dots in Figure 8.5(b) thus indicate stable traveling waves comprising synchronized impulses existing at a higher level of organization than that of their component impulses. Because these ideas can be generalized to several synchronized impulses on many fibers [24, 32], it is interesting to consider whether such groups of impulses might play functional roles on the fiber bundles of motor nerves, sensory nerves, or on the *corpus callosum*, comprising a large number of fibers carrying information between the two hemispheres of the mammalian brain [32]. Such fibers, however, are not smooth but myelinated.

8.5 Ephaptic Coupling of Myelinated Nerves

For many instances of ephaptic coupling, the nerve fibers are myelinated, with impulses jumping from active node to active node as discussed in the previous chapter. Outstanding examples include the motor nerve bundles of vertebrates (see Figure 1.2), optic nerves, and the corpus callosum. Thus, it is of interest to consider models for ephaptic coupling of impulses that are propagating on myelinated nerve fibers [3].

8.5.1 A Numerical Model for Myelinated Interactions

In describing ephaptic interactions between myelinated nerve fibers, it seems important to model the degree to which the locations of active nodes are aligned. To this end, two partially aligned axons are sketched in Figure 8.6(a), and a corresponding circuit diagram is shown in Figure 8.6(b).

In this circuit diagram, the $V_n^{(j)}$ are voltages across the active nodes, where $j = 1, 2$ indicates a particular fiber. Similarly, the $I_n^{(j)}$ define *mesh currents*, the independent variables for an analysis in terms of Kirchhoff's voltage law. (A component of current that circulates about a mesh is called a mesh current. Evidently, all of the mesh currents in Figure 8.6(b) are

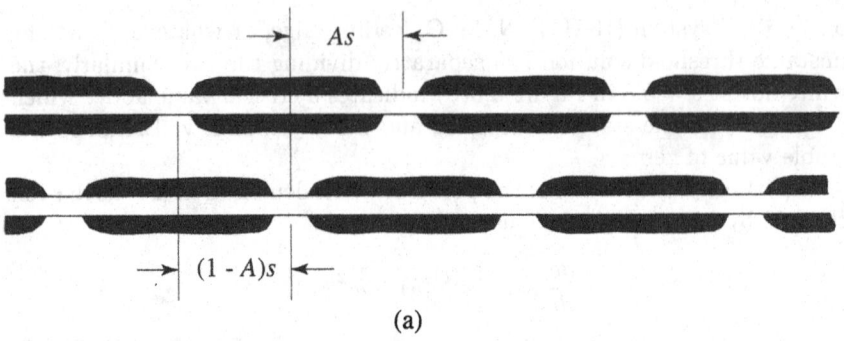

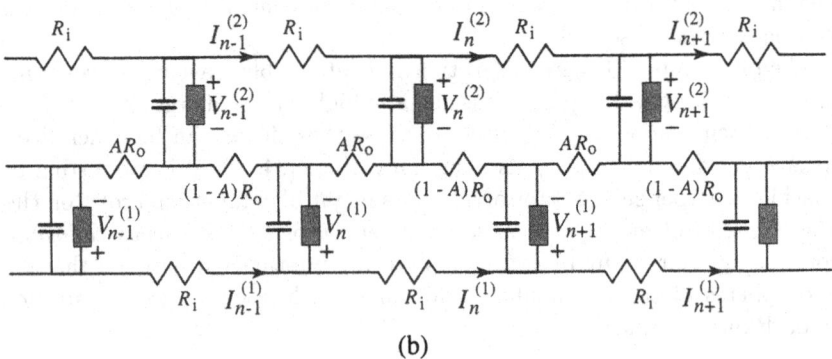

Figure 8.6. (a) Two myelinated nerves on which impulses may be coupled by a linking of their external return currents (not to scale). (b) A circuit diagram of the coupled myelinated nerves.

determined by the $I_n^{(1)}$ and $I_n^{(2)}$.) Equating the voltages about the meshes to zero leads directly to the equations

$$V_n^{(1)} - V_{n+1}^{(1)} = (R_{\mathrm{i}} + R_{\mathrm{o}})I_n^{(1)} + R_{\mathrm{o}}\left[AI_n^{(2)} + (1-A)I_{n-1}^{(2)}\right],$$

$$V_n^{(2)} - V_{n+1}^{(2)} = (R_{\mathrm{i}} + R_{\mathrm{o}})I_n^{(2)} + R_{\mathrm{o}}\left[AI_n^{(1)} + (1-A)I_{n+1}^{(1)}\right],$$

where the voltages across the active nodes are related to the mesh currents by

$$I_{n-1}^{(j)} - I_n^{(j)} = C\frac{dV_n^{(j)}}{dt} + I_{\mathrm{ion},n}^{(j)}.$$

As in the previous chapter, it is analytically convenient to model the ionic current in the cubic approximation

$$I_{\mathrm{ion},n}^{(j)} = \left(\frac{G}{V_2(V_2 - V_1)}\right)V_n^{(j)}(V_n^{(j)} - V_1)(V_n^{(j)} - V_2),$$

which was introduced in Equation (5.9).

In this formulation, an *alignment parameter* A indicates the degree of nodal alignment. With $A = 1$, the active nodes on the two fibers are exactly aligned, whereas they are evenly staggered for $A = 1/2$.

For the present analysis of these equations, the parameter values from the standard frog axon of the previous chapter are used. Furthermore, it is convenient to measure voltages and currents in units of V_2 and V_2/R_f, respectively. Thus, the equations for ephaptic coupling become

$$D\left(\mathcal{V}_n^{(1)} - \mathcal{V}_{n+1}^{(1)}\right) = i_n^{(1)} + \alpha\left[A i_n^{(2)} + (1 - A)i_{n-1}^{(2)}\right],$$

$$D\left(\mathcal{V}_n^{(2)} - \mathcal{V}_{n+1}^{(2)}\right) = i_n^{(2)} + \alpha\left[A i_n^{(1)} + (1 - A)i_{n+1}^{(1)}\right],$$

$$i_{n-1}^{(j)} - i_n^{(j)} = R_f C \frac{d\mathcal{V}_n^{(j)}}{dt} + i_{\text{ion},n}^{(j)}, \qquad (8.11)$$

$$i_{\text{ion},n}^{(j)} = \left(\frac{R_f G}{1 - a}\right)\mathcal{V}_n^{(i)}(\mathcal{V}_n^{(j)} - a)(\mathcal{V}_n^{(j)} - 1),$$

where

$$\alpha \equiv \frac{R_o}{R_i + R_o}$$

and $\mathcal{V}_n^{(j)} \equiv V_n^{(j)}/V_2$, $i_n^{(j)} \equiv R_f I_n^{(j)}/V_2$, $a \equiv V_1/V_2 = 0.205$, $R \equiv (R_i + R_o)$, and $j = 1, 2$. Also, as in Chapter 7,

$$D \equiv \frac{R_f}{R} = \frac{2\text{ mm}}{s}$$

is the discreteness parameter for the myelinated axon, and s is the internodal distance. Large values of D imply the continuum limit, where $D \approx 1$ or less indicates the saltatory limit.

In studies of coupled impulses on myelinated fibers, the following features have been numerically confirmed [3].

The Continuum Limit
For $D \gg 1$, $A = 1$, and $\delta = 0$ (two coupled impulses), all voltages and currents are identical on the two fibers, effectively increasing the current in each fiber by a factor of $1 + \alpha$. This is equivalent to multiplying both C and G by the same factor, and from Equation (7.9) conduction velocity is proportional to $\sqrt{G/C^2}$. Thus, it follows that the speed of two coupled impulses is decreased by a factor of $\sqrt{1 + \alpha}$ below the speed of a single impulse on the same system, in accord with Equation (8.10).

For $A = 1/2$, the numerically observed conduction velocity also decreases by a factor of $\sqrt{1 + \alpha}$, as expected, because the average values of the currents and voltages over several nodes are independent of node alignment in the continuum limit.

Failure
For $D \leq 1$, $A = 1$ and $\delta = 0$ (two coupled impulses), twice the external current flows through the external resistance (R_o); thus, the effective loop resistance seen by each fiber is

$$R_i + 2R_o = R(1 + \alpha).$$

In Equation (7.10), the value of the discreteness parameter at which failure occurs, called D^*, is proportional to the effective loop resistance. It follows that D^* is increased by the factor $1 + \alpha$, or

$$\frac{\text{critical node spacing for failure of two impulses}}{\text{critical spacing for failure of a single impulse}} = \frac{1}{1 + \alpha}.$$

For $A = 1/2$, the dependence of D^* on α is weaker and of the opposite sign. Referring to Figure 8.6, we see that the nodes are evenly staggered for $A = 1/2$, implying an external resistance of $R_o/2$ linking adjacent current loops. Numerical studies show that in the saltatory limit the jumps alternate, the impulse first advancing on one fiber and then on the other. Thus the effective loop resistance can be computed for a single fiber as

$$
\begin{aligned}
R_i + 2\left[\left(\frac{R_o}{2}\right) \| (R_i + R_o/2)\right] &= R_i + R_o - \frac{1}{2}\frac{R_o^2}{R_i + R_o} \\
&= R(1 - \alpha^2/2),
\end{aligned}
$$

where the symbol "$\|$" implies evaluating the parallel combination of the resistors indicated,[4] and contributions of order R_o^3 have been neglected. Therefore,

$$\frac{\text{critical node spacing for failure of two impulses}}{\text{critical spacing for failure of a single impulse}} = 1 + \frac{1}{2}\alpha^2 + O(\alpha^3).$$

Dynamics of Impulse Coupling
Recall from Figures 8.3 and 8.5 that the dynamics of coupled impulses on smooth fibers are governed by the difference in their speeds, which in turn are functions of the impulse separation. Assuming as before that the impulse on fiber #2 leads the impulse on fiber #1 by a distance δ, $v^{(1)}(\delta)$ and $v^{(2)}(\delta)$ have been defined as the numerically computed impulse speeds on fibers #1 and #2, respectively. For sufficiently small values of coupling ($\alpha \leq 0.1$), the difference between these speeds is found to be proportional to the coupling constant α.

Upon analyzing many numerical integrations of Equations (8.11), Stephane Binczak has determined $d\delta/dt$ as a function of δ at various values of the system parameters. These numerical data allow construction of

$$\frac{v^{(1)}(\delta) - v^{(2)}(\delta)}{\alpha} = \left[-\frac{d\delta/dt}{\alpha}\right]_{\text{num.}}$$

---

[4]For resistors R_1 and R_2, $R_1 \| R_2 \equiv R_1 R_2/(R_1 + R_2)$.

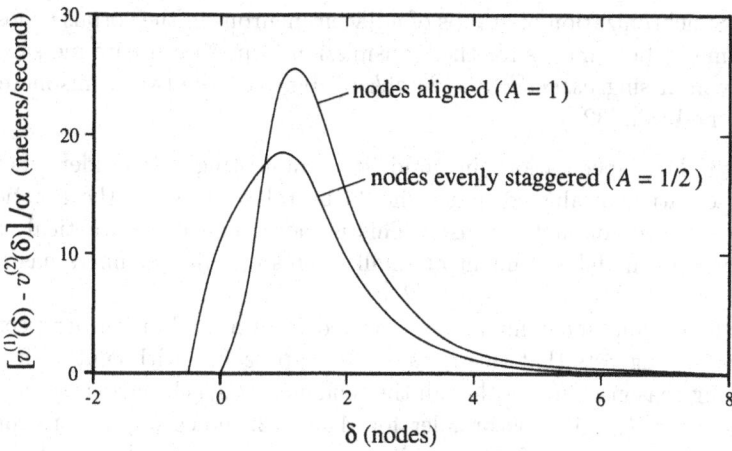

Figure 8.7. Numerical computations of relative speeds of two impulses $[v^{(1)} - v^{(2)}]/\alpha$ as functions of impulse spacing (δ) for the standard frog axon of Chapter 7. (Data courtesy of Stephane Binczak.)

as functions of δ for $A = 1$ and $A = 1/2$, which are plotted in Figure 8.7 for the standard frog axon of Chapter 7 [3].

This figure shows that an alignment of the active nodes ($A = 1$) leads to a somewhat stronger and more localized synchronization of impulses than for the staggered case ($A = 1/2$). From a qualitative perspective, this seems reasonable because it is the external resistance shared by two fibers that induces impulse synchronization, and with $A = 1$ this shared external resistance is entirely situated between two nodes. With $A = 1/2$, on the other hand, the external coupling resistance is shared among three nodes, resulting in a somewhat broader and weaker impulse coupling.

8.5.2 Neurological Implications

Functional significance of coupled impulses on myelinated fibers may arise from at least three considerations [32].

- Synchronization of impulses on bundles of motor neurons might provide a means for adjusting and maintaining timings among coupled impulses, allowing for coordinated stimulations of muscle cells.

- Impulse synchronization on bundles of optical or auditory axons in the central nervous system may help to ensure the timings necessary for computations in the dendritic fields of subsequent neurons [31]. (We consider the nature of such dendritic computations in the following chapter.)

- Synchronization of groups of adjacent neurons in the corpus callosum might be a means for the transmission of more intricate messages—comprising *assemblies of impulses*—between the two hemispheres of the brain [32].

With these speculations in mind, it is interesting to consider whether an observation of aligned nodes should be taken to imply the functional importance of coupled impulses. This is more than a hypothetical question because nodal alignment of small groups of adjacent fibers has been observed in the corpus callosum [33].

Although alignment might be expected from a qualitative perspective, Figure 8.7 suggests that this question be approached with caution for the following reasons. First, although the tendency to synchronize is somewhat stronger for $A = 1$, it is broader for $A = 1/2$, indicating a more robust coupling. Thus, both of these limiting cases lead to impulse synchronization near $\delta = 0$. If the fibers are short (with few nodes), the need for rapid synchronization may favor node alignment, whereas more rapid synchronization would be less important for longer fibers. Second, the node separation at which failure occurs is insensitive to ephaptic coupling for staggered nodes, whereas it decreases linearly with the coupling for aligned nodes.

In any case, Figure 8.7 suggests that ephaptic coupling is to be expected on myelinated fibers for every sort of nodal orientation—aligned, unaligned, or random.

8.6   Recapitulation

The chapter opened with a sketch of the empirical evidence for ephaptic coupling of nerve impulses on parallel fibers, which shows the change in threshold on one fiber as a function of the timing of an impulse on a neighboring fiber. A simple Markin–Chizmadzhev analysis of coupled nerves explains this effect and also the phenomenon of synchronization, in which two coupled impulses travel at exactly the same speed. Assuming that the threshold level is sufficiently high, this tentative conclusion is supported by a perturbation study of coupled impulses in the leading-edge approximation. A simple qualitative analysis is provided for the case in which the firing threshold is much less than the leading-edge amplitude.

Without going into the analytic details, similar results are presented for a corresponding perturbation analysis of the FitzHugh–Nagumo model. Computations of the difference in impulse speed as a function of longitudinal spacing between impulses are found to have the same multiphase behavior as the above-mentioned experimental observations of threshold variation.

Extension of these ideas to myelinated nerves is effected through numerical analyses of coupled pairs of the standard frog axon, which was characterized in the previous chapter. These studies imply that the speeds of coupled impulses decrease below their values on isolated fibers, that with nodes aligned, the critical node spacing for failure decreases with stronger coupling, and that for evenly staggered nodes, the critical node spacing for failure is essentially independent of coupling. Numerical studies also show that synchronization of impulses on coupled myelinated fibers is somewhat stronger or broader, depending on whether the active nodes are aligned or evenly staggered.

From a neurological perspective, it is suggested that impulse synchronization on parallel fibers might provide means for coordinating neural time codes and for transmitting more intricate patterns of information. The computational significance of such time codes is considered in the following chapter.

References

[1] A Arvanitaki, Effects evoked in an axon by the activity of a contiguous one, *J. Neurophysiol.* 5 (1942) 89–108.

[2] MVL Bennett, GD Pappas, E Aljure, and Y Nakajima, Physiology and ultrastructure of electronic junctions, *J. Neurophysiol.* 30 (1967) 180–208.

[3] S Binczak, JC Eilbeck, and AC Scott, Ephaptic coupling of myelinated nerve fibers, *Physica D* 148 (2001) 159–174.

[4] EH Blair and J Erlanger, Interaction of medullated fibers of a nerve tested with electric shocks, *Am. J. Physiol.* 131 (1940) 483–493.

[5] A Bose, Symmetric and antisymmetric pulses in parallel coupled nerve fibers, *SIAM J. Appl. Math.* 55 (1995) 1650–1674.

[6] A Bose and CKRT Jones, Stability of the in-phase travelling wave solution in a pair of coupled nerve fibers, *Indiana Univ. Math. J.* 44 (1995) 189–220.

[7] JW Clark and R Plonsey, A mathematical study of nerve fiber interaction, *Biophys. J.* 10 (1970) 937–957.

[8] JW Clark and R Plonsey, Fiber interaction in a nerve trunk, *Biophys. J.* 11 (1971) 281–294.

[9] JC Eilbeck, SD Luzader, and AC Scott, Pulse evolution on coupled fibers, *Bull. Math. Biol.* 43 (1981) 389–400.

[10] R Granit, LE Leksell, and CR Skoglund, Fiber interaction in injured or compressed regions of nerve, *Brain* 67 (1944) 125–140.

[11] L Goldman and JS Albus, Computation of impulse conduction in myelinated fibers, *Biophys. J.* 8 (1968) 596–607.

[12] H Grundfest and J Magnes, Excitability changes in dorsal roots produced by electrotonic effects from adjacent afferent activity, *Am. J. Physiol.* 164 (1951) 502–508.

[13] E Hering, Beiträge zur allgemeinen Nerven- und Muskelphysiologie. IV. Über Nervenreizung durch den Nervenstrom, *Sitzungsber. k. Akad. Wiss. (Wien)* 85 part 3 (1882) 237–275.

[14] HH Jasper and AM Monnier, Transmission of excitation between excised non-myelinated nerves: An artificial synapse. *J. Cell. Comp. Physiol.* 11 (1938) 259–277.

[15] JGR Jefferys, Nonsynaptic modulation of neuronal activity in the brain: Electric currents and extracellular ions, *Physiol. Rev.* 75 (1995) 689–723.

[16] VB Kazantsev, DV Artykhin, and VI Nekorkin, Dynamics of excitation pulses in two coupled nerve fibers, *Radiophys. Quantum Electron.* 41 (1998) 1079–1086.

[17] B Katz and OH Schmitt, Excitability changes in a nerve fiber during the passage of an impulse in an adjacent fiber, *J. Physiol. (London)* 96 (1939) 9P–10P.

[18] B Katz and OH Schmitt, Electric interaction between two adjacent nerve fibers, *J. Physiol. (London)* 97 (1940) 471–488.

[19] B Katz and OH Schmitt, A note on the interaction between nerve fibers, *J. Physiol. (London)* 100 (1942) 369–371.

[20] JP Keener, Frequency dependent coupling of parallel excitable fibers, *SIAM J. Appl. Math.* 49 (1989) 210–230.

[21] SD Luzader, *Neurophysics of Parallel Nerve Fibers*, PhD thesis, University of Wisconsin, Madison, 1979.

[22] VS Markin, Electrical interactions of parallel nonmyelinated fibers I. Change in excitability of the adjacent fiber, *Biophysics* 15 (1970) 122–133.

[23] VS Markin, Electrical interactions of parallel nonmyelinated fibers II. Collective conduction of impulses, *Biophysics* 15 (1970) 713–721.

[24] VS Markin, Electrical interactions of parallel nonmyelinated fibers III. Interaction in bundles, *Biophysics* 18 (1973) 324–332.

[25] AS Marrazzi and R Lorente de Nó, Interaction of neighboring fibers in myelinated nerve, *J. Neurophysiol.* 7 (1944) 83–101.

[26] T Otani, Über eine Art Hemmung und Bahnung in Folge der Wechselbeziehungen Nervenfasern zueinander, *Jpn. J. Med. Sci.* 4 (1937) 355–372.

[27] F Ramón and JW Moore, Ephaptic transmission in squid giant axons, *Am. J. Physiol.* 234 (1978) C162–C169.

[28] B Renshaw and PO Therman, Excitation of intraspinal mammalian axons by nerve impulses in adjacent axons, *Am. J. Physiol.* 133 (1941) 96–105.

[29] A Rosenbleuth, The stimulation of myelinated axons by nerve impulses in adjacent myelinated axons, *Am. J. Physiol.* 132 (1941) 119–128.

[30] AC Scott, The electrophysics of a nerve fiber, *Rev. Mod. Phys.* 47 (1975) 487–533.

[31] AC Scott, *Nonlinear Science: Emergence and Dynamics of Coherent Structures*, Oxford University Press, Oxford, 1999.

[32] *Phys. Scr.* 20 (1979) 395–401.

[33] SG Waxman, private communications, 1978 and 1999.

9
Neural Modeling

Although we are now familiar with several ways in which mathematical studies have contributed to an understanding of nerve impulse dynamics, much remains to be done before the global behavior of a neuron is considered well-modeled.

In this chapter, attention is focused on two questions, the first being: What are the dendrites doing? Because a compelling response to this question is not currently known, the emphasis is on establishing a context in which credible answers may emerge. Thus, dendritic trees are modeled in three different ways: as linear gatherers of incoming (synaptic) signals, as fully nonlinear computing systems performing logical operations (or switchings) on all synaptic codes, and as power series in the incoming impulse rates, anticipating that the functioning of any particular dendritic tree may lie somewhere between the two idealized limits.

Second, we ask: What might the axons be doing? Here suggestions of unusual computational abilities are supported by observations of impulse blockage at branchings of axonal trees, which leads to the speculation that some trees may translate time codes on their trunks into space and time codes on the distal twigs.

The dynamics of individual neurons comprise a wide variety of axonal, dendritic, and synaptic behaviors that are but briefly surveyed in this chapter. For those who wish to delve more deeply into this area of neuroscience, Michael Arbib's *Handbook of Brain Theory and Neural Networks* [9] and Christof Koch's *Biophysics of Computation* [52] are recommended.

9.1 Linear Dendritic Models

A typical neuron is served by an array of dentrites, which are tree-like
structures that receive incoming signals from synapses and present them
to the outgoing axonal tree for transmission to other neurons or to muscle cells. Some appreciation for the variety of such trees may be obtained
from Figure 9.1, which shows four different classes of hippocampal dendrites.[1] Nowadays, neuroscientists are asking: What are the functions of
these dendritic trees? Are their dynamics linear or nonlinear? Do they process the incoming data streams, or are they merely passive conduits for this
information?

Over the past few decades, the simplest assumption has been that dendritic trees gather a linear weighted sum of the input signals for three
reasons. First, on many neurons, there was no clear evidence to the contrary.
Second, if all-or-nothing propagation is supposed to occur on dendrites,
then the entire tree might be expected to ignite, precluding the opportunity
to integrate incoming information. Third, assuming a linear combination of
the various streams of incoming information makes it easier to sort out various causal influences, easing somewhat the daunting difficulties of studying
neural systems.

Although current empirical evidence supports nonlinear information processing on dendrites [41, 119], it is convenient to begin our discussion by
exploring the assumption of dendritic linearity, not least because it helps
us to understand some key constraints on approaches to the threshold of
nonlinearity [106].

9.1.1 Passive Dendrites

From the discussion of the Hodgkin–Huxley axon in Chapter 4, the basic
example of a passive (or *dissipative*) neural process on a fiber is described
by the linear diffusion equation

$$\frac{1}{r}\frac{\partial^2 V}{\partial x^2} - c\frac{\partial V}{\partial t} = g_{\text{rest}}V,\tag{9.1}$$

where we know something about the parameters. The series resistance per
unit length r and the membrane capacitance per unit length c are as defined
in Chapter 4, but g_{rest} is the *resting* conductance per unit length of the

---

[1]The hippocampus is a seahorse-shaped region of the inner brain that is implicated in short-term memory. Presently, some 200 such images can be viewed at
www.neuro.soton.ac.uk maintained by the Centre for Neuroscience at the University of
Southampton [27]. Complete geometrical data on each neuron are available at this site,
which is organized to accept contributions from around the globe. Another interesting
site is www.dendrites.org.

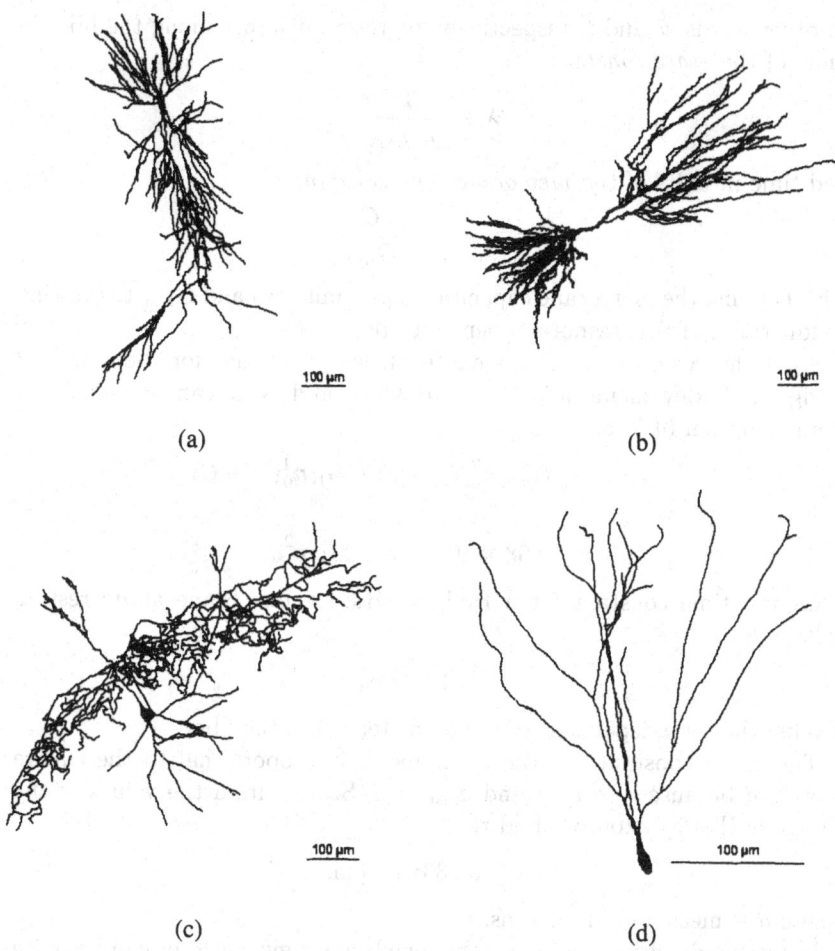

Figure 9.1. Typical dendritic trees in the hippocampus of the rat. (a) CA1 pyramidal cell [90]. (b) CA3 pyramidal cell [127]. (c) Interneuron [76]. (d) Granule cell [128]. (From the Southampton–Duke Public Morphological Archive [27].)

fiber, which is more than an order of magnitude smaller than the active conductance (g) that was considered in Chapter 5.

It is convenient to normalize this equation by measuring time in units of c/g_{rest} and distance along the fiber in units of $1/\sqrt{rg_{\text{rest}}}$. Then Equation (9.1) reduces to the normalized form

$$\frac{\partial^2 V}{\partial \tilde{x}^2} - \frac{\partial V}{\partial \tilde{t}} = V,$$ (9.2)

where

$$\tilde{x} \equiv x\sqrt{rg_{\text{rest}}} \quad \text{and} \quad \tilde{t} \equiv t\frac{g_{\text{rest}}}{c}.$$

In other words, $\tilde{x}$ and $\tilde{t}$, respectively, represent distance along the fiber in units of the *space constant*

$$\lambda \equiv \frac{1}{\sqrt{r g_{rest}}}$$

and time in units of the *membrane time constant*

$$\tau \equiv \frac{c}{g_{rest}} = \frac{C}{G_{rest}},$$

with C being the membrane capacitance per unit area and G_{rest} the resting conductance of the membrane per unit area.

From the considerations of Chapter 4, we recall that for the standard Hodgkin–Huxley membrane $C = 1\,\mu F/cm^2$, and G_{rest} can be calculated from Equation (4.5) as

$$G_{rest} = G_{Na}m_0^3(0)h_0(0) + G_K n_0^4(0) + G_L$$

$$= 0.68 \times 10^{-3} \quad mhos/cm^2.$$

Thus the time constant for a Hodgkin–Huxley membrane at its resting voltage is

$$\tau = 1.5 \quad ms,$$

a value that is independent of the diameter (d) of the fiber.

The space constant, on the other hand, is proportional to the square root of d because $r \propto 1/d^2$ and $g_{rest} \propto d$. Scaling from the values of the Hodgkin–Huxley axon, we find that

$$\lambda = 0.033\sqrt{d} \quad cm,$$

where d is measured in microns.

In resting dendrites, however, the membrane time constant can be quite different from the H–H value, as indicated in Table 9.1 [106, 116]. Because differences in τ stem from variations in the resting conductance, corresponding variations in λ are estimated in the last column of the table.

Return now to Equation (9.2), observing that it has the *exact* solution

$$V(\tilde{x}, \tilde{t}) = \frac{1}{\sqrt{4\pi \tilde{t}}} \exp\left(-\frac{\tilde{x}^2}{4\tilde{t}} - \tilde{t}\right), \tag{9.3}$$

a result that was anticipated in Equation (2.3). From its three-dimensional plot in Figure 9.2, we see the dynamic character of this function in that it spreads out with time into an ever-wider bell-shaped curve of exponentially decreasing area.

Note that Equation (9.3) is normalized (multiplied by the factor $1/\sqrt{4\pi}$) such that

$$\int_{-\infty}^{+\infty} V(\tilde{x}, \tilde{t})d\tilde{x} = e^{-\tilde{t}}.$$

Table 9.1. Time and space constants for resting dendritic membranes compared with Hodgkin–Huxley values, where d is the fiber diameter in microns. (Data from [106] and [116].)

Neuron	τ (ms)	λ (cm)
H–H	1.5	$0.033\sqrt{d}$
Hippocampal CA3	~ 70	$\sim 0.23\sqrt{d}$
Hippocampal CA1	~ 30	$\sim 0.15\sqrt{d}$
Neocortical pyramidal	10–20	$\sim 0.1\sqrt{d}$
Cochlear	~ 0.2	$\sim 0.01\sqrt{d}$

Thus, as $\tilde{t} \to 0$ from positive values, $V(\tilde{x}, \tilde{t})$ approaches a *Dirac delta function* (or *unit impulse function*) defined by two properties: first,

$$\int_{-\varepsilon}^{+\varepsilon} V(\tilde{x}, 0) d\tilde{x} = 1$$

for all $\varepsilon > 0$; and second, $V(\tilde{x}, 0) = 0$ for $\tilde{x}$ not zero. In other words, at $\tilde{t} = 0$ the total area of the function shown in Figure 9.2 is concentrated at $\tilde{x} = 0$, and for positive values of time this area decays exponentially and spreads out into a bell-shaped curve.

Because Equation (9.3) is linear and synaptic inputs are localized in space and time, this delta-function property is useful for constructing a general solution for arbitrary input signals. Thus if a particular synaptic signal delivers Q_0 coulombs of electric charge to a dendritic fiber at location $x = 0$ and time $t = 0$, Equation (9.1) has the corresponding solution

$$V(x, t) = \frac{Q_0}{c\lambda} \sqrt{\frac{\tau}{4\pi t}} \exp\left(-\frac{x^2}{4Dt}\right) \exp(-t/\tau), \qquad (9.4)$$

where

$$D = \frac{\lambda^2}{\tau} = \frac{1}{rc},$$

is the *diffusion constant* for the process. When Equation (9.4) is used to describe the dynamics of membrane charge on a passive dendritic fiber, the following observations should be kept in mind.

- The total charge input from a particular synapse decays with time as

$$Q(t) = Q_0 e^{-t/\tau},$$

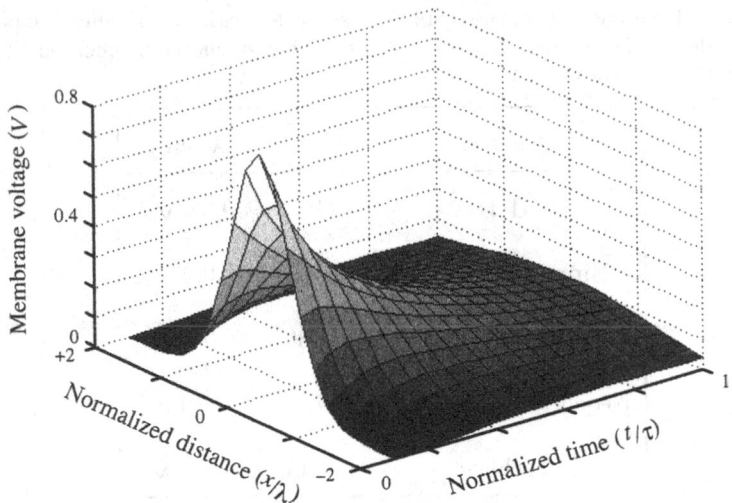

Figure 9.2. A three-dimensional plot of Equation (9.3) showing the diffusion of transmembrane voltage on a passive dendritic fiber. At $t = 0$, all of the area under this curve is concentrated as a "delta-function" (not shown) located at $x = 0$.

so for $t \gg \tau$, the influence of that input on the cell-body voltage is no longer present.

- At time t, the charge from a particular synaptic input spreads out to a distance of about

$$\sqrt{Dt} = \lambda\sqrt{t/\tau}$$

along the fiber, so for a dendritic path large compared with λ, there is little influence of an input on the cell body.

- Although the diameters of fiber segments vary throughout a dendritic tree, each segment can be measured in units of the corresponding length constant λ. Thus, the normalized lengths of typical dendritic trees—from the base of the trunk to tips of the twigs—can be estimated from microscopic observations of segment lengths and diameters. When all branches are added together, this total distance is called the *electrotonic length* (Λ) of an axonal tree. Typical Λs of cat motoneuron dendrites, for example, lie between 1 and 2, with an average of about 1.5 [14].

- Figure 9.2 shows that the voltage at an electrotonic distance of $\Lambda = 1.5$ from a synapse rises until t is about 0.5τ, after which it levels off and begins to decay.

- If the trees have $\Lambda \sim 1.5$ and $t > \tau$, then the synaptic charge is spread out over the entire dendritic structure and cell body, implying that the resulting voltage at the cell body is given by

$$V_{\text{cell body}} \sim \frac{Q_0}{C_{\text{total}}} e^{-t/\tau},$$

where C_{total} is the entire membrane capacitance of the dendritic trees and the cell body. This is the voltage that contributes to igniting an action potential at the initial segment of an axonal tree.

We now have a number of results at hand, but what are they telling us? One insight comes from recognizing that with $g_{\text{rest}} = 0$, Equation (9.1) is a *conservation law* for electric charge (see Appendix A). Thus,

$$i(x,t) = -\frac{1}{r}\frac{\partial V}{\partial x} \tag{9.5}$$

is the longitudinal current through the fiber, which is also the *flow* of the conserved quantity (charge). Similarly, $cV(x,t)$ is the local density of the charge. The term on the right-hand side in Equation (9.1) gives the leakage of this conserved charge across the membrane, which leads to the factor of exponential decay with time—$\exp(-t/\tau)$—in Equation (9.4).

Because the flow of heat along an imperfectly insulated thermal conductor is a formally identical phenomenon, we could imagine making analog models of the dendritic trees in Figure 9.1 from a thermally conducting material (e.g., metallic silver). Analogs of the synapses could then provide inputs of heat at various instants of time and locations along the tree. Each amount of heat input in the thermal analog would correspond to injecting a corresponding amount of charge at a fixed point on the dendritic tree at a particular time. This heat would then diffuse with space constant λ and decay with time constant τ (see Figure 9.2), as does the electric charge in a dendrite.

If a dendritic system is assumed to be linear, one can find the total effect of the synaptic inputs by simply adding together all individual contributions. Called the *superposition theorem*, this additive property is in fact the definition of linearity. Interaction among inputs, on the other hand, is a characteristic of *nonlinear* systems, in which the threads of causality may become interwoven [24].

Under the linear assumption, therefore, total voltage at a cell body can be calculated as a function of time from a sum of integrals of the form

$$V_{\text{cell body}}(t) = \sum_j \int_{-\infty}^{t} i_j(t')\mathcal{G}_j(t-t')dt', \tag{9.6}$$

where $i(t')$ is the current injected by synapse #j and $i_j(t')dt'$ is the amount of charge injected between times t' and $t' + dt'$. Called a *Green function*,[2]

---

[2]After George Green, a self-taught miller's son from Nottingham, England, who devised this method for solving linear electrical problems in 1828.

$\mathcal{G}_j(t - t')$ is a function of the type indicated in Equation (9.4), showing how this differential amount of synaptic charge contributes to the voltage at the cell body at a time $(t - t')$ after it was introduced.

Although the detailed construction of such a Green function for a realistic dendritic tree is difficult, our qualitative analysis of Equation (9.4) suggests the following general observation:

> The $\mathcal{G}_j$ are small for distances from the synapse to the cell body that are large compared with the space constant (λ) or for times that are large compared with the membrane time constant (τ), or both.

Can the linear influence of a synaptic signal be extended beyond these confines?

9.1.2 Decremental Conduction

To increase the region of space and time over which its synaptic inputs can be transmitted, a dendritic tree must restore some of the energy that is lost in a purely passive process. To see how this might be managed, recall the concept of *power balance* for a nerve impulse that was introduced in Section 6.4.

From the perspective of the FitzHugh–Nagumo (F–N) model of a nerve fiber

$$\frac{\partial^2 V}{\partial x^2} - \frac{\partial V}{\partial t} = f(V) + R,$$

$$\frac{\partial R}{\partial t} = \varepsilon V,$$

we noted that the energy carried by a solution is given by the integral

$$\mathcal{E} = \frac{1}{2} \int_{-\infty}^{\infty} \left(\left(\frac{\partial V}{\partial t} \right)^2 + \varepsilon V^2 \right) dx. \tag{9.7}$$

Differentiating $\mathcal{E}$ with respect to time and substituting from the F–N equation yields Equation (6.15), implying that $d\mathcal{E}/dt$ is negative if $dF(V)/dV$ is everywhere positive.

If, on the other hand, there is a range of voltage over which $dF(V)/dt < 0$, then the rate of energy loss is reduced. Diminishing the rate at which a solution loses energy reduces its rate of attenuation, which extends a synapse's range of influence.

To appreciate this effect, refer back to Figure 6.3 showing a *critical point* in the locus of wave speeds versus the parameter ε, past which no traveling waves are possible, and denote the critical value of ε by ε_c. Just beyond this critical point, the rate of energy loss is only slightly negative because almost all of the dissipation of the impulse is being restored by the energy generating processes of the system. Under these conditions, a nerve impulse is only slightly attenuated, and the resulting propagation is called *decremental conduction*.

Because of a tendency to overemphasize all-or-nothing propagation, decremental conduction was controversial in the literature of electrophysiology for many years [65]. From work in the 1960s and 1970s, however, the phenomenon of decremental conduction is now known to have a sound analytic basis [30, 49, 50, 56, 57], and a numerical example for the Hodgkin–Huxley axon is shown in Figure 4.7. Furthermore, there is evidence at both the electrophysiological and behavioral levels suggesting that drones of the honeybee *Apis mellifera* use this means to amplify photoreceptor signals in the course of their primary activity—searching for a queen [130, 131].

How can we get an intuitive grasp of decremental conduction? Suppose that an impulse at the critical point is given by

$$V(x, t) = V_c(x - v_c t),$$

and ε is slightly larger than the critical value. In this case, the amplitude of the wave $a(t)$ is expected to decrease slowly with time because the energy losses of the impulse are almost, but not quite, being restored. An approximate ODE describing the attenuation rate is obtained by substituting

$$V(x, t) = a(t) V_c(x - v_c t) = a(t) V_c(\xi)$$

into Equations (9.7) and (6.15), whereupon

$$\frac{da(t)}{dt} \approx -a(t) \frac{\int \left[\left(d^2 V_c/d\xi^2 \right)^2 + \left(dV_c/d\xi \right)^2 dF(V)/dV|_{V=aV_c} \right] d\xi}{\int \left[\left(dV_c/d\xi \right)^2 + \left(\varepsilon/v_c^2 \right) V_c^2 \right] d\xi}. \tag{9.8}$$

Evidently, the numerator of this expression is exactly zero for $a = 1$. For $a < 1$, on the other hand, the wave spends less time in the region where $F(V)$ has a negative slope, so it absorbs less energy and loses amplitude.

This crude analysis is introduced to make the point that dendritic impulses may decay more slowly than is implied by the linear diffusion solution of Equation (9.4). In this manner, linearity of the dynamics may be preserved, which allows the Green function of Equation (9.6) to remain valid with a wider range of influence for the synaptic inputs. More recent simulations of dynamics on typical dendritic structures support this conjecture [31, 132].

9.1.3 Rall's Equivalent Cylinder

Even under the assumption of linearity, computation of the voltage response at a cell body to a large number of synaptic input signals is a formidable

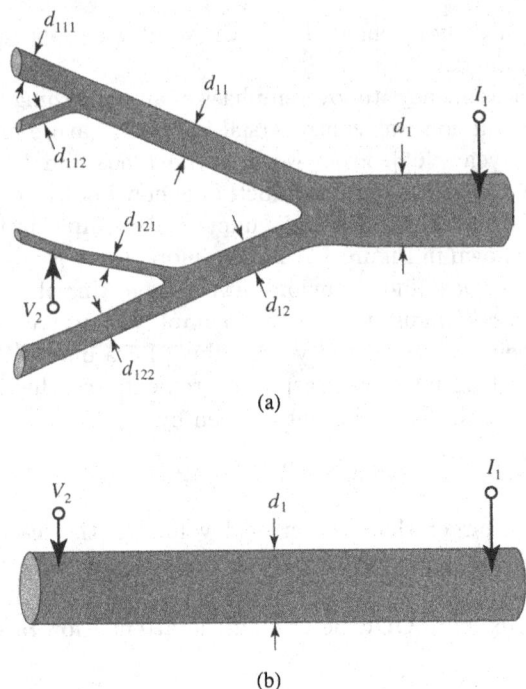

(a)

(b)

Figure 9.3. (a) A branching dendritic structure. (b) Rall's "equivalent cylinder" for the structure in (a).

task; thus, it is of interest to consider an unexpectedly simple case introduced by Wilfred Rall in 1959 [92, 93, 94, 95]. To see how this goes, refer to Figure 9.3(a), which represents an arbitrary dendritic branching region.

Suppose that a steady current I_1 is injected into the large fiber at location #1 on the left-hand side of the diagram from which the resulting steady transmembrane voltage V_2 is to be computed at location #2 on one of the smaller branches. Although time derivatives have been neglected, this remains a difficult calculation because a discontinuity (or *reflection*) in the solution occurs at each branching (or *bifurcation*) in Figure 9.3(a). Dealing with reflections is not a new problem; radio, microwave, acoustic, and optical engineers have long been interested in doing so in order to increase the efficiencies of electromagnetic, sound, or light transmissions. How do they accomplish this?

To minimize reflections, the standard procedure is to make the *characteristic admittance* (Y_0) of the transmission system equal on both sides of a boundary, where

$$Y_0 \equiv \sqrt{\frac{\text{shunt admittance/length}}{\text{series impedance/length}}}.$$

Physically, the characteristic admittance is the reciprocal of the impedance one sees when looking into the open end of a semi-infinite line.[3]

In the case of a nerve fiber operating in the steady-state (time independent) regime, the series impedance per unit length is r, and the shunt admittance per unit length is g_{rest}. Thus the characteristic admittance (Y_0) reduces to a *characteristic conductance*

$$Y_0 \to G_0 = \sqrt{g_{rest}/r}.$$

Now consider two facts:

- The series resistance per unit length (r) is inversely proportional to the square of the fiber diameter (d), and the shunt conductance per unit length (g_{rest}) is proportional to the diameter, implying that

$$G_0 \propto d^{3/2}.$$

- Characteristic conductances on individual fibers add to obtain the total characteristic conductance seen upon entering and leaving a branching region.

To eliminate reflections at the bifurcations shown in Figure 9.3(a), therefore, it is sufficient to require that

$$
\begin{aligned}
d_1^{3/2} &= d_{11}^{3/2} + d_{12}^{3/2}, \\
d_{11}^{3/2} &= d_{111}^{3/2} + d_{112}^{3/2}, \\
d_{12}^{3/2} &= d_{121}^{3/2} + d_{122}^{3/2},
\end{aligned}
\tag{9.9}
$$

ensuring that characteristic conductances are everywhere matched. Equations (9.9) embody the first of four assumptions that Rall made to represent the tree of Figure 9.3(a) by the *equivalent cylinder* shown in Figure 9.3(b).

Rall's second assumption is that the total electrotonic lengths from the end of the main trunk to the ends of the distal twigs are the same. Thus in the particular example of Figure 9.3,

$$
\begin{aligned}
\Lambda &= \ell_1/\lambda_1 + \ell_{11}/\lambda_{11} + \ell_{111}/\lambda_{111} \\
&= \ell_1/\lambda_1 + \ell_{11}/\lambda_{11} + \ell_{112}/\lambda_{112} \\
&= \ell_1/\lambda_1 + \ell_{12}/\lambda_{12} + \ell_{121}/\lambda_{121} \\
&= \ell_1/\lambda_1 + \ell_{12}/\lambda_{12} + \ell_{122}/\lambda_{122},
\end{aligned}
\tag{9.10}
$$

where ℓ_1 and λ_1 are, respectively, the physical length and space constant of the segment of diameter d_1, and so on. The third assumption is that all regions of the dendritic tree have the same cytoplasmic resistivity (ρ) and

---

[3]For those unfamiliar with the jargon of electrical engineering, "impedance" is a generalization of the concept of ohmic resistance that accounts for phase shifts between alternating currents and voltages, and "admittance" is a corresponding generalization of conductance.

specific membrane conductivity (G_{rest}) throughout, and the fourth is that
the boundary conditions at the far ends of the equivalent cylinder are the
same as for the original branching structure.

Under these four hypotheses on the nature of the dendrites, reflections
have been eliminated, and $V(x/\lambda)$ satisfies the same equations in the original branching region and on the equivalent cylinder. Thus the problem
of computing the voltage response V_2 to the injected current I_1 in Figure
9.3(a) reduces to the easier task for the equivalent cylinder of electrotonic
length Λ shown in Figure 9.3(b). Here the voltage V_2 is to be measured
at the same electrotonic distance from the injected current I_1 as in the
original branching structure of Figure 9.3(a).

At this point, the alert reader may be wondering whether I have things
mixed up. In understanding how a passive dendritic structure transmits
synaptic input signals from distant (distal) branches to the cell body, we
are not interested in calculating the voltage response at location #2 that
is caused by current injected at location #1; it is the other way around. To
gauge the efficiency of dendritic transmission, we must know how much voltage is produced at location #1 in response to current injected at location
#2. What do we do?

To find the voltage produced at location #1 (adjacent to the cell body)
to current injected at location #2 (at a distal twig of the dendritic tree), it
is convenient to use a fundamental result from the theory of linear networks
that can be stated as follows [40].[4]

> **Reciprocity theorem:** In a linear network composed of resis-
> tors, capacitors, and inductors, the ratio of voltage measured
> across terminals #1 to current injected into terminals #2 is
> equal to the ratio of voltage measured across terminals #2 to
> current injected into terminals #1.

Thus Rall's results can be summarized as follows. The voltage appearing
at location #1 (a cell body) in response to current injected at location
#2 (through a distal synapse) for the branching structure shown in Figure

---

[4]Proof of this theorem involves showing that the impedance matrix relating voltage
responses to current inputs is symmetric about its main diagonal. If one computes

$$\begin{bmatrix} V_1 \\ V_2 \end{bmatrix} = \begin{bmatrix} Z_{11} & Z_{12} \\ Z_{21} & Z_{22} \end{bmatrix} \begin{bmatrix} I_1 \\ I_2 \end{bmatrix},$$

it follows from symmetries of the algebra that $Z_{12} = Z_{21}$. Thus V_1/I_2 with $I_1 = 0$ is
equal to V_2/I_1 with $I_2 = 0$, Q.E.D. Because the inverse of this impedance matrix is
also symmetric, a similar statement can be made about ratios of measured currents to
impressed voltages. It is important to remember, however, that the reciprocity theorem
does *not* hold for ratios of voltages or for ratios of currents.

9.3(a) is equal to the voltage produced at location #2 in response to current injected at location #1 for the equivalent cylinder shown in Figure 9.3(b).

Importantly, Rall's conclusions are not restricted to the case of steady voltages and currents as has hitherto been assumed for ease of exposition. If the membrane time constant $\tau = c/g_{rest} = C/G_{rest}$ is constant over a branching structure, similar arguments go through when the voltages and currents are allowed to be functions of time.

My aim in this section has been to present the main logic of Rall's work without getting lost in analytic details. For those wishing more specific discussions of representations of dendritic trees by equivalent cylinders, the books by Jack, Noble, and Tsien [46], Keener and Sneyd [47], Koch [52], and Tuckwell [126] are recommended.

Do Rall's assumptions—particularly those expressed in Equations (9.9) and (9.10)—hold for real dendrites? For some trees, these conditions are fulfilled [20], but for others—as one might expect—they seem not to be satisfied [25, 43, 55, 66, 127]. Indeed, recent observations suggest that the effective cylinder diameter decreases as one moves away from the cell body, which makes dendritic propagation more difficult in the direction toward the cell body than away from it [132]. Furthermore, the presence of synapses may alter dendritic transmission properties either by covering some of the active area or by introducing dendritic spines, which can be active [73, 83, 108].

9.2   Inhomogeneous Active Fibers

Attention in this book has so far been limited to nerve fibers with uniform cross sections, but a look at real neurons shows that this assumption is often inappropriate. Fibers can change their diameters both slowly, in gradual taperings, and rapidly at local enlargements, or *varicosities*, of dendrites, as pointed out by Bogoslovskaya et al. [22]. Although varicosities may be artifacts of fixation, they have been widely reported and can have significant effects on the nature of impulse dynamics. Thus the mathematically oriented neuroscientist must appreciate spatial inhomogeneities in order to learn what dendrites are about.

9.2.1   Tapered Fibers

In the late 1960s, Lindgren and Buratti suggested making electronic nerve models (called *neuristors*) having parameters that vary exponentially with distance along the fiber as [58]

$$
\begin{aligned}
r &= r_0 e^{\gamma x}, \\
c &= c_0 e^{-\gamma x}, \\
j_i &= j_0(V) e^{-\gamma x},
\end{aligned}
\tag{9.11}
$$

where r_0, c_0, and j_0 are independent of x and j_0 is a nonlinear function of the voltage V.

The first-order partial differential equations of the system

$$\frac{\partial V}{\partial x} = -ri,$$

$$\frac{\partial i}{\partial x} = -c\frac{\partial V}{\partial t} - j_i,$$

then imply a nonlinear *drift-diffusion* equation of the form

$$\frac{\partial^2 V}{\partial x^2} - \gamma\frac{\partial V}{\partial x} - r_0c_0\frac{\partial V}{\partial t} = r_0j_0(V).$$

For traveling-wave solutions, this reduces to the ODE

$$\frac{d^2V}{d\xi^2} + r_0c_0\left(v - \frac{\gamma}{r_0c_0}\right)\frac{dV}{d\xi} = r_0j_0(V),$$

implying that the traveling-wave speed in the x-direction *increases* by γ/r_0c_0 as a result of the exponential taper.

For biological nerve fibers, the parameter dependence of Equations (9.11) is unrealistic because the parameters vary as $r \propto 1/d^2$, $c \propto d$, and $j_i \propto d$, where d is the diameter of the fiber. If as an exponential tapering of the fiber diameter one assumes $d \propto \exp(-\gamma x)$, then

$$r = r_0e^{2\gamma x},$$

$$c = c_0e^{-\gamma x},$$

$$j_i = j_0e^{-\gamma x},$$

leading to the PDE

$$\frac{\partial^2 V}{\partial x^2} - 2\gamma\frac{\partial V}{\partial x} - r_0c_0e^{\gamma x}\frac{\partial V}{\partial t} = r_0j_0e^{\gamma x}. \tag{9.12}$$

For a nerve impulse that is small compared with $1/\gamma$, the exponential factors in Equation (9.12) remain approximately constant, and the traveling-wave speed is increased in the direction of decreasing fiber diameter as

$$v_0 \longrightarrow v_0 + \frac{2\gamma}{r_0c_0} + O(\gamma^2). \tag{9.13}$$

The dynamics of impulse propagation on tapered fibers have been studied both theoretically and numerically in greater detail by several authors [15, 16, 37, 48, 50, 68, 93, 94], and the results can be summarized as follows.

> If a nerve fiber has a gradual change in its diameter, the impulse velocity in the direction of decreasing (increasing) diameter is increased (decreased). The magnitude of this change in velocity is proportional to the diffusion constant for the fiber $(1/rc)$ (centimeters squared per second) times the spatial rate of diameter change (γ) (fractional change per centimeter).

Table 9.2. Diameter ratios (or fiber widenings) for impulse blocking and passage as functions of taper length over which widening occurs calculated for Hodgkin–Huxley axons [15].

Taper length (cm)	Blocking	Passage
0.088	5.5:1	5:1
0.785	6:1	5.5:1
1.76	8:1	7:1
3.81	>10:1	10:1

An interpretation of this result stems from our discussion of the relationship between threshold and leading-edge charges in Section 5.5. Thus gradually decreasing (increasing) the fiber diameter tends to decrease (increase) the threshold charge, Q_θ, thereby increasing (decreasing) both the safety factor and the impulse velocity.

9.2.2 Varicosities and Impulse Blockage

If the diameter of a fiber increases in the direction of propagation, impulse velocity decreases, and with a sufficiently abrupt increase an impulse can cease to propagate, or become *blocked*. A quantitative description of this phenomenon is provided by the numerical computations on linearly tapered Hodgkin–Huxley axons recorded in Table 9.2, where the first row is for a taper length of 0.088 cm [15]. Close to an abrupt widening, this taper barely passes an isolated impulse at a ratio of 5:1, introducing a time delay of about 0.8 ms [48].

The reason for this delay can be qualitatively appreciated and calculated as the extra time required for the capacitance of the enlarged membrane area to become charged to a voltage that exceeds threshold. Such delays occur whenever nerve impulses encounter varicosities or local enlargements of a fiber, as have been observed on the dendrites of cochlear (auditory) neurons [22].[5] Could the impulse blockages and time delays associated with varicosities play a role in some sort of dendritic computations [5]? The very short membrane time constant for such dendrites (see Table 9.1) suggests that this may be so [116].

In his book *The Problem of Excitation*, Boris Khodorov has assembled the results of many such numerical computations of impulse delay and

---

[5] Although long regarded as fixation artifacts, varicosities are now accepted features of normal nerve fibers [77].

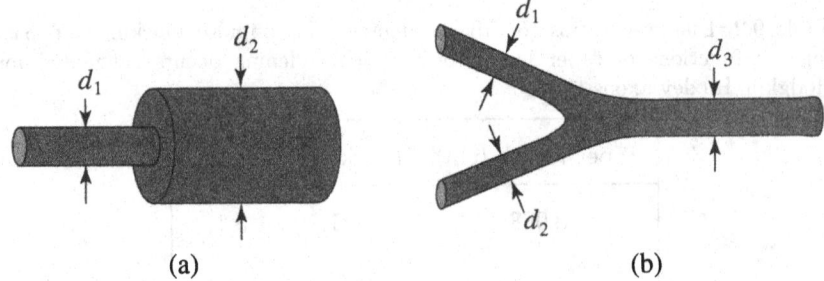

Figure 9.4. (a) Abrupt widening of a nerve fiber. (b) Branching region.

blockage [50]. For abrupt widenings of Hodgkin–Huxley fibers (at 20°C), as shown in Figure 9.4(a), he reports blockage at widening ratios greater than 5.5:1 and passage at 5:1 [48, 50]. Recently, Altenberger, et al. have carried through more refined calculations and found the critical widening ratio for blockage of a standard H–H impulse (at 18.5°C) to be [6]

$$\frac{d_2}{d_1} > 5.43 \,, \tag{9.14}$$

with an error of ±0.05%.

To judge whether impulse blockage could play a role in dendritic or axonal computing, one must understand how to calculate this ratio from measurable fiber parameters. To this end, two analytic approaches have been employed.

Markin–Chizmadzhev (M–C) Analysis
Representing a nerve with the Markin–Chizmadzhev (M–C) model, which was introduced in Section 6.1, Pastushenko and Markin showed that impulse blockage should occur for [80]

$$\left(\frac{d_2}{d_1}\right)^{3/2} > \kappa + 1.11\sqrt{\kappa} - 1.69 \,, \tag{9.15}$$

where

$$\kappa \equiv V_{\max}/V_\theta$$

is the ratio of the maximum level of impulse voltage to the threshold voltage, as indicated in Figure 6.1(b).

Mornev's Analysis
An alternative derivation of the blockage condition was presented by Oleg Mornev in the context of an analysis of spiral waves [74]. Reasoning that an impulse will make it through the enlargement of Figure 9.4(a) if its leading edge can do so, he used a leading-edge model of the type that was developed in Chapter 5. On both sides of the abrupt widening, therefore,

the leading-edge voltage is assumed to obey a nonlinear PDE of the form

$$\frac{1}{r}\frac{\partial^2 V}{\partial x^2} - c\frac{\partial V}{\partial t} = gf(V),\tag{9.16}$$

where r, c, and g have different values in the two regions, and $f(V)$ is a general "cubic-shaped" function such as those given in Equations (5.9) and (5.12).

As a necessary condition for blockage at a widening, it must be possible to construct a time-independent solution with the boundary conditions

$$V \to 0 \quad \text{as} \quad x \to +\infty,$$
$$V \to 1 \quad \text{as} \quad x \to -\infty.$$

For such a solution, the time-derivative terms are neglected, and the preceding PDE reduces to the nonlinear ODE

$$\frac{1}{rg}\frac{d^2 V}{dx^2} = \frac{d^2 V}{d\tilde{x}^2} = f(V),$$

where $\tilde{x} \equiv \sqrt{rg}\,x$. Evidently, this equation can be exactly integrated for any $f(V)$. From Equation (9.5), the internal boundary condition of current continuity at the widening is

$$\frac{1}{r_1}\frac{dV}{dx}\bigg|_{x=0-} = \frac{1}{r_2}\frac{dV}{dx}\bigg|_{x=0+},$$

implying

$$\sqrt{\frac{g_1}{r_1}}\frac{dV}{d\tilde{x}}\bigg|_{\tilde{x}=0-} = \sqrt{\frac{g_2}{r_2}}\frac{dV}{d\tilde{x}}\bigg|_{\tilde{x}=0+}.$$

In other words, current must flow into the widening region at the same rate that it flows out; otherwise, charge would accumulate at the discontinuity. Because $\sqrt{g/r} \propto d^{3/2}$, this boundary condition can be written as

$$d_1^{3/2}\frac{dV}{d\tilde{x}}\bigg|_{\tilde{x}=0-} = d_2^{3/2}\frac{dV}{d\tilde{x}}\bigg|_{\tilde{x}=0+},$$

where d_1 (d_2) is the diameter of the fiber at $\tilde{x} < 0$ ($\tilde{x} > 0$).

With this formulation, Mornev showed that a necessary condition for a static solution to exist at the widening is

$$\left(\frac{d_2}{d_1}\right)^{3/2} > \sqrt{\frac{A^-}{A^+}},\tag{9.17}$$

where A^- (A^+) is the negative (positive)-going area under the curve $f(V)$. That this condition is also sufficient for stopping an impulse has been checked by solving Equation (9.16) numerically with various degrees of widening for the special case $f(V) = V(V - V_1)(V - V_2)$ [18]. If the preceding inequality is satisfied, an impulse approaching the widening along the smaller fiber will not be able to get through, and impulse blockage will occur.

These expressions of blocking conditions—Equations (9.14), (9.15), and (9.17)—are brought to the reader's attention for several reasons. If we are dealing with abrupt widenings of Hodgkin–Huxley fibers, then Equation (9.14) is really all that is needed, but many fibers are not well-modeled by the H–H parameters. Thus Equation (9.15) is useful for cases in which the safety factor can be estimated from experimental data. Equation (9.17), on the other hand, may be more useful when the function $f(V)$ is known. Additionally, Equations (9.15) and (9.17) are alternative derivations of the same qualitative result, reinforcing one's confidence in the form of the equations.

Why should these results be of interest to neuroscientists?

9.2.3 Branching Regions

Because a branching region of an active nerve fiber is an abrupt widening of the structure, a qualitative formulation the dynamics of the bifurcation sketched in Figure 9.4(b) follows from three previously developed concepts. First, as discussed in Section 5.5, the stimulating current—brought to the branch on the incoming fiber—must be sufficient to raise the outgoing fibers above threshold. Second, if an impulse is incoming to the branch on fiber #1, for example, and outgoing on fibers #2 and #3, its stimulating current will divide in proportion to the characteristic admittances of these branches, which are proportional to $d_2^{3/2}$ and $d_3^{3/2}$, respectively, summing to $(d_2^{3/2} + d_3^{3/2})$. Third, from the previous section, the stimulations required to achieve threshold on fibers #2 and #3 are also proportional to $d_2^{3/2}$ and $d_3^{3/2}$. Thus, a blocking condition on $(d_2/d_1)^{3/2}$ in Figure 9.4(a) should correspond to the same condition on $(d_2^{3/2} + d_3^{3/2})/d_1^{3/2}$ in Figure 9.4(b). Assuming this estimate holds, blocking conditions can be expressed in several ways, including the following.

(1) *Hodgkin–Huxley model.* From the numerical studies leading to Equation (9.14) [48, 50, 79], the blocking condition for a single H–H impulse (at 18.5°C) is

$$\frac{d_2^{3/2} + d_3^{3/2}}{d_1^{3/2}} > 5.43^{3/2} = 12.7 .$$

From the considerations discussed in Section 4.6, this condition is expected to become less severe (i.e., the right-hand side is less than 12.7) under narcotization, increased external potassium ion concentration, increased temperature, and for multiple impulses.

(2) *Markin–Chizmadzhev model.* From Equation (9.15), the blocking condition for the M–C model is [80]

$$\frac{d_2^{3/2} + d_3^{3/2}}{d_1^{3/2}} > \frac{V_m}{V_\theta} + 1.11\sqrt{\frac{V_{max}}{V_\theta}} - 1.69,$$

where V_{max} and V_θ are defined in Figure 6.1(b). Guided by the parameter values used to construct Figure 6.2, this implies blocking for $(d_2^{3/2} + d_3^{3/2})/d_1^{3/2} > 5.2$, which is less severe than for the H–H model.

(3) *General leading-edge model.* Describing a nerve impulse in the leading edge approximation of Chapter 5 and appealing to Mornev's analysis, Equation (9.17) implies blockage for [74]

$$\frac{d_2^{3/2} + d_3^{3/2}}{d_1^{3/2}} > \sqrt{\frac{A^-}{A^+}}.$$

Here, A^- and A^+ are, respectively, the negative- and positive-going areas under a general function with the form indicated in Figures 5.1, 5.2, and 5.3.

(4) *Cubic leading edge model.* Choosing a leading-edge model with

$$f(V) \propto V(V - V_1)(V - V_2)$$

as in Figure 5.3(a) and Equation (5.9), the area ratio A^-/A^+ can be computed as a function of $a \equiv V_1/V_2$. Thus Mornev's analysis implies blocking for

$$\frac{d_2^{3/2} + d_3^{3/2}}{d_1^{3/2}} > \sqrt{1 + \frac{1 - 2a}{a^3(2 - a)}}. \tag{9.18}$$

For this model, Equation (5.10) tells us that $20 = 49(1 - 2a)/\sqrt{2}$ at the wave speed for an H–H axon, implying $a = 0.21$, with a corresponding blocking condition of $(d_2^{3/2} + d_3^{3/2})/d_1^{3/2} > 5.9$. In approximate accord with the preceding result of M–C analysis, this is again less severe than the full H–H model.

(5) *Piecewise linear leading-edge model.* Assuming a piecewise leading-edge model with $f(V)$ as in Figure 5.3(b) and Equation (5.12), the area ratio A^-/A^+ can be computed as a function of the parameters V_2, V_1, and β. Mornev's analysis then implies blocking for

$$\frac{d_2^{3/2} + d_3^{3/2}}{d_1^{3/2}} > \frac{V_2 - V_1}{V_1\sqrt{\beta}}.$$

Equation (5.13) implies that the wave speed equals that of an H–H axon at $\beta = 1/40$ and $V_1/V_2 = 0.64$ with blocking at $(d_2^{3/2} + d_3^{3/2})/d_1^{3/2} > 3.6$.

The central importance of the expression on the left-hand side of these inequalities was recognized by Goldstein and Rall, who called it the *geo-*

metric ratio (GR) [37]. In general, for a branching region with coincident impulses incoming on M fibers and outgoing on N fibers, the geometric ratio is defined as

$$\text{GR} \equiv \frac{d_{\text{out}1}^{3/2} + \cdots + d_{\text{out}N}^{3/2}}{d_{\text{in}1}^{3/2} + \cdots + d_{\text{in}M}^{3/2}}. \tag{9.19}$$

As we have seen from the discussion of Rall's equivalent cylinder in Section 9.1.3, the GR measures the degree to which reflections are eliminated at a branch. Thus the condition

$$\text{GR} = 1$$

indicates that the characteristic admittances of the corresponding incoming and outgoing branches are "matched" in engineering jargon. In other words, the sum of admittances of the incoming fibers is equal to the sum of the admittances of the outgoing fibers, thereby eliminating reflections from the branching region.

Quantitative studies of biological branching go back to Leonardo da Vinci, who claimed for botanical trees that: "All the branches of a tree at every stage of its height when put together are equal in thickness to the trunk [below them]" [67]. This observation suggests a branching law—which we may call *Leonardo's law*—for parent and daughter diameters (d_p and $d_1, d_2, \cdots, d_N$) of the form

$$d_p^\Delta = \sum_{j=1}^N d_j^\Delta, \tag{9.20}$$

where Δ is the *branching exponent*. As Mandelbrot [67] and Thompson before him [123] have noted, there are several examples of Leonardo's law in the natural world, including: (1) the lung's bronchial tree, where $\Delta = 3$; (2) rivers, for which $\Delta \sim 2$; (3) human arterial branchings, with $\Delta \sim 2.7$; and (4) the first branching of the abdominal aorta, with $\Delta = 2$. But we are interested in the influence of dendritic and axonal branchings on the dynamics of a neuron.

9.3 Information Processing in Dendrites

Among the reasons for supposing dendrites to be linear transmission systems is that the ignition of a single all-or-nothing impulse might fire the whole dendritic tree, making it difficult to integrate the effects of incoming signals, but the results of the previous section reveal a flaw in this logic. If spikes can be blocked at branching regions, their collective activity need not rage out of control like a forest fire or a sparked heap of gunpowder. On the contrary, impulse blocking would seem to provide bases for at least two types of information processing: dendritic logic and multiplicative interactions among the pulse rates of incoming signals.

Although interest in dendritic information processing has been growing among Western neuroscientists [3, 35, 41, 42, 51, 52, 53, 71, 73, 83, 105, 106, 107, 108, 109, 110, 119], the concept is not new. Since the 1960s, the possibilities for dendritic computations have been pursued by a number of researchers, many in the former Soviet Union [10, 16, 21, 50, 81, 82, 101, 122, 133]. In this section, some formulations are introduced to help the reader evaluate these ideas.

9.3.1 Dendritic Logic

In speculating on the possibility of information processing on dendrites, the first question to consider is whether there is experimental evidence for dendritic action potentials. Interestingly, such evidence was provided in the late 1960s by Llinás and his colleagues from observations on the Purkinje cell of the alligator cerebellum [59, 61, 62]. In the mid-1970s, spikes on Purkinje cell dendrites were shown to arise from voltage dependence of calcium ions rather than sodium ions as in the squid nerve [60, 63, 64]. More recently, evidence has been presented for spikes on the dendrites of pyramidal cells in the hippocampus [8, 17, 42, 69, 97, 86, 129, 136] and the neocortex [7, 71, 85, 118]. Presently, there is little doubt that dendritic spikes are a real neural phenomenon stemming from a variety of active channels [52, 64, 119, 132]. (Those with a taste for numerical studies will enjoy Chapter 15 of Wilson's *Spikes, Decisions, and Actions*, which includes several MATLAB codes for computing dendritic responses from synaptic inputs under various assumptions for active sodium and calcium channels [135].)

Located near the base of the mammalian brain (just above the nape of your neck), the cerebellum is a neural structure with surprisingly regular organization that coordinates arm and leg motions. Within this structure are a large number of Purkinje cells having planar dendritic fields and receiving many synaptic inputs. The human Purkinje cell shown in Figure 9.5, for example, receives some 160,000 synaptic inputs from parallel fibers [34], which are oriented perpendicular to the plane of the dendrites. Because action potentials are known to form on Purkinje dendrites [66], we are faced with the question: What is the function of this intricate structure?

One answer to this question is suggested by the numerical modeling of De Schutter and Bower [31], which shows that more distal input signals are amplified by factors of up to 5 over the purely passive calculations of Section 9.1.1. Thus all signals arrive at the cell body with about the same amplitude, easing constraints on the locations of particular inputs.

Another response emerges from detailed analyses of the dendritic branchings using the concepts of impulse blockage that were developed in the previous section [101]. Thus, a dendritic branching region can be viewed as a switch that either stops or passes an impulse according to whether a

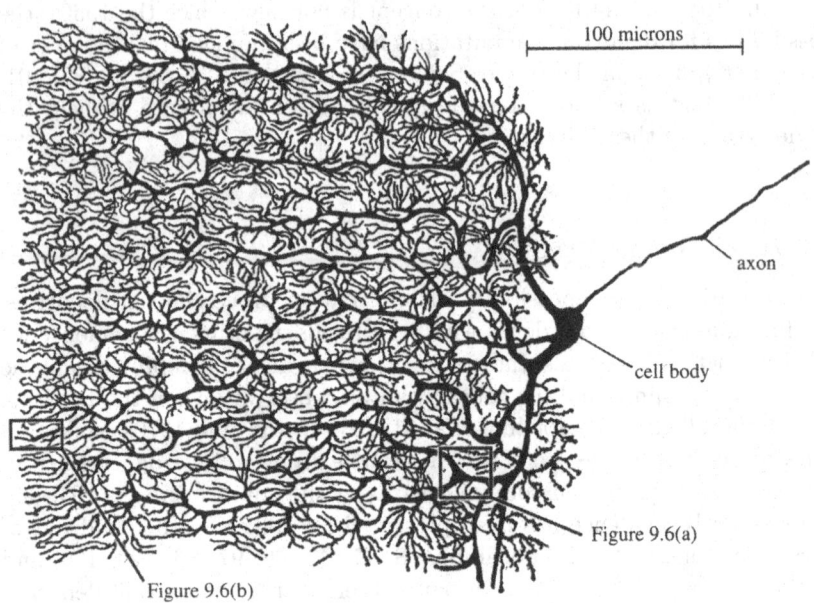

Figure 9.5. Ramón y Cajal's classic image of a Purkinje cell from the human cerebellum [96].

blocking condition is satisfied. To this end, let us consider the bifurcation shown in Figure 9.4(b) with the notation that d_1 and d_2 are daughter diameters and d_3 is the diameter of the parent branch. Extracted and enlarged from Figure 9.5, two possibilities are indicated in Figure 9.6.[6]

OR *Bifurcations*

For the simple branch shown in Figure 9.6(a), it is seen that $d_1 \approx d_2 \approx d_3$. Supposing that an impulse arrives at the branch from (say) daughter #1,

$$\frac{d_2^{3/2} + d_3^{3/2}}{d_1^{3/2}} \approx 2.$$

All of the models treated in the preceding section imply that this GR is too small for blocking of an impulse to occur. Thus incoming impulses on either of the two daughters are able to ignite the parent. Using the jargon of computer engineering, this can be described as an OR junction because

---

[6]The examples given in this section are for illustration only because the Golgi stain technique used by Ramón y Cajal to obtain Figure 9.5 may not record all of the dendritic structures.

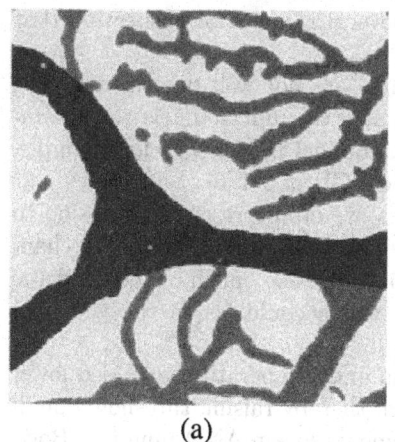

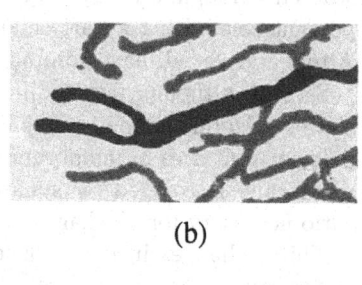

(b)

(a)

Figure 9.6. Details of the Purkinje cell branchings indicated in Figure 9.5. (a) An OR bifurcation. (b) A possible AND bifurcation.

an input on one "or" the other daughter is sufficient to ignite the parent fiber.

In evaluating the computational utility of this OR bifurcation, one should note that an incoming impulse on one daughter will launch an outward-going impulse on the other daughter, disabling that daughter's segment of the dendritic tree for a certain interval of time [107, 132].

AND *Bifurcations*
Computer engineers use the term "AND junction" to describe an element for which inputs on both the first input "and" the second input acting together are required to produce an output signal, implying that one input acting alone is insufficient to produce an output.

If it is assumed that the dendritic trees are composed of Hodgkin–Huxley fibers that support fully developed impulses, the condition for failure of a single incoming impulse is

$$\frac{d_2^{3/2} + d_3^{3/2}}{d_1^{3/2}} > 12.7 \,.$$

From an examination of the various geometric configurations in the dendritic trees of the Purkinje cell in Figure 9.5, it is difficult to find branchings that satisfy this condition. One of the more promising candidates is shown in Figure 9.6(b), from which it is seen that the parent branch diameter (d_3) is about 2.5 times those of the incoming daughter branches (d_1 and d_2). Thus

$$\frac{d_2^{3/2} + d_3^{3/2}}{d_1^{3/2}} \approx 5 \,,$$

which is insufficient to satisfy the preceding condition (GR > 12.7) for an AND bifurcation. There are, however, several reasons for suspecting that this condition is too severe.

First, the action potentials on dendrites are not well described by the standard Hodgkin–Huxley equations, because calcium channels play an important role. Using a particular calcium channel model [75], for example, Altenberger et al. have computed a critical GR of 3.4 [6].

Second, although fast sodium channels are often present in addition to slower calcium channels [60, 63], they have lower density (number of channels per unit area of membrane) [66, 106, 116]. Also, much of the dendritic membrane is covered by synapses [34], which could lessen the widening ratio necessary for blockage.

Third, changes in ionic concentrations and temperature can also lower the safety factor for impulse propagation, thereby raising threshold conditions and easing the geometrical requirements for an AND junction. Body temperatures of mammals, for example, are typically larger than the value of 18.5°C used in H–H calculations of critical widening and close to the critical temperature at which active propagation fails.

Fourth, the fiber length required for an impulse to grow from threshold to its full amplitude is the order of the *active space constant*, $\lambda_a = 1/\sqrt{rg}$. The lengths of some dendritic segments in Figure 9.5 are not large compared with λ_a, implying that voltage amplitudes of impulses arriving at a branch may be less than their full values. This effect also lowers the geometric ratio (GR) needed for blockage.[7]

Fifth, inspection of Figure 9.5 reveals several "delta-shaped" enlargements at bifurcations, which increase the total membrane capacitance and impede impulse transmission.

Sixth, dendrites are tapered, becoming smaller as the distance from the cell body increases [34, 132]. As we have seen in Section 9.2.1, this tends to reduce the safety factor of an incoming spike.

Finally, the incoming impulses may not be isolated but spaced with intervals as small as a few milliseconds. Khodorov reports numerical calculations for H–H impulses (at 20°C) separated by an interval of $T = 2.5$ ms corresponding to a normalized impulse interval $T/T_1 = 0.38$, where T_1 is defined as in Figure 4.8. In this study, the second impulse is blocked at a widening ratio that is less than 3:1 and greater than 1.5:1 [50], implying a critical GR within the range

$$1.8 < \frac{d_2^{3/2} + d_3^{3/2}}{d_1^{3/2}} < 5.2 \,.$$

---

[7]To estimate these space constants from microscope observations, note that for the H–H parameters given in Section 5.2, the active space constant of a fiber is given by $\lambda_a = 2.2\sqrt{d}$, where both λ_a and d are measured in microns.

In the context of the double impulse experiments on the squid giant axon that were introduced in Section 4.7 (see Figure 4.8), it is not difficult to demonstrate the switching action of an active fiber branch [104]. To see this, refer to Figure 9.7, which shows a pair of incoming impulses recorded at point B on a branch of diameter 381μm. Because the outgoing branches are of diameter 218μm and 544μm, the geometric ratio is

$$\text{GR} = \frac{218^{3/2} + 544^{3/2}}{381^{3/2}} = 2.14\,.$$

Figures 9.7(b) and 9.7(c) show that the second impulse (recorded at point A) becomes blocked at a critical impulse spacing of $T = 2.1$ ms (corresponding to $T/T_1 = 0.36$) in approximate accord with the numerical results of Khodorov. To appreciate the implications of these observations, note that there was no setting of the incoming impulse spacing leading to a response between those of Figures 9.7(b) and 9.7(c)—the second impulse either appeared or was blocked in an all-or-nothing manner. In other words, the branch was observed to act as a logical switch.

To get an idea of the GRs to be expected in real dendrites, consider Table 9.3, where branching exponents (Δ) for Equation (9.20) (Leonardo's law) are recorded for a variety of mammalian dendrites [12]. Assuming that the two daughter branch diameters (d_1 and d_2) are equal implies a ratio of parent diameter to either one of the daughter diameters of

$$\frac{d_3}{d_1} = \frac{d_3}{d_2} = 2^{1/\Delta}\,;$$

thus the corresponding geometric ratio is

$$\text{GR} = \frac{d_2^{3/2} + d_3^{3/2}}{d_1^{3/2}} = 2^{3/2\Delta} + 1\,. \tag{9.21}$$

In the last column of Table 9.3 are recorded values of GR calculated from this equation, that suggest a range of values for which blockage might or might not occur. (If the daughters were not assumed to be of equal diameter, this range of GR values would be greater.)

Taking all of these considerations together, it seems reasonable to speculate that two basic elements of the computer engineer—OR and AND switches—may be found at the branchings of real dendritic trees. A third element of computer design is the NOT function, which Koch and his colleagues have shown to be achieved through inhibitory synapses that are located closer to the cell body than the signals they aim to inhibit [52, 53].

It is a fundamental theorem of the algebra of classes that all Boolean functions (or logical statements) can be constructed from the three elements

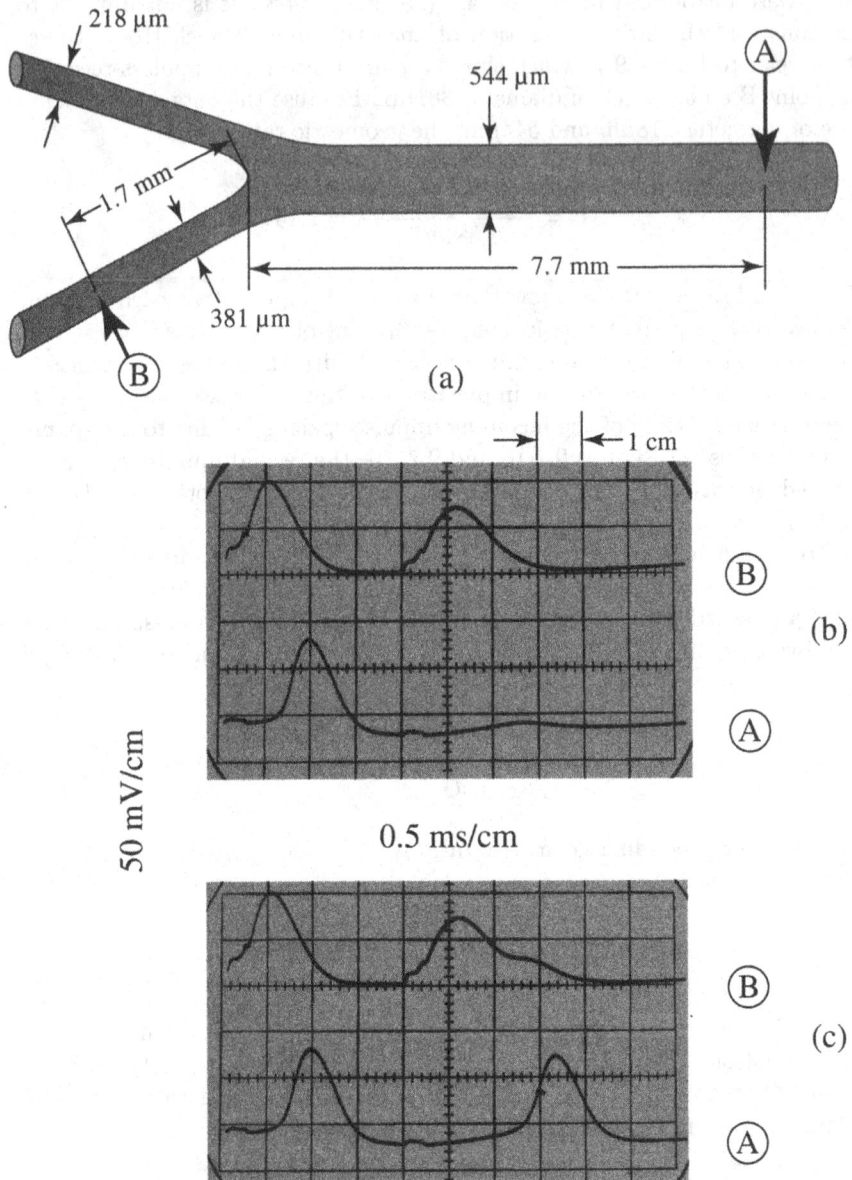

Figure 9.7. Switching action in the branching region of a squid giant axon at 20.3°C [104]. (a) Geometry of the preparation, showing the point of upstream recording of a pair of incoming impulses at B and the point of downstream recording at A (not to scale). (b) Blocking of the second impulse. (c) Passage of the second impulse.

Table 9.3. The GR range for some typical dendrites calculated from observations of branching exponents using Equation (9.21). (Apical dendrites consist of a single tree, whereas basal dendrites comprise several trees.) (Branching exponents are from [12].)

Cell type	Branching exponents (Δ)	GR range
Purkinje	2.36 ± 1.2	2.3–3.5
Stellate	2.24 ± 1.12	2.4–3.5
Granule	2.58 ± 1.8	2.3–4.8
Motoneuron	1.69 ± 0.48	2.6–3.4
Pyramidal (apical)	1.99 ± 0.79	2.5–3.4
Pyramidal (basal)	2.28 ± 0.89	2.4–3.1

AND, OR, and NOT[8] [19]. Thus, one is led to speculate that dendritic trees might realize the most general logical functions of their synaptic inputs. Far from being mere passive channels for delivering synaptic messages to the cell body (or initial segment of the axon), in other words, dendrites may have the ability to compute all functions that are possible in the context of Boolean (computer) algebra. Could this really be so?

Bartlett Mel suggests that such a sweeping conclusion be approached with caution because of the unrealistic requirements that the construction of such dendritic computers would impose on the processes of embryonic growth [72]. How would a developing brain know exactly where to place the excitatory and inhibitory synapses, thereby determining the NOT elements? On the other hand—as Koch points out—synapses may act in functional groups rather than as individuals, easing the task of developmental organization [52].

9.3.2 Multiplicative Nonlinearities

The primary difficulty in confirming or rejecting speculations about dendritic logic is empirical. Because of their small size, it is difficult to measure the internal voltages at selected locations along dendritic fibers [124];

---

[8]As discussed in Chapter 10, Boolean functions are defined on the two-element number system comprising "0" and "1." Thus a function of N variables will specify either "0" or "1" for each of the 2^N combinations.

thus neuroscientists are currently considering other means for dendritic information processing that can be more readily observed.

One such approach is to suppose that dendritic trees do not respond to the precise Boolean codes presented to their synaptic inputs but to average impulse rates. How might this assumption simplify analytic formulations?

Indicating these incoming rates as $F_j(t)$ $(j = 1, 2, \ldots, n)$, where n is the number of synaptic inputs and t is time, the linear analyses of Section 9.1 imply an input to the cell body of

$$I_L(t) = \alpha_1 F_1(t) + \alpha_2 F_2(t) + \cdots + \alpha_n F_n(t) = \sum_{j=1}^{n} \alpha_j F_j(t). \qquad (9.22)$$

The output pulse rate on the axonal tree might then be given by an expression of the form

$$O_L(t + \tau) = S[I_L(t)],$$

a *sigmoid* function, rising smoothly from 0 to 1 as its argument increases from 0 to ∞. Of several possible expressions, a sigmoid function might take the form

$$S[I] = \frac{I^2}{I^2 + \theta^2}.$$

In this formula, θ acts like a threshold in the sense that $S \approx 1$ for $I^2 \gg \theta^2$ and $S \approx 0$ for $I^2 \ll \theta^2$.

Equation (9.22), however, fails to represent the nonlinear aspects of dendritic logic, which were discussed in the previous section. A straightforward way to include such effects is to augment the input variable to

$$I_{\Sigma\Pi} = \sum_{j=1}^{n} \alpha_j F_j + \sum_{j,k} \beta_{jk} F_j F_k + \sum_{j,k,l} \gamma_{jkl} F_j F_k F_l + \cdots, \qquad (9.23)$$

where only one permutation of the indices is counted. Called the "sigma-pi" (or sum of products) model by neuroscientists [52, 114], Equation (9.23) is recognized as a power series in the n inputs that is capable of representing any smooth (analytic) function of those inputs [134]. Thus a rather general expression for the dependence of the outgoing impulse rate on the incoming rates is

$$O_{\Sigma\Pi}(t + \tau) = S[I_{\Sigma\Pi}(t)],$$

but one must bear in mind that this formula includes a rather large number of parameters: n of the αs, $n(n+1)/2$ of the βs, $n(n+1)(n+2)/3!$ of the γs, and so on.[9]

---

[9] For n synaptic inputs and a summation of rth-order products, a general formula for the number of parameters is $(n+r-1)!/(n-1)!r!$, which is the number of ways that r beans can be put into n jars.

For a *quadratic model*

$$I_2(t) = \sum_{j=1}^{n} \alpha_j F_j(t) + \sum_{j,k} \beta_{jk} F_j(t) F_k(t)$$

$$(9.24)$$

$$O_2(t + \tau) = S[I_2(t)],$$

there are a total of $n(n+3)/2$ parameters to be specified, which is computationally feasible up to $n \sim 100$. For higher-order models, the number of parameters grows with correspondingly higher powers of n.

Under the linear model of Equation (9.22), the response of a neuron depends only on the n values of the αs, remaining insensitive to the relative locations of these n input synapses. Among other phenomena, the quadratic model of Equation (9.24) predicts *cluster sensitivity*, in which interactions between pairs of synaptic inputs are taken into account. At least two nonlinear effects in dendritic trees can lead to cluster sensitivity: interactions among neighboring synapses and the presence of AND bifurcations. Mel has tested these predictions of the quadratic model against the numerical behavior of model nonlinear dendrites [71].

In this study, the dendritic trees investigated were those of a neocortical pyramidal cell. Because it is difficult to record from several locations within these dendrites, a numerical model was needed, and a *compartmental model* was chosen [23, 32, 44, 45, 125].

The motivating idea of a compartmental model is to simplify the full nonhomogeneous PDE system describing a dendritic tree with a network of membrane patches (compartments) interconnected by resistors. The membrane patches are like the space-clamped ODEs considered in Section 4.2.3, and upon interconnecting these patches with resistors, the overall network is much like the myelinated systems described in Chapter 7.

Based on some 3000 measurements of dendritic branch lengths and branch diameters on a single pyramidal cell, Mel constructed a dendritic model comprising about 500 compartments [71]. The membrane dynamics of each compartment included the following components, any of which could be turned on or off during a particular computation.

- A. Excitatory passive and active synapses. In the passive synapses, the postsynaptic conductance $G(t)$, defined in Equation (2.4), was assumed to be proportional to $te^{-t/\tau}$, as in Equation (2.6), independent of the transmembrane voltage. For active synapses, on the other hand, the postsynaptic conductance was proportional to

$$\frac{e^{-t/\tau_1} - e^{-t/\tau_2}}{1 + Ke^{-\gamma V}},$$

with $\tau_1 \gg \tau_2$ as in Equation (2.7). Here, the dependence on membrane voltage (V) represents the fact that active postsynaptic channels are

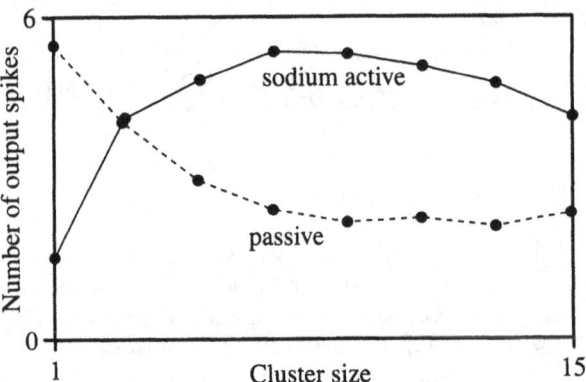

Figure 9.8. Response of passive (dashed line) and sodium-active (solid line) pyramidal cell dendrites to synaptic inputs of varying cluster size. (From data in reference [71].)

blocked (by magnesium ions) at voltages near and more negative than the resting voltage, becoming unblocked at positive values.

- B. Fast sodium channels, leading to Hodgkin–Huxley spikes similar to those described in Section 4.5.

- C. Two types of slow calcium spikes, with impulse durations of about 10 ms [52]. (Because the precise dynamics of the calcium spikes are uncertain, different models were used to check whether the overall dendritic behavior is sensitive to the details of this effect.)

For each numerical run, incoming trains of 100 Hz (impulses/s) were applied to 100 randomly selected synapses, and the number of output spikes generated by the cell body during the first 100 ms of stimulation was recorded. Although selected randomly, the 100 synapses were constrained to lie in contiguous "clusters" with sizes ranging from 1 (unclustered) to 15. (If the cluster size did not divide evenly into 100, a single smaller cluster was stimulated.) Figure 9.8 shows the qualitative behavior of this model, where the recorded number of output spikes is averaged over 50 to 100 different computations.

In this figure, two different assumptions are made: *passive*, implying passive synapses with effects B and C turned off, and *sodium active*, implying passive synapses with effect B on and C off.

As the cluster size is increased, the number of output spikes observed on the passive membrane model decreases. According to Mel, this is because inputs from nearby synapses increase the membrane permeability, thereby shunting away some of the injected input current. In this passive case, corresponding to the assumptions of Section 9.1.1, the number of output spikes is maximum when the input synapses are as widely dispersed as possible.

The opposite effect is observed when the dendritic membrane is sodium-active. In this case, the number of output spikes initially increases with cluster size, as expected if the dendritic model included a significant number of the AND bifurcations described in the previous section. (The decrease of output spiking at larger cluster sizes is attributed to the opposing influence of interacting synapses.)

These qualitative observations were quite robust in the numerical studies of Mel. Thus any or all of the nonlinear effects (active synapses or effects B and C) considered led to numerical observations of cluster sensitivity qualitatively similar to that of the "sodium-active" curve in Figure 9.8.

The implications of this numerical work for the functioning of pyramidal cells in the neocortex are twofold. First, as Mel suggests, cluster sensitivity may be employed as a means for neocortical information processing that is based on average impulse rates. Second, the numerical evidence for cluster sensitivity under a rather wide variety of neurologically plausible assumptions implies that AND bifurcations are to be expected in the dynamics of typical dendritic trees.

Although numerical evidence for neuronal multiplication remains sparse, recordings from specific neurons in the owl's auditory map are suggestive [84]. In these observations, a significant component of a neuron's output signal can be computed as the product of two factors: the time difference between signals at the two ears and corresponding intensity differences.

9.4 Axonal Information Processing?

The theoretical and numerical results presented in Section 9.3.1 suggest that blockage of impulse propagation should occur less readily on axonal than dendritic trees. To see why, consider the branching notation of Figure 9.4(b) in which d_3 is the diameter of the parent and d_1 and d_2 are the daughter diameters. In the course of normal dendritic dynamics, impulses enter a branch along a daughter fiber (say fiber #1), so the relevant geometric ratio (GR) for calculating blockage is

$$\text{GR(dendritic)} = \frac{d_2^{3/2} + d_3^{3/2}}{d_1^{3/2}},$$

as we have seen in the previous section.

Under normal axonal operation (so-called *orthodromic* conduction, where the axonal impulse propagates away from the cell body), on the other hand, impulses arrive at a branching junction along the parent fiber (fiber #3), for which the relevant GR is

$$\mathrm{GR(axonal)} = \frac{d_1^{3/2} + d_2^{3/2}}{d_3^{3/2}}.$$

Because the parent diameter (d_3) is typically greater than those of the daughters $(d_1$ and $d_2)$, these equations imply a lower value of GRs for axonal than for dendritic transmissions, suggesting that impulse blockage is less likely at axonal branchings.

Indeed, for axonal branchings it is not uncommon for the relevant GR to be unity, leading to a form of Leonardo's law [67, 123]

$$d_3^{3/2} = d_1^{3/2} + d_2^{3/2}$$

with branching exponent $\Delta = 3/2$.

As an example of this law, the present author has measured the diameters of the parent and daughter fibers of the first branch on 109 giant axons of the squid (*Loligo vulgaris*) taken from the Bay of Naples over a period of five months from December to May [104]. The mean of the GRs was found to be 1.017 with an rms variation of ± 0.029.[10]

There is an evolutionary explanation for the empirical observation that GR ≈ 1 for branchings of squid giant axons, with the logic as follows. As you may have noticed at your favorite Italian restaurant, a squid is tubular in shape, having its rear open to the sea. When danger is perceived, the function of the giant axon is to send an "escape!" message from the animal's brain to its posterior muscles [137]. These muscles rapidly contract, squirting water out the back and enabling the squid to literally rocket away from a presumed predator. For such a mechanism to be effective, the escape signal must arrive at all parts of the posterior muscle as rapidly as possible. This requirement is eased by having the GRs at axonal branchings equal unity, implying no impulse blockage.

At variance with these ideas are the results of several experimental studies demonstrating that impulse blockage does occur on axonal trees, reinforcing previous speculation that blockage occurs on dendrites [13, 29, 38, 78, 113, 115, 117, 122]. How might such empirical evidence be understood?

- If the frequency of an impulse train on the trunk of an axonal tree becomes greater than about 100 impulses per second, then each impulse necessarily propagates in the refractory zone of the previous impulse [103]. As we have seen in Figure 4.8 and the previous section, the safety factor is thereby diminished so each impulse becomes more susceptible to blockage or extinction.

---

[10]Similar observations (GR $= 1.01 \pm 0.15$) on branchings of excitor motor axons in the crayfish (*Orconectes virilis*) have been reported by Dean Smith [112].

One way to relieve the tension in this situation is to have the first impulse go down daughter #1, the next down daughter #2, the next down daughter #1, and so on. In the course of such *impulse steering*, the frequencies would be halved on the daughter branches, with the interpulse intervals increasing from 10 ms to a more comfortable 20 ms.

More generally, impulse steering might provide a means whereby an axonal tree translates a time code on its main trunk into time and space codes on its distal (distant) branches [133]. Because such behavior might be useful in neural processing of information, it should be studied numerically using realistic models for the branching axon.

- Most theoretical studies to date suppose that the sodium and potassium channels are distributed uniformly over the fiber membrane, but this simplifying assumption must be verified in real neurons. If the ion channels are not distributed uniformly but are more sparse near the crotches of branching junctions ("hot spots") [41, 66, 88, 89, 110, 121, 132], a means for impulse blockage and impulse train steering becomes available.

- Adelman and FitzHugh have shown that the safety factor of Hodgkin–Huxley impulses can be substantially reduced through the buildup of potassium ion concentration in restricted regions surrounding the axon, altering the potassium equilibrium potential (V_K in Equation (4.3)) [4]. In the course of their studies of impulse propagation through branchings of a lobster axon, Grossman, Parnas, and Spira suggest this mechanism as an explanation for differential blockage [38, 39]. The same explanation has been advanced by Smith for observed blockages in crayfish axons [111, 112].

- Many axons are not uniform structures like that of the squid but myelinated (see Chapter 7); thus Khodorov and his colleagues have considered what happens at transitions between smooth and myelinated regions [98]. Near branchings of myelinated cat fibers, studies by Quick et al. [91] show that the ratio of internode spacing to fiber diameter (β in Table 7.2) becomes much smaller near the twigs of axonal trees, which indicates that the continuum limit of Section 7.2.1 is being approached and the safety factor increased. Also the nearest active nodes of the two daughters may not be equidistant from the crotch of a branch, providing yet another means for impulse steering at higher rates of data transmission.

- Synaptic contacts have been observed on the active nodes of myelinated axons of cat motoneurons [87], suggesting additional opportunities for axonal information processing. Might each active node on a myelinated nerve act as an individual switch that processes incoming synaptic codes as it decides whether to ignite the next node?

- As discussed in the previous chapter, impulses on parallel fibers can interact through ephaptic couplings of their external current loops. This phenomenon is well established both theoretically and experimentally and provides opportunities for impulse steering at branchings, offering possibilities for information processing on bundles of axons such as vertebrate motor nerves and the corpus callosum [102].

Considering the repertory of interesting dynamics that is currently being recognized on real neurons [35, 51, 52, 71], it seems prudent to remain open to the possibility of axonal information processing.

9.5 Numerical Models

A moment's reflection on the foregoing should convince the reader that the task of computing the dynamic behaviors of dendritic and axonal trees is formidable. Even under the linear assumption of Section 9.1, there remains the task of describing the trees, and in the study of Mel about 3000 measurements of branch lengths and diameters were used to characterize the dendrites of a single pyramidal cell [71]. How is mathematical neuroscience to make progress if such an effort precedes every computation?

One answer is the establishment of an *Archive of Neuronal Morphology* by the Centre of Neuroscience at the University of Southampton, from which the dendritic trees in Figure 9.1 were obtained [27]. Neuroscientists can now obtain specific geometrical data (branch lengths, diameters, branching angles, and so on) from real neurons upon which subsequent analyses can be based. Because this website is organized to accept new data, it will continue to grow and become an invaluable resource in future years.

Another approach is to gather a number of measurements on similar neurons (perhaps from the above *Archive*) from which a statistical characterization of that class can be extracted [26, 43, 120]. For particular examples of the various classes of neurons in the mammalian central nervous system, consider the four dendritic trees from the hippocampus of the rat shown in Figure 9.1 [27]. As Giorgio Ascoli and his colleagues have shown, such characterizations allow the generation of thousands of statistically equivalent neurons per second on commonly available personal computers [11]. Using a computational package called "L-Neuron" (LN),[11] a geometrical description of each dendritic or axonal tree is available within the computer, ready for analysis.

Like biological trees, these dendrites grow by successive branchings; thus it is necessary to have histograms of the following features [11, 12]: num-

---

[11]Information on various versions, databases, and neuroanatomical structures grouped by morphological classes is available on the Web at www.krasnow.gmu.edu/L-Neuron.

ber of trees, length of parent branch, taper of parent branch, branching exponent, branch angle, relative diameters of daughters, and minimum diameter, all of which may depend on distance (or number of branchings) from the cell body. At the formation of each branch during the growth process, values for these parameters are selected randomly from the respective histograms, resulting in sets of model trees that are statistically identical to the original neurons.

Given data files on tree structures, the linear passive theory sketched in Section 9.1.1 provides a theoretical basis for computer codes causally relating synaptic inputs to voltage response at the nerve body. Such codes employ either the Green function of Equation (9.6), discussed in detail by Rinzell and Rall [99] and more recently by Abbott and his colleagues [1, 2, 28], or the Fourier transform approach proposed by Koch and Poggio [54]. Upon development, these linear codes can be compared with existing compartmental codes such as NEURON [44, 45] or GENESIS [23].

In a compartmental code, each compartment comprises three components: the membrane capacitance (in farads), the resistance (in ohms) from the center of one compartment to the center of the next, and a linear or nonlinear formulation of the ionic current (in amperes) crossing the compartment membrane. In Mel's code, for example, 163 dendritic segments were represented by about 500 compartments, for an average of about three compartments per segment of dendritic fiber. As the number of compartments per section increases, solutions of the compartmental model are expected to approach exact solutions of the partial differential system describing the tree.

On the positive side, compartmental models have two main advantages:

- The transmembrane ionic current is readily changed from passive to active, or from fast sodium current to slower calcium current, without altering the structure of the numerical code. (Codes based on Green functions or Fourier/Laplace transforms, on the other hand, cannot easily be generalized to include nonlinear membranes.)

- By reducing the number of compartments per section, the computational load is decreased.

In using compartmental models, however, it is always necessary to consider the accuracy of the ultimate numerical computation, and this will depend on what is being computed. For Mel's studies of cluster sensitivity (see Figure 9.8), using three compartments to describe one section of dendritic fiber seems sufficient. If one were to study details of the dendritic logic described in Section 9.3.1, on the other hand, more compartments might be required, thereby increasing the computational difficulties.

9.6 Some Outstanding Research Problems

In Chapter 2, we introduced two neural models: the single-switch representation of McCulloch and Pitts [70] and Waxman's multiplex neuron, which supposes that more intricate information takes place in the dendritic and axonal trees of a single nerve cell [133]. Throughout this survey, it has been emphasized that finding an accurate description of real neurons remains the ultimate objective. We are now, I hope the reader agrees, in a better position to see how the insights of mathematics might help neuroscientists proceed toward this goal.

Because so much remains to be done before real neurons are adequately described, this chapter closes with some suggestions for future research activities in the area of mathematical neuroscience. No claim of completeness is made; these are merely some ideas that I might offer to applied mathematics students who are interested in neuroscience or to neuroscience students with a taste for mathematics.

Statistical Models of Neurons

To construct realistic neuron models within a computer, one can divide them into classes (perhaps genetically determined) and then find histograms of the statistical probabilities for various growth parameters [11]. Much of the data for such statistical characterizations currently lies buried in collections of photographs or could be recorded in the course of ongoing experiments. As "computational anatomy" becomes a more important aspect of neuroscience research, it is of increasing interest to gather all available statistical data on the classes of neurons that are being studied in each laboratory.

In assembling such statistics, it would be helpful to have an efficient means for automatically transferring geometrical information on neural structures from photographs or electron micrograms of axonal and dendritic trees (e.g., branchings, segment lengths, diameters, branching exponents) into computer data files. This is a daunting task, in part because crossovers in visual images must be distinguished from true branchings, and checking that a code works properly in all cases is tedious. Nonetheless, available computer power is growing year by year, and such a development would be of great value to the neuroscience community.

Studies of this type respond to Walter Elsasser's appeal for biological studies that go beyond measurements of magnitudes and seek "correlations between various phenomena observable either on a given object or perhaps only on a class of objects" [33].

Evaluating Compartmental Models

Although compartmental codes (like NEURON [44, 45] and GENESIS [23]) are popular for computational studies of neural systems, concerns about their accuracy remain, particularly in their ability to correctly model the dendritic switching phenomena described in Section 9.3.1. In the course of numerical checks of the blocking conditions predicted by Equation (9.18), for example, a rather fine spatial step was needed to confirm the theoretical predictions [18]. Additionally, the compartmental model used by Goldfinger (to compare blocking conditions at abrupt widenings and at branchings) [36] gave results at variance with the finite element studies of Altenberger et al. [6].

In using compartmental analyses, therefore, neuroscientists must consider how well the behaviors of real neural structures are represented. Might tests based on linear assumptions (Green functions or Fourier transforms) provide useful benchmarks for such evaluations? Perhaps the M–C model, introduced in Chapter 6, could serve to bridge the computational gap between compartmental codes and a full PDE description of branching fibers [80, 81, 82]. Could compartmental approximations mask dendritic logic? Or might they overestimate threshold phenomena by introducing the possibility of failure between poorly selected compartments? Do the answers to these questions depend on the nature of the active membrane process (fast sodium or slow calcium) that is assumed? Such numerical studies are expected to become ever more feasible in coming years.

Inhomogeneous Fibers

Although exploratory numerical studies have been carried out on the dynamic effects of changing the cross section of a Hodgkin–Huxley fiber [15, 16, 22, 37, 48, 49, 50, 79, 108, 36, 6, 138], work remains to be done, and the necessary computing power is now widely available. In particular, it should be interesting to check the assumption (underlying Section 9.2.3) that a blocking condition on $(d_2/d_1)^{3/2}$ in Figure 9.4(a) is equivalent to the same condition on $(d_2^{3/2} + d_3^{3/2})/d_1^{3/2}$ in Figure 9.4(b). Additionally, relations between Equations (9.15) and (9.17) and numerical studies of widening on the H–H model can be explored. How good are these approximate formulations? What might they be missing? What is the best way to account for sodium turn-on delay? How can the time delay generated at a varicosity be conveniently described?

Decremental Conduction

The concept of a *critical point* in active propagation was introduced in Chapter 4 and discussed in Chapter 6 as the region of parameter space beyond which action potentials cannot be supported by the nerve, but the nonlinear dynamics in this region are not well understood. Numerical computations based on the Hodgkin–Huxley (H–H) and FitzHugh–Nagumo (F–N) models might be helpful in clarifying behavior near the critical point, providing bases for improved analytic descriptions and theoretical understanding. In this context, the phenomenon of decremental conduction, discussed in Sections 4.6 and 9.1.2, merits careful theoretical investigation aimed at understanding its nonlinear features and providing guidelines for electrophysiologists.

Impulse Steering
H–H, F–N, M–C, and compartmental models could also be used to study
the phenomenon of impulse steering at axonal branchings, which was
mentioned in Section 9.4 [133]. Is this a realizable phenomenon or mere
theoretical speculation? Assuming it is real, what are the experimental
conditions for getting steering started? Might it stem from dynamic insta-
bilities related to a high impulse rate on the axonal trunk, or must it be
induced by (threshold) perturbations of finite amplitude? How is impulse
steering influenced by the locations of active nodes on myelinated fibers?
How far up the tree can impulse steering occur?

Impulse Dynamics on Short Segments
Motivated by Ramón y Cajal's classic image of Purkinje cell dendrites in
Figure 9.5, H–H and F–N models might be used to study more intricate
branchings for which interbranch segments are of the order of the active
space constant ($\lambda_a = 1/\sqrt{rg}$), providing realistic corrections to the idealized
estimates in Section 9.3.1. How do the finite lengths of branch segments
influence the input–output behaviors of real trees? Can one describe these
behaviors in terms of Boolean functions (as was suggested in Section 9.3.1),
or are more or less intricate representations needed?

Second Impulse Blockage
Although the observations of double impulse switching shown in Figure 9.7
[103, 104] are in approximate accord with the exploratory computations
reported by Khodorov [50], wide ranges of numerical uncertainty wait to
be resolved. Using a combination of H–H, F–N, M–C, and compartmental
models, it should now be possible to compute more precisely the ways in
which the critical impulse spacing for the second impulse block depend
upon the GR and the nature of the active channels (sodium or calcium)
for realistic dendritic models. In such calculations, it would be interesting
to include degrading effects of temperature, external ionic concentrations,
narcotization, and so on as outlined in Section 4.6.

9.7 Recapitulation

Linear diffusion of transmembrane voltage on passive models of dendritic
fibers was discussed first in this chapter, emphasizing key experimental
parameters and simplifications arising from the assumption of linearity.
The phenomenon of decremental conduction was suggested as a means by
which synaptic inputs can be amplified without giving up the powerful superposition properties of linear models, and the theoretical bases for Rall's
"equivalent cylinder model" were presented.

Active conduction of fully nonlinear impulses was then considered on
a variety of inhomogeneous fibers, emphasizing the formulation of simple

conditions for impulse blockage. Although the insights from these studies underlie speculations about information processing on the dendritic trees, real dendrites may operate somewhere between the regimes of linear diffusion and fully nonlinear switching. Thus, the possibility of multiplicative response, in which a dendritic tree responds nonlinearly to the average pulse rates of the incoming signals was also considered. Information-processing possibilities were then discussed for axonal branchings, including the speculation that time codes on the main trunk may be translated into time and space codes on the distal twigs of an axonal tree.

As the computing power available to individual researchers continues to increase, it is always important to reassess the spectrum of numerical strategies. Presently, databases of neuronal structures are becoming available that make it possible to grow model neurons within a computer that are statistically equivalent to classes of biological neurons. Thus detailed computations of dendritic behaviors are now feasible under a variety of dynamic assumptions—an exciting opportunity.

Finally, a brief list of theoretical, numerical, and experimental problems was presented that, it is hoped, may encourage research in the dynamics of individual neurons over the next few years.

References

[1] LF Abbott, Simple diagrammatic rules for solving dendritic cable problems, *Physica A* 185 (1992) 343–356.

[2] LF Abbott, E Farhi, and S Gutmann, The path integral for dendritic trees, *Biol. Cybern.* 66 (1991) 49–60.

[3] PR Adams, The platonic neuron gets the hots, *Curr. Biol.* 2 (1992) 625–627.

[4] WJ Adelman, Jr and R FitzHugh, Solutions of the Hodgkin–Huxley equations modified for potassium accumulation in a periaxonal space, *Fed. Proc. (Fed. Am. Soc. Exp. Biol.)* 34 (1975) 1322–1329.

[5] H Agmon-Snir, CE Car, and J Rinzel, The role of dendrites in auditory coincidence detection, *Nature* 393 (1998) 268–272.

[6] R Altenberger, KA Lindsay, JM Ogden, and JR Rosenberg, The interaction between membrane kinetics and membrane geometry in the transmission of action potentials in non-uniform excitable fibres: a finite element approach, *J. Neurosci. Meth.* 112 (2001) 101–117.

[7] Y Amitai, A Friedman, B Connors, and M Gutnick, Regenerative electrical activity in apical dendrites of pyramidal cells in neocortex, *Cereb. Cortex* 3 (1993) 26–38.

[8] R Andersen, J Storm, and HV Wheal, Thresholds of action potentials evoked by synapses on the dendrites of pyramidal cells in the rat hippocampus in vitro, *J. Physiol. (London)* 383 (1987) 509–526.

[9] M Arbib (ed), *The Handbook of Brain Theory and Neural Networks*, MIT Press, Cambridge, Mass., 1995.

[10] YI Arshavskii, MB Berkinblit, SA Kovalev, VV Smolyaninov, and LM Chailakhyan, The role of dendrites in the functioning of nerve cells, *Dokl. Akad. Nauk SSSR* 163 (1965) 994–997.

[11] GA Ascoli, Progress and perspectives in computational neuroanatomy, *Anat. Rec.* 257 (1999) 195–207.

[12] G Ascoli and JL Krichmar, L-Neuron: A modeling tool for the efficient generation and parsimonious description of dendritic morphology, *Neurocomputing* 32–33 (2000) 1003–1011.

[13] DH Barron and BHC Matthews, Intermittent conduction in the spinal cord, *J. Physiol. (London)* 85 (1935) 73–103.

[14] JN Barrett and WE Crill, Specific membrane properties of cat motoneurons, *J. Physiol. (London)* 239 (1974) 301–324.

[15] MB Berkinblit, ND Vvedenskaya, LS Gnedenko, SA Kovalev, AV Kholopov, SV Fomin, and LM Chailakhyan, Computer investigation of the features of conduction of a nerve impulse along fibers with different degrees of widening, *Biophysics* 15 (1970) 1121–1130.

[16] MB Berkinblit, ND Vvedenskaya, LS Gnedenko, SA Kovalev, AV Kholopov, SV Fomin, and LM Chailakhyan, Interaction of nerve impulses in a node of branching (investigation of the Hodgkin–Huxley model), *Biophysics* 16 (1972) 105–113.

[17] LS Bernardo, LM Masukawa, and DA Prince, Electrophysiology of isolated hippocampal dendrites, *J. Neurosci.* 2 (1982) 1614–1622.

[18] S Binczak, personal communication.

[19] G Birkhoff and S MacLane, *A Survey of Modern Algebra,* Macmillan, New York, 1953.

[20] SA Bloomfield, JE Hamos, and SM Sherman, Passive cable properties and morphological correlates of neurons in the lateral geniculate nucleus of the cat, *J. Physiol. (London)* 383 (1987) 653–668.

[21] KY Bogdanov and VB Golovchinskii, Extracellularly recorded action potentials and the possibility of the spread of excitation over the dendrites, *Biophysics* 15 (1970) 672–681.

[22] LS Bogoslovskaya, IA Lyubinskii, NV Pozin, YV Putsillo, LA Shmelev, and TM Shura-Bura, Spread of excitation along a fiber with local inhomogeneities (results of modelling), *Biophysics* 18 (1973) 944–948.

[23] JM Bower and D Beeman, *The Book of Genesis: Exploring Realistic Neural Models with the General Neural Simulation System,* second edition, Springer-Verlag, New York, 1998.

[24] M Bunge, *Causality and Modern Science,* third revised edition, Dover, New York, 1979.

[25] RE Burke, Spinal cord: Ventral horn. In *Synaptic Organization of the Brain,* GM Shepherd (ed), Oxford University Press, New York, 1998, pp 77–120.

[26] RE Burke, WB Marks, and B Ulfhake, A parsimonious description of motoneuronic dendritic morphology using computer simulation, *J. Neurosci.* 12 (1992) 2403–2416.

[27] RC Cannon, DA Turner, GK Pyapali, and HV Wheal, An on-line archive of reconstructed hippocampal neurons, *J. Neurosci. Methods* 84 (1998) 49–54. (www.neuro.soton.ac.uk)

[28] B Cao and LF Abbott, A new computational method for cable theory problems, *Biophys. J.* 64 (1993) 303–313.

[29] SH Chung, SA Raymond, and JY Lettvin, Multiple meaning in single visual units, *Brain. Behav. Evol.* 3 (1970) 72–101.

[30] JW Cooley and FA Dodge, Digital computer solutions for excitation and propagation of the nerve impulse, *Biophys. J.* 6 (1966) 583–599.

[31] E De Schutter and JM Bower, Simulated responses of cerebellar Purkinje cells are independent of the dendritic location of granule cell inputs, *Proc. Natl. Acad. Sci. USA* 91 (1994) 4736–4740.

[32] Ö Ekeberg, P Wallén, A Lansner, H Travén, L Brodin, and S Grillner, A computer based model for realistic simulations of neural networks, I: The single neuron and synaptic interaction, *Biol. Cybern.* 65 (1991) 81–90.

[33] WM Elsasser, *Reflections on a Theory of Organisms: Holism in Biology*, The Johns Hopkins University Press, Baltimore, 1998 (first published in 1987).

[34] JC Fiala and KM Harris, Dendritic structure. In reference [119].

[35] F Gabbiani, W Metzner, R Wessel, and C Koch, From stimulus encoding to feature extraction in weakly electric fish, *Nature* 384 (1996) 564–567.

[36] MD Goldfinger, Computation of high safety factor impulse propagation at axonal branch points, *NeuroReport* 11 (2000) 449–456.

[37] SS Goldstein and W Rall, Changes of action potential, shape and velocity for changing core conductor geometry, *Biophys. J.* 14 (1974) 731–757.

[38] Y Grossman, I Parnas, and ME Spira, Differential conduction block in branches of a bifurcating axon, *J. Physiol. (London)* 295 (1979) 283–305.

[39] Y Grossman, I Parnas, and ME Spira, Ionic mechanisms involved in differential conduction of action potentials at high frequency in a branching axon, *J. Physiol. (London)* 295 (1979) 307–322.

[40] EA Guillemin, *Introductory Circuit Theory*, John Wiley & Sons, New York, 1953.

[41] M Haüsser, N Spruston, and GJ Stuart, Diversity and dynamics of dendritic signalling, *Science* 290 (2000) 739–744.

[42] O Herreras, Propagating dendritic action potential mediates synaptic transmission in CA1 pyramidal cells *in situ*, *J. Neurophysiol.* 64 (1990) 1429–1441.

[43] DE Hillman, Neuronal shape parameters and substructures as a basis for neuronal form. In *The Neurosciences: Fourth Study Program*, FO Schmitt and FG Worden (eds), MIT Press, Cambridge, MA, 1979, pp 477–498.

[44] M Hines, A program for simulation of nerve equations with branching geometries, *Int. J. Biomed. Comput.* 24 (1989) 55–68.

[45] M Hines, NEURON—a program for simulation of nerve equations. In *Neural Systems: Analysis and Modelling*, FH Eckman (ed), Kluwer Academic, Boston, 1993.

[46] JJB Jack, D Noble, and RW Tsien, *Electric Current Flow in Excitable Cells,* Oxford University Press, Oxford, 1975.

[47] J Keener and J Sneyd, *Mathematical Physiology,* Springer-Verlag, New York, 1998.

[48] BI Khodorov, YN Timin, SY Vilenkin, and FB Gul'ko, Theoretical analysis of the mechanisms of conduction of a nerve impulse over an inhomogeneous axon. I. Conduction through a portion with increased diameter, *Biophysics* 14 (1969) 323–335.

[49] BI Khodorov, YN Timin, SY Vilenkin, and FB Gul'ko, Theoretical analysis of the mechanisms of conduction of a nerve pulse over an inhomogeneous axon. II. Conduction of a single impulse across a region of the fiber with modified functional properties, *Biophysics* 15 (1970) 145–152.

[50] BI Khodorov, *The Problem of Excitability,* Plenum, New York, 1974.

[51] C Koch, Computation and the single neuron, *Nature* 385 (1997) 207–210.

[52] C Koch, *Biophysics of Computation: Information Processing in Single Neurons,* Oxford University Press, New York, 1999.

[53] C Koch, T Poggio, and V Torre, Nonlinear interaction in a dendritic tree: Localization timing and role in information processing, *Proc. Natl. Acad. Sci. USA* 80 (1983) 2799–2802.

[54] C Koch and T Poggio, A simple algorithm for solving the cable equation in dendritic trees of arbitrary geometry, *J. Neurosci. Methods* 12 (1985) 303–315.

[55] AU Larkman, G Major, KJ Stratford, and JJB Jack, Dendritic morphology of pyramidal neurones of the visual cortex of the rat. IV: Electrical geometry, *J. Comp. Neurol.* 323 (1992) 137–152.

[56] KN Leibovic and NH Sabah, On synaptic transmission, neural signals and psychophysiological phenomena. In *Information Processing in the Nervous System,* KN Leibovic (ed), Springer-Verlag, New York, 1969.

[57] KN Leibovic, *Nervous System Theory,* Academic Press, New York, 1972.

[58] AG Lindgren and RJ Buratti, Stability of waveforms on active non-linear transmission lines, *Trans. IEEE Circuit Theory* 16 (1969) 274–279.

[59] R Llinás, C Nicholson, JA Freeman, and DE Hillman, Dendritic spikes and their inhibition in alligator Purkinje cells, *Science* 160 (1968) 1132–1135.

[60] R Llinás and R Hess, Tetrodotoxin-resistant dendritic spikes in avian Purkinje cells, *Soc. Neurosci. Abstr.* 2 (1976) 112.

[61] R Llinás, C Nicholson, and W Precht, Preferred centripital conduction of dendritic spikes in alligator Purkinje cells, *Science* 163 (1969) 184–187.

[62] R Llinás and C Nicholson, Electrophysiological properties of dendrites and somata in alligator Purkinje cells, *J. Neurophysiol.* 34 (1971) 532–551.

[63] R Llinás and M Sugimori, Electrophysiological properties of *in vitro* Purkinje cell dendrites in mammalian cerebellar slices, *J. Physiol. (London)* 305 (1980) 197–213.

[64] RR Llinás, The intrinsic electrophysiological properties of mammalian neurons: Insights into central nervous system function, *Science* 242 (1988) 1654–1664.

[65] R Lorente de Nó and GA Condouris, Decremental conduction in peripheral nerve: Integration of stimuli in the neuron, *Proc. Natl. Acad. Sci. USA* 45 (1959) 593–617.

[66] JC Magee, Voltage-gated ion channels in dendrites. In [119].

[67] BB Mandelbrot, *The Fractal Geometry of Nature,* WH Freeman and Company, San Francisco, 1982, Chapter 17.

[68] VS Markin and VF Pastushenko, Spread of excitation in a model of an inhomogeneous fiber. I. Slight change in dimensions of fiber, *Biophysics* 14 (1969) 335–344.

[69] H Mayakawa and H Kato, Active properties of dendritic membrane examined by current source density analysis in hippocampal CA1 pyramidal neurons, *Brain Res.* 399 (1986) 303–309.

[70] WS McCulloch and WH Pitts, A logical calculus of the ideas immanent in nervous activity, *Bull. Math. Biol.* 5 (1943) 115–133.

[71] BW Mel, Synaptic integration in an excitable dendritic tree, *J. Neurophysiol.* 70 (1993) 1086–1101.

[72] BW Mel, Information processing in dendritic trees, *Neural Comput.* 6 (1994) 1031–1085.

[73] JP Miller, W Rall, and J Rinzel, Synaptic amplification by active membranes in dendritic spines, *Brain Res.* 325 (1985) 325–330.

[74] OA Mornev, Elements of the "optics" of autowaves. In *Self-Organization of Autowaves and Structures Far from Equilibrium,* VI Krinsky (ed), Springer-Verlag, Berlin, 1984.

[75] C Morris and H Lecar, Voltage oscillations in the barnacle giant muscle fibre, *Biophys. J.* 35 (1981) 198–213.

[76] DD Mott, DA Turner, MM Okazaki, and DV Lewis, Interneurons of the dentate–hilus border of the rat dentate gyrus: Morphological and electrophysiological heterogeneity, *J. Neurosci.* 17 (1997) 3990–4005.

[77] S Ochs, R Pourmand, RA Jersild, Jr, and RN Friedman, The origin and nature of beading: A reversible transformation of the shape of nerve fibers, *Prog. Neurobiol.* 52 (1997) 391–426.

[78] I Parnas, Differential block at high frequency at branches of a single axon innervating two muscles, *J. Neurophysiol.* 35 (1972) 903–914.

[79] I Parnas and I Segev, A mathematical model for conduction of action potentials along bifurcating axons, *J. Physiol. (London)* 295 (1979) 323–343.

[80] VF Pastushenko and VS Markin, Propagation of excitation in a model of an inhomogeneous nerve fiber. II. Attenuation of pulse in the inhomogeneity, *Biophysics* 14 (1969) 548–552.

[81] VF Pastushenko, VS Markin, and YA Chizmadzhev, Propagation of excitation in a model of an inhomogeneous nerve fiber. III. Interaction of pulses in the region of the branching node of a nerve fibre, *Biophysics* 14 (1969) 929–937.

[82] VF Pastushenko, VS Markin, and YA Chizmadzhev, Propagation of excitation in a model of an inhomogeneous nerve fiber. IV. Branching as a summator of nerve pulses, *Biophysics* 14 (1969) 1130–1138.

[83] DH Perkel and DJ Perkel, Dendritic spines: Role of active membrane in modulating synaptic efficiency, *Brain Res.* 325 (1985) 331–335.

[84] JL Peña and M Konishi, Auditory spatial receptive fields created by multiplication, *Science* 292 (2001) 249–252.

[85] H Pockberger, Electrophysiological and morphological properties of rat motor cortex neurons in vivo, *Brain Res.* 539 (1991) 181–190.

[86] NP Poolos and JD Kocsis, Dendritic action potentials activated by NMDA receptor-mediated EPSPs in CA1 hippocampal pyramidal cells, *Brain Res.* 524 (1990) 342–346.

[87] R Poritsky, Two and three dimensional ultrastructure of boutons and glial cells on the motoneuronal surface in the cat spinal cord, *J. Comp. Neurol.* 135 (1969) 423–452.

[88] RR Pozanski and J Bell, A dendritic cable model for the amplification of synaptic potentials by an ensemble average of persistent sodium channels, *Math. Biosci.* 166 (2000) 101–121.

[89] RR Pozanski and J Bell, Theoretical analysis of the amplification of synaptic potentials by small clusters of persistent sodium channels in dendrites, *Math. Biosci.* 166 (2000) 123–147.

[90] GK Pyapali and DA Turner, Increased dendritic extent in CA1 hippocampal pyramidal cells from aged F344 rats, *Neurobiol. Aging* 17 (1996) 601–611.

[91] DC Quick, WR Kennedy, and L Donaldson, Dimensions of myelinated nerve fibers near the motor and sensory terminals in cat tenuissimus muscles, *Neuroscience* 4 (1979) 1089–1096.

[92] W Rall, Branching dendrite trees and motoneuron resistivity, *Expl. Neurol.* 1 (1959) 491–527.

[93] W Rall, Theory of physiological properties of dendrites, *Ann. N.Y. Acad. Sci.* 96 (1962) 491–527.

[94] W Rall, Electrophysiology of a dendrite neuron model, *Biophys. J.* 2 (1962) 145–167.

[95] W Rall and J Rinzel, Branch input resistance and steady attenuation to one branch of a dendritic neuron model, *Biophys. J.* 13 (1973) 648–688.

[96] S Ramón y Cajal, *Histologie du Système Nerveux de l'Homme et des Vertébrés*, Malaine, Paris, 1909.

[97] WG Regehr and DW Tank, Postsynaptic NMDA receptor-mediated calcium accumulation in hippocampal CA1 pyramidal cell dendrites, *Nature* 345 (1990) 807–810.

[98] SV Revenko, YE Timin, and BI Khodorov, Special features of the conduction of nerve impulses from the myelinated part of the axon into the non-myelinated terminal, *Biophysics* 18 (1973) 1140–1145.

[99] J Rinzel and W Rall, Transient response in a dendritic neuron model for current injected at one branch, *Biophys. J.* 14 (1974) 759–790.

[100] AC Scott, *Active and Nonlinear Wave Propagation,* Wiley-Interscience, New York, 1970.

[101] AC Scott, Information processing in dendritic trees, *Math. Biosci.* 18 (1973) 153–160.

[102] AC Scott and SD Luzader, Coupled solitary waves in neurophysics, *Phys. Scr.* 20 (1979) 395–401.

[103] AC Scott and U Vota-Pinardi, Velocity variations on unmyelinated axons, *J. Theor. Neurobiol.* 1 (1982) 150–172.

[104] AC Scott and U Vota-Pinardi, Pulse code transformations on axonal trees, *J. Theor. Neurobiol.* 1 (1982) 173–195.

[105] I Segev, Dendritic processing. In [9].

[106] I Segev and M London, A theoretical view of passive and active dendrites. In [119].

[107] I Segev and M London, Untangling dendrites with quantitative models, *Science* 290 (2000) 744–750.

[108] I Segev and W Rall, Computational study of an excitable dendritic spine, *J. Neurophysiol.* 60 (1988) 499–523.

[109] I Segev and W Rall, Excitable dendrites and spines: Earlier theoretical insights elucidate recent direct observations, *Trends Neurosci.* 21 (1998) 453–460.

[110] GM Shepherd and RK Brayton, Logic operations are properties of computer simulated interactions between excitable dendritic spines, *Neuroscience* 21 (1987) 151–166.

[111] DO Smith, Mechanisms of action potential propagation failure at sites of axon branching in the crayfish, *J. Physiol. (London)* 301 (1980) 243–259.

[112] DO Smith, Morphological aspects of the safety factor for action potential propagation at axon branch points in the crayfish, *J. Physiol. (London)* 301 (1980) 261–269.

[113] DO Smith, Axon conduction failure under *in vivo* conditions in crayfish, *J. Physiol. (London)* 344 (1983) 327–333.

[114] W Softky and C Koch, Single-cell models. In reference [9].

[115] DC Spray, ME Spira, and MV Bennett, Peripheral fields and branching patterns of buccal mechanosensory neurons in the opisthobranch mollusc, *Navanax intermis, Brain Research* 182 (1980) 253–270.

[116] N Spruston, G Stuart, and M Häusser, Dendritic integration. In [119].

[117] SD Stoney, Jr, Limitations on impulse conduction at the branch point of afferent axons in frog dorsal root ganglion, *Exp. Brain Res.* 80 (1990) 512–524.

[118] G Stuart and B Sakmann, Active propagation of somatic action potentials into neocortical pyramidal cell dendrites, *Nature* 367 (1994) 69–72.

[119] G Stuart, N Spruston, and M Häusser, *Dendrites,* Oxford University Press, Oxford, 1999.

[120] Y Tamori, Theory of dendritic morphology, *Phys. Rev. E* 48 (1993) 3124–3129.

[121] DW Tank, M Sugimori, J Conner, and R Llinás, Spatially resolved calcium dynamics of mammalian Purkinje cells in the cerebellar slice, *Science* 242 (1988) 773–777.

[122] L Tauc and GM Hughes, Modes of initiation and propagation of spikes in the branching axons of molluscan central neurons, *J. Gen. Physiol.* 46 (1963) 533–549.

[123] DW Thompson, *On Growth and Form*, abridged edition, Cambridge University Press, Cambridge, 1961.

[124] AM Thomson and J Deuchars, Temporal and spatial properties of local circuits in neocortex, *Trends Neurosci.* 17 (1994) 119–126.

[125] RD Traub, RKS Wong, R Miles, and H Michelson, A model of the CA3 hippocampal pyramidal neuron incorporating voltage-clamp data on intrinsic conductances, *J. Neurophysiol.* 66 (1991) 635–650.

[126] HC Tuckwell, *Introduction to Theoretical Neurobiology*, Cambridge University Press, Cambridge, 1988.

[127] DA Turner, XG Li, K Pyapali, A Ylinen, and G Buzaki, Morphometric and electrical properties of reconstructed hippocampal CA3 neurons recorded in vivo, *J. Comp. Neurol.* 356 (1995) 580–594.

[128] DA Turner, submitted.

[129] YW Turner, DER Meyers, and JL Barker, Localization of tetrodotoxin-sensitive field potentials of CA1 pyramidal cells in the rat hippocampus, *J. Neurophysiol.* 2 (1991) 1375–1387.

[130] AM Vallet and JC Coles, Is the membrane voltage amplifier of drone photoreceptors useful at physiological light intensities? *J. Comp. Physiol. A* 173 (1993) 163–168.

[131] AM Vallet, JC Coles, JC Eilbeck, and AC Scott, Membrane conductances involved in amplification of small signals by sodium channels of drone honeybee, *J. Physiol. (London)* 456 (1992) 303–324.

[132] P Vetter, A Roth, and M Häusser, Propagation of action potentials in dendrites depends on dendritic morphology, *J. Neurophysiol.* 85 (2001) 926–937.

[133] SG Waxman, Regional differentiation of the axon, a review with special reference to the concept of the multiplex neuron, *Brain Res.* 47 (1972) 269–288.

[134] N Wiener, *Nonlinear Problems in Random Theory*, Technology Press, Cambridge, MA, 1958.

[135] HR Wilson, *Spikes, Decisions, and Actions: The Dynamical Foundations of Neuroscience*, Oxford University Press, Oxford, 1999.

[136] RKS Wong, DA Prince, and A Busbaum, Intradendritic recordings from hippocampal neurons, *Proc. Natl. Acad. Sci. USA* 76 (1979) 986–990.

[137] JZ Young, *Doubt and Certainty in Science*, Oxford University Press, Oxford, 1951.

[138] Y Zhou and J Bell, Study of propagation along nonuniform excitable fibers, *Math. Biosci.* 119 (1994) 169–203.

10

Constructive Brain Theories

Given some modest appreciation for the dynamics of individual neurons, it is natural to ask how they might act in concert, which is a central question of neuroscience. In accord with the constructive perspectives of modern science, this problem can be phrased: How does a collection of interacting neurons manage to behave like a brain? The aim of this chapter is to sketch some answers to this question.

We begin with a brief review of the first McCulloch–Pitts paper, in which brain modeling was approached by stripping real neurons down to their essential features: all-or-nothing response and a threshold for firing (see Section 2.4.1) [36]. Interestingly, this simple "M–P neuron" survives to the present day as a workhorse of neural network modeling.

McCulloch and Pitts also suggested the division of neural network models into two broad classes: "nets with circles" and "nets without circles." Their quaint jargon distinguishes between networks possessing internal feedback loops and those simpler networks for which the information flows in one direction only, from input to output terminals, providing a basis for organizing this chapter.

A key property of biological brains is their ability to learn, which was modeled in 1958 by Rosenblatt through a significant modification of the M–P neuron [44]. In a class of neural networks called the "perceptron," information is constrained to flow only in one direction (a net without circles), but the input weightings of each model neuron are allowed to change during the course of a *training period* as required by an appropriate *learning algorithm*.

Nets *with* circles are of central interest in neuroscience because biological brains—even those of the most simple creatures—do indeed have many internal loops of positive feedback threading through their constituent neurons. As we have seen in previous chapters of this book, such closed causal loops (or "re-entry") lead to the emergence of new dynamic entities, the nerve impulse being an outstanding example. With the emergence of novel coherent states arises the need for describing their dynamics, compounding the difficulties of mathematical formulation and analysis. Such matters are addressed in the following two chapters.

If each model neuron in a network is allowed to compute the most general Boolean function of its inputs, as suggested in the previous chapter, it is straightforward to compute the number of nets with circles that can be created from a given number of neurons and to sketch the various types of behavior. The number of such systems grows very rapidly with the number of constituent neurons, however, soon becoming unmanageable; thus, some guiding perspectives are needed.

As a simple brain model that includes closed loops of causal implication (positive feedback), Hopfield's "spin-glass" model is presented in the context of previously noted concepts of phase-space analysis of nonlinear systems [27]. The number of stable stationary states in this model is considered as an estimate for the information storage capacity of real brains.

The chapter closes with a brief introduction to cortical field theories, the dynamics of which are in accord with observations of Gestalt psychology, suggesting means for communication among the emergent states of real brains.

10.1   Nets Without Circles

In this section, attention is restricted to nets *without* circles for two reasons. First, it is evident that such network models are easier to analyze and understand just because they do not give rise to emergent entities. (There is an adage, no less true for being ancient, that one should learn to walk before trying to run.) Second, from a mathematical perspective, there are several rather simple results on the geometric interpretation of the pattern-classification problem and on procedures for learning that are of general interest and may play supporting roles in the information-processing activities of real brains.

Although it was proposed back in the 1950s that the trainable properties of nets without circles offer a basis for understanding the human brain [4, 44, 45], this view has not been widely held since the demise of behaviorism as a credible psychological theory. Nonetheless, nets without circles do comprise a class of *learning machines* that have been of engineering interest since the late 1950s for a variety of tasks, including automatic sorting of photographs, converting handwritten characters to digitally defined letters, recognizing speech, generating suggestions for medical diagnoses, making weather predictions directly from atmospheric data, analyzing aerial photographs for economic data, and so on [23, 35].

However such systems fare in the realms of engineering, the peculiar properties of nets without circles may be employed for special purposes in certain restricted regions of the human brain, such as processing information on the way from the retina to the primary areas of the visual cortex or from the ears to the temporal lobes. Thus it seems prudent for neuroscientists to be aware of what nets without circles can do.

10.1.1 McCulloch–Pitts (M–P) Networks

In their 1943 paper, McCulloch and Pitts began by assuming a class of neural networks with the following properties: the activity of any constituent "neuron" is an all-or-nothing process; a fixed number of synapses must be stimulated within the period of latent addition in order to ignite a "neuron," and this number is independent of previous activity; the only significant delay occurs at synapses; ignition of a "neuron" is prevented by activation of a single inhibitory synapse; and the network structure does not change with time [36]. The term "neuron" is used here with quotation marks to emphasize that real neurons are more intricate than the model. Although this indication will be dropped in subsequent discussions, the reader should keep the caveat in mind.

McCulloch and Pitts were under no illusion that their assumptions are physiologically correct; indeed, they specifically mention that *facilitation and extinction* ("in which antecedent activity temporarily alters responsiveness to subsequent stimulation") and *learning* have been ignored. They defended their approach, however, as a way to establish baseline estimates of what neural networks can do.

A key aspect of the M–P formulation was their recognition that the all-or-nothing property of a neuron (an impulse is either present or it is not on a certain nerve at a certain time) can be viewed as a logical proposition (this statement is either true or false), so Boolean algebra (the algebra of classes) can be invoked to describe their model networks [3]. Thus they obtained two main results.

First, M–P showed that their model neuron could represent the three fundamental circuit elements of the computer engineer—the AND, OR, and NOT gates—which we met in the preceding chapter. Second, they appealed to the algebra of classes to show that any Boolean function can be modeled by one or more of their networks, and each such network corresponds to one or more Boolean functions. What is a Boolean function?

Written in the two-element number system "1" and "0" (which indicates that a statement is true or false or that an all-or-nothing impulse is present or absent), the three basic operations of Boolean arithmetic are:

$$
\begin{bmatrix}
1 \text{ AND } 1 & = & 1 \\
1 \text{ AND } 0 & = & 0 \\
0 \text{ AND } 1 & = & 0 \\
0 \text{ AND } 0 & = & 0
\end{bmatrix}
,
\begin{bmatrix}
1 \text{ OR } 1 & = & 1 \\
1 \text{ OR } 0 & = & 1 \\
0 \text{ OR } 1 & = & 1 \\
0 \text{ OR } 0 & = & 0
\end{bmatrix}
, \text{ and }
\begin{bmatrix}
\text{NOT } 1 & = & 0 \\
\text{NOT } 0 & = & 1
\end{bmatrix}.
$$

In the context of this arithmetic, a Boolean function specifies the output variable for each combination of input variables. Thus a particular Boolean function of three inputs A, B, and C might be denoted as $F(A, B, C)$ and defined as in the following table.

A	B	C	$F(A, B, C)$
0	0	0	0
0	0	1	0
0	1	0	0
0	1	1	0
1	0	0	0
1	0	1	1
1	1	0	0
1	1	1	1

A Boolean expression for this particular function is

$$
\begin{aligned}
F(A, B, C) &= (A \text{ AND } B \text{ AND } C) \text{ OR} (A \text{ AND NOT } B \text{ AND } C) \\
&= A \text{ AND } C \qquad\qquad\qquad\qquad\qquad\qquad (10.1)
\end{aligned}
$$

indicating in ordinary English that an output impulse will appear if either of two input conditions occurs: there are impulses at A, B, and C, or there are impulses at A and at C but not at B. In this formulation, "at" refers to a location in space-time because the AND operation requires temporal coincidence.

Because a Boolean function of N inputs has 2^N input combinations for which the corresponding output is either 0 or 1, there are evidently

$$
2^{2^N}
$$

distinct functions of N inputs. Each of these Boolean functions can be defined as in the preceding table and expressed as in Equation (10.1).

In retrospect, the demonstration by McCulloch and Pitts that any possible dependence on the output of a neural network can be realized (as engineers like to say) through a suitable combination of model neurons may seem modest. These results are now well known to computer engineers, and techniques for designing networks with a minimum number of switching elements (AND, OR, and NOT functions) have been available for decades [25]. In the early 1940s, however, engineers were striving to construct telephone switching stations with networks of magnetomechanical relays, and the modern digital computer was but a dream. In its day, therefore, the M–P paper was strikingly original.

More to the point in evaluating McCulloch–Pitts networks is the recognition that each nerve cell is modeled by a *single switch* represented by the Heaviside step function $H(I)$ in Equation (2.10), an assumption with two implications.

- This is a *convenient* assumption to make because the linear summation of input variables to the jth neuron

$$I_j = \sum_{k=1}^{N} \alpha_{jk} V_k(t) - \theta_j \qquad (10.2)$$

  in Equation (2.10) keeps the threads of causality distinct, facilitating analysis of the system [7].

- In the context of neuroscience, however, it is a *dangerous* assumption because causal relations among input signals to real neurons are far more intricate than is indicated in Equation (10.2).

10.1.2 Learning Networks

Although M–P networks can "in principle" be arranged to do whatever can be done without circles, their design is not straightforward and requires selection of the weighting parameters α_{jk} and θ_j in Equation (10.2) for all neurons in the net. How might a neuron manage to solve this problem?

In 1958, Rosenblatt suggested that the α_{jk} and θ_j could be changed incrementally if a particular neuron is not responding correctly [44, 45]. His *training algorithm* led to a class of learning networks composed of M–P neurons with adjustable weights, which he called the *perceptron* [4, 5, 37].

At about the same time, an identical idea arose within the engineering community [23, 50]. Here, the class of networks was dubbed "ADALINE" (for ADAptive LInear NEtworks), and the constituent element was called a "linear threshold unit" (LTU). In this stream of activity, the aim was not to understand brain dynamics but to design computing machines that could be trained to recognize patterns in data sets.

To be specific, let us suppose that the Boolean function of Equation (10.1) is to be used for predicting the weather, where $A = 1$ indicates that

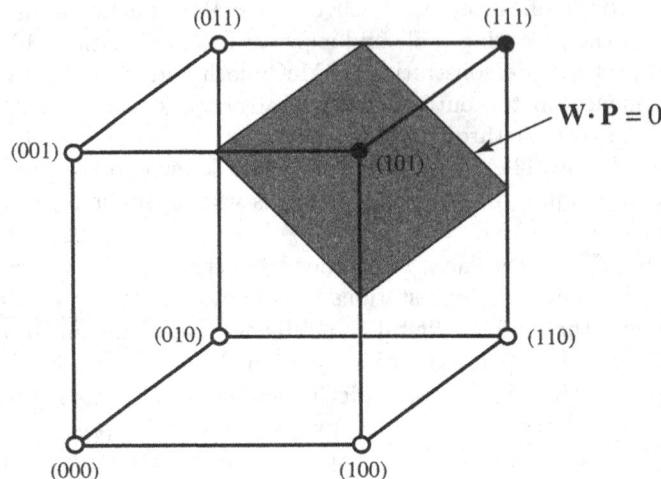

Figure 10.1. The geometrical interpretation of the pattern-recognition task indicated by Equations (10.1) and (10.3).

the barometer is rising and $A = 0$ that it is falling, $B = 1$ implies daytime and $B = 0$ night, and $C = 1$ indicates that it is clear and $C = 0$ indicates cloudiness. With $F(A, B, C)$ defined as in Equation (10.1), it is reasonable to expect that $F = 1$ implies that no rain is to be expected within the next few hours.

To understand how the training algorithm works, it helps to view pattern-recognition problems in a geometrical context. Thus, the eight values of these three input variables can be taken as vertices of a cube, as indicated in Figure 10.1, with the black dots indicating where $F = 1$ and the open dots where $F = 0$. The shaded area indicates a *linear discriminant plane* in pattern space on one side of which $F = 1$ and on the other $F = 0$.

Suppose that we wish to realize the logical function of Equation (10.1) with the M–P model neuron

$$\tilde{F} = H\left(\sum_{k=1}^{3} \alpha_k V_k(t) - \theta\right), \tag{10.3}$$

where $V_1 \equiv A$, $V_2 \equiv B$, and $V_3 \equiv C$. (Recall that $H(x)$ is the Heaviside step function, which equals 0 when x is negative and 1 otherwise.)

Two questions arise: (1) How do we choose α_1, α_2, α_3, and θ? (2) If these weighting parameters are incorrectly chosen, how can they be altered so that the functions computed from Equations (10.1) and (10.3) are the same?

To answer these questions, it is convenient to define a four-dimensional *weight vector* as

$$\mathbf{W} \equiv (\alpha_1, \alpha_2, \alpha_3, -\theta)$$

and a four-dimensional *augmented pattern vector* as

$$\mathbf{P} \equiv (V_1, V_2, V_3, 1).$$

Then the inner product of the weight vector and the augmented pattern vector,[1]

$$\mathbf{W} \cdot \mathbf{P} = \sum_{k=1}^{3} \alpha_k V_k(t) - \theta,$$

is just the argument of the Heaviside step function in Equation (10.3). Thus to realize the Boolean expression of Equation (10.1) with the M–P neuron of Equation (10.3), it suffices to choose the three α_js and θ so that the condition

$$\mathbf{W} \cdot \mathbf{P} = 0$$

corresponds to a discriminant plane lying between the vertices where $F = 1$ (the dark circles) and those where $F = 0$ (the open circles), as shown in Figure 10.1. This answers question (1).

To answer question (2), suppose that we have mistakenly chosen the components of the weight vector ($\mathbf{W_1}$) such that

$$\mathbf{W_1} \cdot \mathbf{P} < 0$$

for (say)

$$\mathbf{P} = (1, 1, 1, 1),$$

but all of the other vertices in Figure 10.1 lie on the correct side of the discriminant plane. Then Equation (10.3) tells us that $\tilde{F} = 0$ for $V_1 = V_2 = V_3 = 1$. In other words, if the barometer is rising, it is daytime, and the sky is not cloudy, we should expect rain. Clearly, this is not a correct prediction and the weight vector must be changed, but how?

If the weight vector were altered by adding an increment in a direction orthogonal (at right angles) to $\mathbf{P}$, the inner product $\mathbf{W} \cdot \mathbf{P}$ would not change; thus, it is necessary to alter the weight vector in the direction of $\mathbf{P}$. To accomplish this, assume

$$\mathbf{W_2} = \mathbf{W_1} + c\mathbf{P}, \tag{10.4}$$

where c is a positive real constant that must be determined. Taking the inner product of both sides of Equation (10.4) with $\mathbf{P}$ and requiring that $\mathbf{W_2} \cdot \mathbf{P} > 0$ shows that for

$$c > -\frac{\mathbf{W_1} \cdot \mathbf{P}}{\mathbf{P} \cdot \mathbf{P}} \tag{10.5}$$

the inner product $\mathbf{W_2} \cdot \mathbf{P} > 0$.

---

[1]The inner (or "dot") product of two vectors is the sum of the products of their components.

If $\mathbf{W_1} \cdot \mathbf{P} > 0$ gives an incorrect result for some $\mathbf{P}$, on the other hand, it is necessary to decrease $\mathbf{W} \cdot \mathbf{P}$, so making

$$c < -\frac{\mathbf{W_1} \cdot \mathbf{P}}{\mathbf{P} \cdot \mathbf{P}} \qquad (10.6)$$

will give the correct response.

When the inequality in Equation (10.5) or (10.6) is barely satisfied, then the weight vector has been readjusted with a minimum of change. In our example of the weather predictor, this ensures that $\tilde{F} = F$ for $V_1 \equiv A = 1$, $V_2 \equiv B = 1$, and $V_3 \equiv C = 1$. Because these changes in the weight vector may have caused some of the other inputs to give erroneous results, it is necessary to check all of the other input conditions and make minimal corrections corresponding to Equations (10.4) and (10.5) wherever necessary.

In more general cases, it may be that no discriminant plane exists, for example, with a function defined as 1 at two diagonally opposite vertices—(000) and (111)—and 0 otherwise. If such cases are excluded, the Boolean function is said to be *linearly separable*.

In other words, an M–P (or LTU) representation for a linearly separable Boolean function exists by definition, leading to the following theorem.

> **Training theorem:** If a Boolean function (F) is linearly separable and the weight vectors of an M–P neuron $(\tilde{F})$ are successively modified as indicated in Equations (10.4) and (10.5) or (10.6), then the sequence
>
> $$\mathbf{W_1} \to \mathbf{W_2} \to \mathbf{W_3} \to \mathbf{W_4} \to \cdots$$
>
> converges in a finite number of steps to a weight vector for which $\tilde{F}$ is identical to F [37, 38].

This result is biologically interesting because the information needed to make such successive weight modifications is just what a neuron has available at the tips of its dendrites. From Equations (10.5) and (10.6), this information comprises the current values of the synaptic strengths and the threshold (given by $\mathbf{W_j}$) and the current values of the input signals (given by $\mathbf{P}$).

Put differently, if a neuron were informed that its response to a particular pattern is undesired, it could correct that behavior by increasing or decreasing its synaptic weights in amounts proportional to the current input signals, which suggests the following question for neuroscientists: Are there biologically credible means through which a real neuron might come to know that a certain response is unwanted by its organism?

Table 10.1. The number of Boolean networks ($\mathcal{N}$) for various numbers of switches (N).

N	$\mathcal{N} = 2^{N2^N}$
1	$2^2 = 4$
2	$4^4 = 256$
3	$8^8 \doteq 1.7 \times 10^7$
4	$16^{16} \doteq 1.8 \times 10^{19}$
5	$32^{32} \doteq 1.5 \times 10^{48}$
6	$64^{64} \doteq 3.9 \times 10^{115}$

10.2 Nets with Circles

Because the human brain is threaded through with myriad closed loops of causal implication, any serious study of its dynamics must deal with the many new entities that emerge. This section presents two constructive theories of such networks. The first indicates the degree of intricacy to be expected, and the second suggests ways in which methods of statistical physics may lead to understanding.

10.2.1 General Boolean Networks

Let us begin by imagining the most general class of networks that can be constructed from N model neurons (or switches), each of which is allowed to compute an arbitrary Boolean function of its N inputs. Because there are

$$2^{2^N}$$

Boolean functions of N inputs and each of the N neurons is chosen to be one of these, there are

$$\mathcal{N} = \left(2^{2^N}\right)^N = 2^{N2^N}$$

different systems in this class of general Boolean networks. For modest numbers of neurons, the number of possible systems soon becomes very large, as is seen from Table 10.1. To deal with such large numbers, combinatoric mathematicians have whimsically defined the *googol* $\equiv 10^{100}$ as a finite number above which arithmetic becomes problematic [11]. To see why

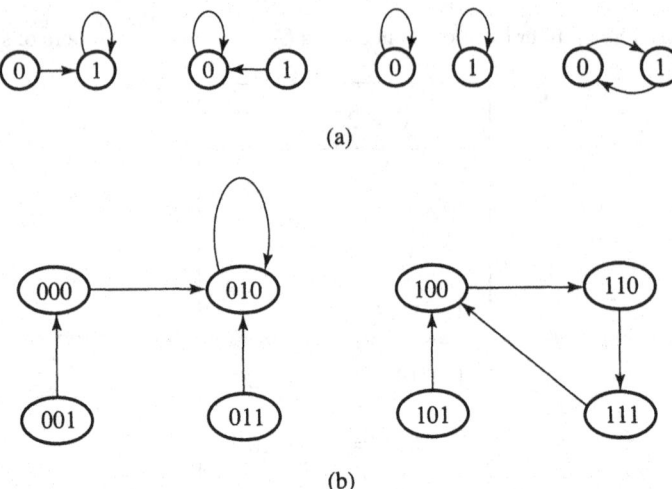

(a)

(b)

Figure 10.2. (a) The four Boolean systems that can be constructed from a single switch. (b) One of the more than sixteen million systems that can be constructed from three switches.

they have introduced such a definition, let us go down the table, making references along the way to Figure 10.2.

In this figure are indicated *state diagrams* of certain Boolean systems. Each such diagram shows the 2^N states of some system, which advances from one state to the next in a fixed time interval. Each state, therefore, has one outgoing arrow, indicating which of the 2^N states the system will go to in the next increment of time. Thus, the 2^N arrows can be chosen in 2^N different ways, leading once again to

$$\mathcal{N} = \left(2^N\right)^{2^N} = 2^{N2^N}$$

different systems composed of N switches.

For $N = 1$, there are $2^2 = 4$ Boolean systems that can be constructed from a single switch. These four systems are shown in Figure 10.2(a), where "0" indicates that the switch is off and "1" implies on. Reading from left to right, the first system turns on if it is off and then stays on. The second system turns off if it is on and then stays off. The third system stays off if it is off and stays on if it is on, as expected for a light switch. Finally, the last system—which electrical engineers call a "free running multivibrator"— turns off if it is on and turns on if it is off, thereby generating a periodic signal.

For $N = 2$, implying two switches or neurons, there are $4^4 = 256$ possible Boolean systems, which could be worked out in a few hours. For $N = 3$, this number has increased to $8^8 = 16,777,216$ systems, one of which is shown in Figure 10.2(b). Presumably, a computer code could be written to generate diagrams for all of these systems. For $N = 4$ and 5, it would be imprudent to attempt such a code.

For $N \geq 6$, interestingly, it can be asserted with confidence that no computing system will ever be constructed that generates and records the diagrams for all possible systems. This is because the atomic weight (or the total number of protons and neutrons) of the universe is only about 10^{80}; thus there is not enough paper—or memory storage of any sort—for the task.

To emphasize the importance of this point in theoretical biology, physicist Walter Elsasser has proposed that the term *immense* be used to describe finite numbers larger than a googol [12]. Typically, the number of possible members of a biological species is immense, whereas the number of actual members—past, present, and future—is not.

Dealing with sets in which the number of possible members is much larger than the number of actual members, Elsasser suggests, helps to make the biological and social sciences fundamentally different from the physical sciences. Thus, a physicist studying (say) hydrogen can perform as many experiments as desired on identical atoms, leading to generalizations formulated as laws of physics. Similarly, the chemist can study as many (say) benzene molecules as are needed to formulate reliable laws of chemistry.

The biologist or psychologist studying (say) *Homo sapiens*, on the other hand, faces quite a different challenge. There are a great many more possible humans than will ever actually exist—past, present, or future. To see this, note first that the total number of actual human beings is certainly less than a googol (not immense). Then, consider what $\mathcal{N}$ would be in Table 10.1 if N were anything like the 10^{10} neurons in a human brain. It is clear from such a comparison that psychological observations are necessarily made on very limited subsets of the possible members of our species.[2]

In Elsasser's terminology, biologists, psychologists, and anthropologists study *heterogeneous sets*, the members of which exhibit substantial differences. (We have seen an example of biological heterogeneity in Table 4.1 showing the variability of membrane data on giant axons of the squid.) Physicists and chemists, on the other hand, deal with *homogeneous sets*, for which members are essentially identical. To emphasize this distinction, consider the case of alloys.

The conducting wires of electric circuits, for example, are often connected together by an alloy of tin and lead called solder, for which the melting point depends on the ratio of the components. Given two small

---

[2]It may be objected that the brain's neurons do not compute arbitrary Boolean functions of their inputs, so the estimate of $\mathcal{N}$ may be high. From work of Yajima et al. [54], the number of systems composed of N threshold (M–P) neurons with n inputs each is greater than $2^{n^2 N/2}$, but this estimate is low. Recently, Poirazi and Mel [42] have studied memory capacity of systems of model neurons in which individual dendritic trees respond nonlinearly to their inputs. From both combinatoric and numerical calculations, these authors find that capacities of nonlinear neurons exceed those of linear models by "orders of magnitude."

blobs of solder with the same tin-to-lead ratio, the detailed arrangements of the tin and lead atoms will differ, and it is not difficult to show that the total number of possible arrangements is immense. Nonetheless, the average properties of solder (melting point, electrical resistivity, specific heat, ductility, and so on) are almost exactly the same for most arrangements; thus all possible solder blobs form a homogeneous set, falling comfortably within the purview of physical science.

In a heterogeneous set, on the other hand, small variations count, so members have very different global properties. Thinking about it, there are many heterogeneous sets in our ordinary experience, including the number of possible natural languages, protein molecules, musical compositions, English sonnets, chess games, people, and so on. Thus there will always be many different languages, useful proteins, beautiful melodies and poems, interesting chess games, and exciting human personalities that have not been realized and never will be.

10.2.2   Attractor Neural Networks

Although the large numbers of the previous section may discourage some who would develop a constructive theory of the brain's dynamics, one should not give up altogether. With reference to the state diagram of Figure 10.2(b), three qualitatively different sorts of behavior are observed.

(1) First, we note *transients* such as

$$(001) \rightarrow (000) \rightarrow (010)$$

in which the system passes through a sequence of states, never to return.

(2) Amid these transients emerge *stable attractors* such as (010) with the *basin of attraction* (000), (001), and (011).

(3) Finally, there are *limit cycles* such as

$$(100) \rightarrow (110) \rightarrow (111) \rightarrow (100) \rightarrow (110) \rightarrow (111) \rightarrow \cdots$$

having a period of three time units and the basin of attraction (101).

It comes as no surprise to find emergent entities in switching networks; such behavior arises directly from the positive feedback associated with closed causal loops, as we have often seen. The problem with nets comprising a realistic number of neurons is that there is no direct way of accounting for all of the emerging states. Some limit cycles may be as short as one time unit; others might approach 2^N time units, snaking through most states of the system, and one cannot proceed by looking at all possible systems.

In 1982, Hopfield made some progress in circumventing such combinatoric difficulties by assuming a modified version of the neural network we have previously considered [27]. Motivated by the McCulloch–Pitts formulation of Equation (2.10), his basic "neuron" obeys the dynamic equation

$$s_j(t + \tilde{\tau}) = \text{sign}\left(\sum_{k=1}^{N} J_{jk}\, s_k(t)\right),\qquad (10.7)$$

where $J_{kk} = 0$ and $\text{sign}(\cdot)$ is the "sign" function with properties

$$\text{sign}(y) = \begin{cases} +1 & \text{for } y \geq 0 \text{ and} \\ -1 & \text{for } y < 0, \end{cases}$$

and the "spin" variables s_j take the values $+1$ and -1 instead of the Boolean numbers 1 and 0.

In comparing Hopfield's *attractor neural network* with the McCulloch–Pitts (M–P) model, the following points of difference should be noted.[3]

- In M–P, the variables V_j (with $j = 1, 2, \ldots, N$) indicate the instantaneous voltages of the N neurons in the network, whereas the s_j in an attractor neural network represent the average firing rate of the jth neuron measured on a linear scale from *quiescent* (-1) to *fully active* $(+1)$.

- In an attractor neural network, the states of individual switches are not changed all at once but successively altered in a randomly selected order. Thus $\tau = N\tilde{\tau}$ is the time required for updating average firing rates for the entire net, corresponding to about 1–2 s in the human brain.

- The attractor neural network threshold is assumed to be the same for all N model neurons, which are joined by a symmetric $N \times N$ interconnection matrix

$$J = [J_{jk}],$$

  where $J_{kk} = 0$ and $J_{jk} = J_{kj}$. (This symmetry condition means that the coupling from neuron k to neuron j is equal to the coupling from neuron j to neuron k, which is not neurologically realistic.)

- When numerically convenient, the "$\text{sign}(\cdot)$" function in Equation (10.7) can be replaced by a *sigmoid* function, qualitatively like $\tanh(\cdot)$, which rises in a monotone manner from -1 to $+1$ as x increases from $-\infty$ to $+\infty$ [28].

This attractor neural network is convenient for analysis because it has a *Lyapunov functional* (E) possessing the following pair of properties [29, 32]:

---

[3] Although attractor neural networks are sometimes referred to as "ANNs" [1], others use these initials for the broader class of "artificial neural networks"; thus the acronym is avoided here.

first, E must be bounded from below, and second, E must either decrease or remain constant with each time step.[4]

For the dynamics described by Equation (10.7), a Lyapunov functional is

$$E = -\frac{1}{2}\sum_{j=1}^{N}\sum_{k=1}^{N} J_{jk}\, s_j s_k\,, \tag{10.8}$$

with the proof as follows.

(1) Note first that for finite J_{jk}, E is evidently bounded from below.

(2) To see how E changes under the dynamics of Equation (10.7), suppose first that the ith switch changes from $s_i = +1$ to -1. The corresponding change in E is

$$\begin{aligned}\Delta E &= s_i\sum_{j=1}^{N} J_{ij}\, s_j + s_i\sum_{j=1}^{N} J_{ji}\, s_j \\ &= 2\sum_{j=1}^{N} J_{ij}\, s_j\end{aligned}$$

because $J_{ij} = J_{ji}$. Because the summation must be negative for s_i to decrease, ΔE is negative.

(3) Next suppose that the ith switch changes from $s_i = -1$ to $+1$. The corresponding change in E is

$$\Delta E = -2\sum_{j=1}^{N} J_{ij} s_j\,.$$

Because the summation must be positive for s_i to increase, ΔE is again negative.

(4) Finally, if s_i does not change, ΔE is zero.

Because E is bounded from below and ΔE is either negative or zero after each time step ($\bar{\tau}$), E eventually ceases to decrease, and the system either moves around on a limit cycle at constant E or sits at a stable attractor, which may be found numerically. In Hopfield's formulation, each stable attractor is viewed as a *pattern* stored nonlocally by the net.

Each such pattern will have a basin of attraction into which the system can be forced by sensory inputs. In other words, if external stimulations nudge an attractor neural network into the basin of attraction for one of its patterns, the system will move to that attractor and remain there, providing

---

[4]Further motivation for Hopfield's model is the fact that Equation (10.8) is an expression for the total energy of interacting magnets (or atomic spins), which have been of interest to physicists for decades. Thus several standard results from condensed-matter physics translate directly into statements concerning the dynamics of nerve systems, drawing physical scientists into neuroscience [1, 26].

a model for the brain's ability to recall intricate memory patterns under the influence of sensory information.

Assume a pattern of the form $\mathbf{X_m} = (x_1^m, x_2^m, \ldots, x_N^m)$, where the components are either $+1$ or -1 with equal probability. To learn this pattern, the interconnection matrix can then be constructed by the rules

$$\Delta J_{ij} \propto x_i^m x_j^m,$$
$$\Delta J_{ij} = 0,$$

which has the effect of increasing interconnection strengths where both neurons are in the same state and reducing them where the states are different. If p patterns are learned in this manner,

$$J_{ij} = \frac{1}{N} \sum_{m=1}^{p} x_i^m x_j^m,$$

where the normalization by N is chosen to keep the components of J at a uniform level as the number of patterns is increased.[5]

How many such patterns can be stored by this system?

Supposing that p patterns are randomly chosen, each will seem like noise to the others. Every time a new pattern is added to the store, in other words, the elements of J are adjusted, effectively introducing noise into the task of recovering formerly stored patterns. Because the elements of J are normalized to N, the total noise amplitude seen by each stored pattern will grow as $\sqrt{p/N}$.

If p remains constant while $N \to \infty$, the noise amplitude goes to zero and the storage system works well. With N held constant while p is increased, on the other hand, a maximum number of patterns (p_m) is eventually reached that is proportional to N [1, 26].

Allowing the network to have stable patterns close to those learned (where "close" means that no more than 1% of the neurons deviate), it appears from a combination of theoretical arguments and numerical evidence that

$$p_m \approx 0.14\,N. \tag{10.9}$$

---

[5]For example, if $\mathbf{X} = (+1, -1, +1, -1)$, then

$$\Delta J = \begin{bmatrix} 0 & -1 & +1 & -1 \\ -1 & 0 & -1 & +1 \\ +1 & -1 & 0 & -1 \\ -1 & +1 & -1 & 0 \end{bmatrix},$$

and $\Delta J X \propto X$, which is a stationary point of Equation (10.7).

Interestingly, this is a sharp boundary. If one attempts to store more than the critical number of patterns, the probability of retrieval falls rapidly to zero.

Assuming this spin-glass model of the brain bears some relation to neurological reality, a human neocortex of 10^{10} to 10^{11} neurons might be expected to store something like 10^9 to 10^{10} intricate patterns. We will consider another derivation of this important number in the following chapter.

10.3   Field Theories for the Neocortex

Another way to deal with the immense number of possible systems into which the brain's neurons may become organized is to develop a *field theory* for neural activity. Such an approach was first proposed in 1956 by Beurle [2], who assumed that the neural mass of the neocortex can be locally described by the fraction $F(x,t)$ of cells at position x that are firing at time t, and the probability $p(x)$ of two cells being interconnected is an exponentially decreasing function of the distance between them [47]; thus,

$$p(x) \propto e^{-|x|/\sigma} .$$

In this theory, activity at a particular region of the cortex induces activity at neighboring regions, which leads to a *wave of information* propagating through the neural mass. Because Beurle supposed that all of the cortical neurons are excitatory, the waves described by his theory correspond roughly to the leading-edge formulations of Chapter 5, albeit with the activity averaged over many neurons rather than localized on a single fiber. Salient features of his study include the following.

- A wave of information may involve the activity of only a small fraction of the local neurons, allowing waves to pass through each other with little interference. Thus many different messages may propagate throughout the neocortex and carry information from one region to another.[6]

- If the neural interconnections (synapses) are supposed to increase in strength upon exposure to the activity of a particular wave (a learning mechanism), one can imagine a means for holographic-like recall [2, 6, 16, 17, 34, 43]. Thus, waves induced by external (sensory) stimulation would become coupled to subsequent internal waves, leading to the possibility that but a fragment of the original stimulation is required to trigger the related internal response.

---

[6]That cortical information waves pass through one another leads some to confuse them with *solitons*, but the two phenomena are quite different. Solitons conserve energy, whereas waves of information do not, being dynamically akin to waves of activity in the heart [46].

- Although neuroscientists often fret over the "binding problem" of relating activities in different parts of the cortex—combining, for example, the voice, image, and personality of a friend into a single perception—Beurle's information waves may provide a means of achieving such coupling.

Following Beurle's lead, Griffith (among others) developed a field theory of neural activity in which time and space dependencies are brought in through their lowest derivatives [19, 20, 21, 24, 39, 40, 41]. In this theory, the probability of a neuron firing in the next time interval (S) is a sigmoid function of the present firing rate (F), say

$$S(F) = \frac{F^2}{F^2 + \theta^2},$$

with S and F necessarily positive and θ a threshold parameter.

Thus, without spatial variations (i.e., "space-clamped"), the first time derivative can be approximated as

$$\frac{dF}{dt} \approx \frac{S(F) - F}{\tau},$$

implying

$$\frac{dF}{dt} \approx -f(F),$$

where

$$f(F) = -\frac{F^3 - F^2 + \theta^2 F}{\tau(F^2 + \theta^2)}$$

is a cubic function qualitatively like those sketched in Figures 5.2 and 5.3.

To introduce spatial dependence, Griffith reasoned that the connectivity would be the same in both the $+x$ and the $-x$ directions; thus, the lowest space derivative is the second, implying a nonlinear diffusion equation

$$D\frac{\partial^2 F}{\partial x^2} - \frac{\partial F}{\partial t} = f(F). \tag{10.10}$$

For a mean interconnection distance indicated by σ and a neural response time of τ, the diffusion constant is of order

$$D \sim \frac{\sigma^2}{\tau}.$$

Although Equation (10.10) is formally identical to Equation (5.5), its interpretation is quite different. Equation (5.5) describes the leading edge of a nerve impulse, traveling along an unmyelinated axon, whereas Equation (10.10) represents a wave of activity propagating through a neural medium such as the neocortex.

To bring *recovery* into the picture, Wilson and Cowan took advantage of the fact that some of the neocortical neurons are inhibitory; thus, they developed a theory in two dependent variables [51, 52]:

- $E(x, t)$: the fraction of excitatory neurons that are firing as a function of x and t, and

- $I(x, t)$: the corresponding fraction of inhibitory neurons,

which are assumed to interact as [53]

$$\tau \frac{dE}{dt} = S_E \left(\int [w_{EE}(x, x')E(x') - w_{IE}(x, x')I(x')]dx' + P(x, t) \right) - E,$$

(10.11)

$$\tau \frac{dI}{dt} = S_I \left(\int [w_{EI}(x, x')E(x') - w_{II}(x, x')I(x')]dx' + Q(x, t) \right) - I.$$

In these coupled integro-differential equations, the nonlinearity of neural response is introduced through the sigmoid functions

$$S_E(y) \equiv \frac{100y^2}{\theta_E^2 + y^2} \quad \text{and} \quad S_I(y) \equiv \frac{100y^2}{\theta_I^2 + y^2},$$

diffusion stems from the interconnection probabilities between neurons at x and x',

$$w_{ij}(x, x') = b_{ij} \exp \left(-\frac{|x - x'|}{\sigma_{ij}} \right),$$

and external (sensory) inputs to the excitatory and inhibitory cells are represented by $P(x, t)$ and $Q(x.t)$, respectively.

It is, of course, difficult to fix the many parameters of such a model, but Wilson suggests the following values as reasonable "guesstimates" for the human neocortex: $\sigma_{EE} = 40\,\mu$m, $\sigma_{EI} = \sigma_{IE} = 60\,\mu$m, $\sigma_{II} = 30\,\mu$m, $\theta_E = 20$, and $\theta_I = 40$, and he has made available several MATLAB codes for exploring the resulting dynamics [53]. Those familiar with MATLAB are encouraged to play with these codes and explore the following spectrum of behaviors.

- *Stationary patterns of activity:* With $b_{EE} = 1.95$, $b_{IE} = b_{EI} = 1.4$, and $b_{II} = 2.2$, a short pulse of stimulation ($P = 1.0$ for 10 ms over a range of 100 μm, and $Q = 0$) induces a stationary pattern in which the longer-ranging inhibitory activity surrounds and contains the more localized excitatory activity. Qualitatively, this behavior is similar to *Turing patterns*, which are found in studies of nonlinear reaction-diffusion systems of more than one dimension [46].

- *Transient activity:* With $b_{EE} = 1.5$, $b_{IE} = b_{EI} = 1.3$, and $b_{II} = 1.5$, a brief stimulation ($P = 2.0$ over 5 μm for 5 ms, and $Q = 0$) causes a transient response, with E rising to a maximum value of about 28 in about 30 ms and then relaxing back to zero.

- *Localized oscillations:* With $b_{EE} = 1.9$, $b_{IE} = b_{EI} = 1.5$, and $b_{II} = 1.5$, a constant stimulation over a spatial range of 100 μm or more results in a variety of spatially localized oscillations.

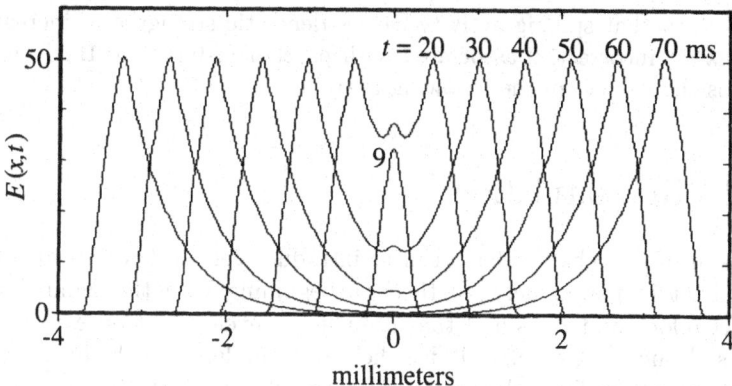

Figure 10.3. Outgoing wave solutions of the Wilson–Cowan equations (10.11) generated by a brief pulse of excitation near the origin.

- *Waves of activity:* With $b_{EE} = 1.9$, $b_{IE} = b_{EI} = 1.5$, and $b_{II} = 1.5$, and a strongly inhibitory input ($Q = -90$ while P is applied briefly over 100 μm), subsequent outgoing waves of activity are shown in Figure 10.3. Traveling at a speed of about 0.06 mm/ms, these waves are qualitatively similar to the impulses with recovery that were discussed in Chapter 6 for single fibers.

In his book, Wilson offers many more examples of such dynamics, discussing ways in which neural field theories similar to Equations (10.11) can model a variety of mental phenomena, including phase transitions, hallucinations, and epileptic seizures [8, 13, 15, 18, 22, 30, 31, 48, 53]. Recently, Ermentrout and Kleinfeld have developed a simple model of cortical wave motion through a network of weakly coupled oscillators in which only the phase of the oscillators is influenced by interactions [14].

On a much longer time scale, note that nonlinear field effects may also play a role in the development of mesoscopic cortical structure along lines suggested by Alan Turing in 1952 [49]. Such structure occurs in the visual cortices of mammals, where alternating bands of neurons, receiving inputs dominated by one or the other eye, are seen. Starting from this observation, Martha Constantine-Paton and Margaret Law have used experiments on "three-eyed frogs" to show that these cortical "stripes" are a form of "Turing pattern" [9, 10, 33]. Interestingly, the active nonlinearity driving the pattern-formation process stems not from biochemistry as was originally proposed by Turing. Rather, it is a positive feedback phenomenon with a closed causal loop having the structure

Increased spiking activity

↓ ↑

Dendritic strength

Thus, the basic driving force is neural activity.

The idea that spiking activity causes dendritic strengths to increase is (somewhat incorrectly) associated with psychologist Donald Hebb, whose work is the subject of the following chapter.

10.4   Recapitulation

The aim of this chapter has been to introduce certain key threads of development in neural network theories beginning with the seminal work of McCulloch and Pitts and the subsequent development of learning machines. Although geometrical ideas help us to understand the fundamental learning theorem for such systems, the immense number of possible neural arrangements precludes exhaustive searches of particular networks.

To deal with this difficulty, physicists have proposed "spin-glass" models of the brain in which neural behavior is idealized in order to obtain a Lyapunov (or energy) functional governing global dynamics. Following this approach, it has been estimated that the number of complex concepts that the human brain can store is of the order of 10^9 to 10^{10}.

Finally, some nonlinear field theories for cortical dynamics were sketched that display a wide variety of qualitative behaviors, including stationary patterns, transients, localized oscillations, and waves of information. Although it is difficult to fix the parameters of these nonlinear diffusion models with precision from neurological data, they can be studied numerically with currently available tools.

References

[1] DJ Amit, *Modeling Brain Function: The World of Attractor Neural Networks*, Cambridge University Press, Cambridge, 1989.

[2] RL Beurle, Properties of a mass of cells capable of regenerating pulses, *Philos. Trans. R. Soc. London* A240 (1956) 55–94.

[3] G Birkhoff and S MacLane, *A Survey of Modern Algebra*, Macmillan, New York, 1953.

[4] HD Block, The Perceptron: A model for brain functioning, *Rev. Mod. Phys.* 34 (1962) 123–135.

[5] HD Block, BW Knight, Jr, and F Rosenblatt, Analysis of a four-layer series coupled Perceptron, *Rev. Mod. Phys.* 34 (1962) 135–142.

[6] A Borsellino and T Poggio, Holographic aspects of temporal memory and optomotor responses, *Kybernetik* 10 (1972) 58–60.

[7] M Bunge, *Causality and Modern Science*, third revised edition, Dover, New York, 1979.

[8] PS Churchland and TJ Sejnowski, *The Computational Brain*, MIT Press, Cambridge, MA, 1994.

[9] M Constantine-Paton and MI Law, Eye-specific termination bands in tecta of three-eyed frogs, *Science* 202 (1978) 639–641.

[10] M Constantine-Paton and MI Law, The development of maps and stripes in the brain, *Sci. Am.* December 1982.

[11] RE Crandall, The challenge of large numbers, *Sci. Am.* February 1997, 72–78.

[12] WM Elsasser, *Reflections on a Theory of Organisms: Holism in Biology,* The Johns Hopkins University Press, Baltimore, 1998 (first published in 1987).

[13] GB Ermentrout and JD Cowan, A mathematical theory of visual hallucination patterns, *Biol. Cybern.* 34 (1979) 137–150.

[14] GB Ermentrout and D Kleinfeld, Traveling electrical waves in cortex: Insights from phase dynamics and speculation on a computational role, *Neuron* 29 (2001) 33–44.

[15] A Fuchs, JAS Kelso, and H Haken, Phase transitions in the human brain: Spatial mode dynamics, *Int. J. Bifurcation Chaos* 2 (1992) 917–939.

[16] D Gabor, Holographic model of temporal recall, *Nature* 217 (1968) 584.

[17] D Gabor, Improved holographic model of temporal recall, *Nature* 217 (1968) 1288.

[18] J Glanz, Mastering the nonlinear brain, *Science* 277 (1997) 1758–1760.

[19] JS Griffith, A field theory of neural nets: I. Derivation of field equations, *Bull. Math. Biophys.* 25 (1963) 187–195.

[20] JS Griffith, A field theory of neural nets: II. Properties of field equations, *Bull. Math. Biophys.* 27 (1965) 111–120.

[21] JS Griffith, *Mathematical Neurobiology: An Introduction to the Mathematics of the Nervous System,* Academic Press, New York, 1971.

[22] H Haken, *Principles of Brain Functioning: A Synergetic Approach to Brain Activity, Behavior and Cognition,* Springer-Verlag, Berlin, 1996.

[23] J Hawkins, Self-organizing systems: A review and commentary, *Proc. IRE* 49 (1961) 31–48.

[24] CE Hendrix, Transmission of electric fields in cortical tissue: A model for the origin of the alpha rhythm, *Bull. Math. Biophys.* 27 (1965) 197–213.

[25] FC Hennie, *Finite-State Models for Logical Machines,* John Wiley & Sons, New York, 1968.

[26] J Hertz, A Krogh, and RG Palmer, *Introduction to Neural Computation,* Addison-Wesley, Reading, MA, 1991.

[27] JJ Hopfield, Neural networks and physical systems with emergent collective computational abilities, *Proc. Nat. Acad. Sci. (USA)* 79 (1982) 2554–2558.

[28] JJ Hopfield, Neurons with graded response have collective computational properties like those of two-state neurons, *Proc. Natl. Acad. Sci. USA* 81 (1984) 3088–3092.

[29] EA Jackson, *Perspectives of Nonlinear Dynamics,* Cambridge University Press, Cambridge, 1990.

[30] VK Jirsa, R Friedrich, H Haken, and JAS Kelso, A theoretical model of phase transitions in the human brain, *Biol. Cybern.* 71 (1994) 27–35.

[31] JAS Kelso, SL Bressler, S Buchanan, GC De Guzman, A Fuchs, and T Holroyd, A phase transition in human brain and behavior, *Phys. Lett. A* 169 (1992) 134–144.

[32] J La Salle and S Lefschetz, *Stability by Liapunov's Direct Method*, Academic Press, New York, 1961.

[33] MI Law and M Constantine-Paton, Right and left eye bands in frogs with unilateral tectal ablations. *Proc. Natl. Acad. Sci. USA* 77 (1980) 2314–2318.

[34] HC Longuet-Higgins, Holographic model of temporal recall, *Nature* 217 (1968) 104.

[35] K Mainzer, *Thinking in Complexity: The Complex Dynamics of Matter, Mind, and Mankind*, Springer-Verlag, Berlin, 1994.

[36] WS McCulloch and WH Pitts, A logical calculus of the ideas immanent in nervous activity, *Bull. Math. Biophys.* 5 (1943) 115–133.

[37] M Minsky and S Papert, *Perceptrons*, MIT Press, Cambridge, MA, 1969.

[38] NJ Nilsson, *Learning Machines: Foundations of Trainable Pattern-classifying Systems*, second edition, Morgan Kaufmann, San Mateo, CA, 1990.

[39] PL Nuñez, The brain wave equation: A model for the EEG, *Math. Biosci.* 21 (1974) 279–297.

[40] PL Nuñez, Wave-like properties of the alpha rhythm, *Trans. IEEE Biomed. Eng.* BME-21 (1974) 473–482.

[41] PL Nuñez, *Electric Fields of the Brain: The Neurophysics of EEG*, Oxford University Press, New York, 1981.

[42] P Poirazi and BW Mel, Impact of active dendrites and structural plasticity on the memory capacity of neural tissue, *Neuron* 29 (2001) 779–796.

[43] KH Pribram, The neurophysiology of remembering, *Sci. Am.*, January 1969, 73–85.

[44] F Rosenblatt, The Perceptron: A probabilistic model for information storage and organization in the brain, *Psychol. Rev.* 65 (1958) 386–408.

[45] F Rosenblatt, *Principles of Neurodynamics*, Spartan Books, New York, 1962.

[46] AC Scott, *Nonlinear Science: Emergence and Dynamics of Coherent Structures*, Oxford University Press, Oxford, 1999.

[47] DA Sholl, *The Organization of the Cerebral Cortex*, Methuen, London, 1956.

[48] P Tass, Cortical pattern formation during visual hallucinations, *J. Biol. Phys.* 21 (1995) 177–210.

[49] AM Turing, The chemical basis of morphogenesis, *Philos. Trans. R. Soc. London* B237 (1952) 37–72.

[50] B Widrow and JB Angell, Reliable, trainable networks for computing and control, *Aerospace Eng.* September, 1962, 78–123.

[51] HR Wilson and JD Cowan, Excitatory and inhibitory interactions in localized populations of model neurons, *Biophys. J.* 12 (1972) 1–24.

[52] HR Wilson and JD Cowan, A mathematical theory of the functional dynamics of cortical and thalamic nervous tissue, *Kybernetik* 13 (1973) 55–80.

[53] HR Wilson, *Spikes, Decisions, and Actions: The Dynamical Foundations of Neuroscience*, Oxford University Press, Oxford, 1999.

[54] S Yajima, T Ibaraki, and I Kawano, On autonomous logic nets of threshold computers, *Trans. IEEE on Comp.* 17 (1968) 385–391.

11
Neuronal Assemblies

Although the suggestion that neurons in the human brain may act in functional groups reaches back at least to the beginning of the twentieth century (when Charles Sherrington published his *The Integrative Action of the Nervous System* [85]), it was in Donald Hebb's classic *Organization of Behavior* that the cell-assembly concept was first carefully formulated. Largely neglected for several decades [13], Hebb's theory of neural assemblies has more recently begun to attract broad interest from the neuroscience community. Why, one wonders, was such a reasonable suggestion so long ignored? Several answers come to mind.

First, Hebb was far ahead of his time. As a psychologist, moreover, he was telling electrophysiologists and neurologists what they should be doing when these people had much on their collective plate. Throughout most of the twentieth century, electrophysiologists were facing numerous difficulties in recording from single neurons. Adequate impulse amplifiers needed to be designed and suitable microelectrodes fabricated before voltages could be measured from even a single cell. If mere hit-or-miss recordings were to be avoided, it was necessary to position accurately the tips of these electrodes, knowing what cells are located where. As the levels of the observed signals became smaller, means for shielding measurements from ambient electromagnetic noise were ever more in demand. With single-neuron recording being the primary experimental focus, therefore, it is not surprising that theoreticians refrained from embracing more complicated formulations that required simultaneous recordings from many neurons for which empirical support was not soon expected.

Second, as we have seen in Chapter 9, it is difficult enough to describe properly the dynamics of individual neurons; thus, a theory that assumed interacting assemblies of neurons would be venturing even further out onto the thin ice of speculation.

A third reason for the tendency to simplify the theoretical picture—in North America, at least—was the unfortunate domination of psychology by the beliefs of behaviorism, which focused attention on the conditioning of stimulus–response reflexes, thereby ignoring much that comprises mental reality. From the behaviorist perspective, the concept of internal cerebral states was rightly shouldered into the background because the simpler ideas of "connection theory" seemed adequate to explain acceptable psychological data.

With all of these strikes against it, how did Hebb's theory ever manage to see the light of day?

11.1 Birth of the Cell-Assembly Theory

During the 1940s, Hebb became impressed with several sorts of evidence that cast doubt on behaviorist assumptions and suggested that more subtle theoretical perspectives were needed to explain psychological facts [34]. Among such facts is the surprising robustness of the brain's dynamics, a well-known example of which was provided by railroad workman Phineas Gage, who survived having a piece of iron rod go through his brain [56]. With characteristic directness, Hebb put the matter thus: How is it that a person can register an IQ of 160 after the removal of a prefrontal lobe [32]?

His first publication on the cell assembly stemmed from observations of chimpanzees raised in a laboratory where, from birth, every stimulus was under experimental control. Such animals, Hebb noted, exhibited spontaneous fear upon seeing a clay model of a chimpanzee's head [33]. The chimps in question had never witnessed decapitation, yet some of them "screamed, defecated, fled from their outer cages to the inner rooms where they were not within sight of the clay model; those that remained within sight stood at the back of the cage, their gaze fixed on the model held in my hand" [35, 36, 38].

Such responses are clearly not reflexes; nor can they be explained as conditioned responses to stimuli, for there was no prior example in the animals' repertory of responses. Moreover, they earned no behavioral rewards by acting in such a manner. But the reactions of the chimps do make sense as disruptions of highly developed and meaningful internal configurations of neural activity according to which the chimps somehow recognized the clay head as a mutilated representation of beings like themselves.

Another contribution to the birth of his theory was Hebb's rereading of Marius von Senden's *Space and Sight* [84], which was originally published in Germany in 1932. In this work, von Senden gathered records on 65 patients who had been born blind due to cataracts up to the year 1912. At ages varying from 3 to 46 years, the cataracts were surgically removed, and a variety of reporters had observed the patients as they went about handling the sudden and often maddeningly novel influx of light.

One of the few generalizations over these cases, von Senden noted, was that the process of learning to see "is an enterprise fraught with innumerable difficulties, and that the common idea that the patient must necessarily be delighted with the gifts of light and colour bequeathed to him by the operation is wholly remote from the facts." Not every patient rejoiced upon being forced to make sense of incoming light that was all but incomprehensible, and many found the effort of learning to see to be so difficult that they simply gave up.

That such observations are not artifacts of the surgery or uniquely human was fortuitously established through observations on a pair of young chimpanzees that had been reared in the dark by a colleague of Hebb [81]. After being brought out into the light, these animals showed no emotional reactions to their new experiences. They seemed unaware of the stimulation of light and did not try to explore visual objects by touch. Hebb conjectured that the chimps showed no visual response because they had not yet formed the neural assemblies needed for perception.

Finally, Hebb pointed out that the learning curve for an individual subject in a behavioral experiment is not the smoothly rising curve shown in psychology textbooks. This is because the textbook curves are averages over many learning experiments, whereas the observations in a particular experiment are influenced by whether the subject is paying attention to the task. Thus the factor of *attention* (otherwise called attitude, expectancy, hypothesis, intention, vector, need, perseveration, or preoccupation), Hebb felt, must somehow be included in any satisfactory theory of learning.

As was noted in Chapter 1, these considerations led Hebb to propose that nerve cells do not necessarily act as individuals in the dynamics of the brain but often as functional groups, which he called cell assemblies, with the following properties.

- Each complex assembly comprises a "three-dimensional fishnet" of many thousands of interconnected cells sparsely distributed over much of the brain.

- The interconnections among the cells of a particular assembly grow slowly in numbers and strength as a person matures in response to both external stimuli and internal dynamics that are tailored to the particular experiences of the organism.

- One mechanism suggested for the growth of neuronal interconnections postulated the strengthening of dendritic contacts through use. (That this feature has become widely known among nerve network mavens as a "Hebbian synapse" amused Hebb because it was one of

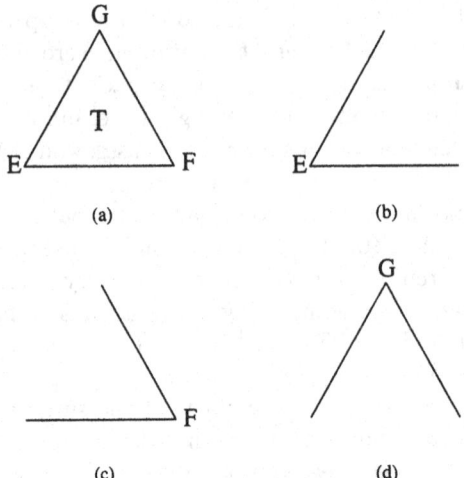

Figure 11.1. Diagrams related to the process of learning to see a triangle.

the few aspects of the theory that he did *not* consider to be original
[64].) In Chapter 9, we saw that a real neuron has several means for
altering its behavior, including changes in the geometry of dendritic
spines or branching, variations in the distributions of ionic channels
over the dendritic and axonal membranes, development of dendrodendritic interactions, changes in amplification levels of decremental
conduction, and so on.

- Upon ignition—effected through some combination of external stimuli
  and the partial activities of other assemblies—a particular assembly
  remains briefly active, yielding in a second or so to partial exhaustion
  of its constituent neurons.

- During the period of time that an assembly is active, the attention
  of the brain is focused on the concepts embodied in that assembly.

- As one assembly ceases its activity, another ignites, then another, and
  so on, in a temporal series of events called the *phase sequence*, which
  is experienced by each of us as a train of thought.

As a simple example of assembly formation, consider how an infant might
learn to perceive the triangle T shown in Figure 11.1(a). The constituent
sensations of the vertices are first supposed to be centered on the retina by
eye movement and mapped onto the primary visual area (V1) of the optical
lobes of the neocortex (located in the back of your head). Corresponding cell
assemblies E, F, and G then develop in the secondary visual area through
nontopological connections with area V1. The process of examining the
triangle involves elementary phase sequences in which E, F, and G are
sequentially ignited. Gradually, these subassemblies are supposed to fuse
together into a common assembly for perception of the triangle T.

With further development of the assembly T—which reduces its threshold for ignition through the strengthening of the internal connections among E, F, and G—a glance at one corner, with a few peripheral cues, serves to ignite the entire assembly representing T. At this point in the learning process, T is established as a second-order cell assembly for perception of a triangle, including E, F, and G among its constituent subassemblies.

Is there empirical evidence supporting Hebb's theory?

11.2   Early Evidence for Cell Assemblies

Upon formulating the cell-assembly theory for brain dynamics, Hebb and other psycholigists began the process of empirical evaluation that is central to science. By the mid-1970s, these efforts had produced the following results.

Robustness

In Chapter 1, we considered a social analogy for the cell-assembly concept in which the brain is likened to a community and the neurons to its individual citizens. From this perspective, the remarkable robustness of the brain to physical damage can be understood. If a motorcycle club gets into a fight, losing several of its members, the strength of the club is not permanently reduced because new members can be added. Similarly, a damaged cell assembly can recruit additional neurons to participate in its activities. (Such recruitment of new assembly members may occur during rehabilitation from a stroke, a lobotomy, or other forms of neurological damage.)

Furthermore, because the cells of an assembly may be widely dispersed over much of the brain, partial destruction of the brain does not completely destroy any of the assemblies. Thus, the cell-assembly theory offers the same sort of robustness under physical damage as a hologram but is more credible because it does not require a regular structure that can reinforce scattered waves of neural activity.

Learning a New Language

As a graduate student in the "post-Sputnik" days of the late 1950s, I had the experience of learning to read Russian, having no prior knowledge of the language whatsoever. This effort proceeded in stages, commencing with the task of recognizing Cyrillic letters and associating these new shapes with novel sounds. Upon mastering the alphabet, it became possible to learn words comprising these letters, and with enough words, sentences and then paragraphs could eventually be understood. Thus it appears to me an empirical observation that language learning is a step-by-step process, during which a hierarchically organized memory is slowly constructed.

Interestingly, the full perception of a letter or word involves the melding of visual, auditory, and motor components, which underscores the concept

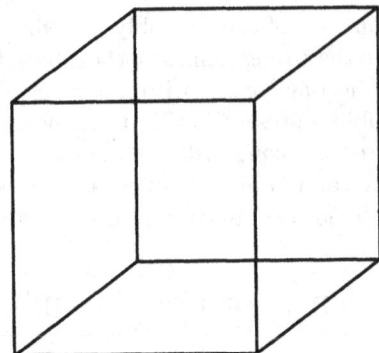

Figure 11.2. The Necker cube.

of subassemblies being distributed widely over the brain, a point to which we will return in the following chapter.

The general idea of hierarchical learning and memory has been rather carefully formulated by Braitenberg and Pulvermüller [8]. Although the acquisition of most of our basic skills lies buried in the forgotten past, most learning seems layered, with each stage necessarily mastered before it becomes possible to move on to the next. In the context of Hebb's theory, these stages involve the formation of subassemblies from which assemblies of higher order will subsequently emerge.

Ambiguous Perceptions
No discussion of the brain can neglect the mention of ambiguous figures, which have fascinated Gestalt psychologists for generations, and my favorite example—the Necker cube—is shown in Figure 11.2. Attempting to "bridge the long gap between the facts of neurology and those of psychology," Hebb's theory provides an explanation for the properties of such figures [34]. Gestalt phenomena are thus understood in a visceral manner by supposing that an assembly is associated with the perception of each orientation. Upon regarding Figure 11.2, I sense something switching inside my head every few seconds as the orientations change.

From the several cases of people learning to see that were cited by von Senden [84], it is clear that the ability to perceive an object in three spatial dimensions is itself learned, and the Necker cube is particularly interesting because perceptions of its two possible orientations would seem to be of equal likelihood. In the following section, we model the dynamics of switching between perceptions of two such orientations, where the overall symmetry of the situation suggests that the parameters of the two assemblies are identical, thereby simplifying analysis.

Stabilized Images
In Hebb's view, some of the strongest evidence in support of the cell-assembly theory was obtained from *stabilized-image* experiments, which

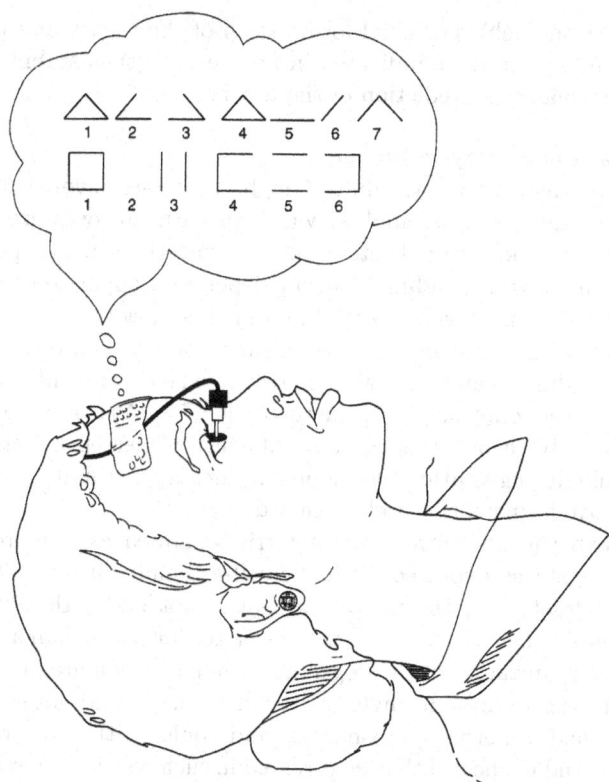

Figure 11.3. Sketch of contact lens and optical apparatus mounted on the eyeball of a reclining observer. The wire is connected to a small lamp that illuminates the target. The thought balloon shows sample sequences of patterns perceived by the subject with images that are stabilized on the retina by the apparatus. In the upper row a triangle is the target, and in the lower row, the target is a square (after a photograph in Pritchard [77]).

were carried out at McGill University in the early 1960s [35, 36, 64, 76, 77]. The experimental setup is sketched in Figure 11.3, where a simple geometric figure (e.g., a triangle or a square) is projected as a fixed image onto the retina. The subjects are asked to relax and simply report what they see, and because this is an introspective experiment, typical results are displayed in a thought balloon.

At first, subjects report seeing the entire figure, but after a few moments the figures change. Habituation effects (perhaps electrochemical changes in the stimulated retinal neurons) cause entire parts of the figures to disappear or to fall out of perception. It is the manner in which perceptions of the figures alter that is of particular interest. Subjects reported that the component lines or angles (i.e., subassemblies) of a triangle and a square would jump in and out of perception all at once. These observations are as expected from Hebb's original formulation of the theory and the learning sequence for a triangle indicated in Figure 11.1; thus stabilized-image experiments confirm a prediction of the theory.

Learning Environments for Animals

According to Hebb's theory, adult thought processes involve continuous interactions among cell assemblies, which in turn are organized by sensory stimulation and internal interactions during the learning period of a young animal. How does adult behavior depend on opportunities for percept formation during development? Experiments show that rats reared in a rich perceptual environment—a "Coney Island for rats"—are notably more intelligent as adults than those raised in restricted environments, which provides yet another confirmation of the theory [64, 78]. As is anticipated from the cell-assembly theory, this positive influence of perceptual stimulation occurred only during youthful development; increased stimulation of adults is less effective in increasing rodent smarts.

Similar experiments with Scottish terriers showed even more striking differences, again as expected from the cell-assembly theory [89]. This is because the fraction of the neocortex that is not under the influence of sensory inputs—the *associative cortex*—is larger for a dog than a rat. Thus, the internal organization of the dog's brain should play a greater role in its behavior. Terriers reared in single cages, where they could not see or touch other dogs, had abnormal personalities and could neither be trained nor bred. Other studies showed that dogs reared in such restricted environments did not respond to pain, as if they were lobotomized [62].

Sensory Deprivation of Humans

In his original formulation of the cell-assembly theory [34], Hebb speculated that perceptual isolation would cause emotional problems because the phase sequence needs the guidance of meaningful sensory stimulation to remain organized in an intelligible manner. To test this aspect of the theory, experiments on perceptual isolation were performed by Heron and his colleagues in the 1950s [37, 64]. In these studies, the subjects were college students who were paid to do nothing. Each subject lay quietly on a comfortable bed wearing soft arm cuffs and translucent goggles, hearing only a constant buzzing sound for several days. During breaks for meals and the toilet, the subjects continued to wear their goggles, so they averaged about 22 hours a day in total isolation.

Many subjects took part in the experiment intending to plan future work or prepare for examinations. According to Hebb [35], the main results were that a subject's ability to solve problems in his or her head declined rapidly after the first day as it became increasingly difficult to maintain coherent thought, and for some it was difficult to daydream. After about the third day, hallucinations became increasingly complex. One student said that his mind seemed to be hovering over his body like a ball of cotton wool. Another reported that he seemed to have two bodies but did not know which was really his. Such observations are in accord with a variety of anecdotal reports from truck drivers, shipwreck survivors, solitary sailors, long-distance drivers, and the like that extended periods of monotony breed hallucinations. (Reporting on his famous solo flight across the Atlantic Ocean, for example, U.S. aviator Charles Lindbergh noted "vapor-like shapes crowding the fuselage, speaking with human voices, giving me advice and important messages" [50].)

After the perceptual isolation experiments were concluded, subjects experienced difficulties with visual perception lasting for several hours and were found to have a significant slowing of their electroencephalograms or brain waves. They also seemed more vulnerable to propaganda. Although the specific results of these experiments were not predicted by the cell-assembly theory, the disorganizing effect of sensory deprivation on coherent thought had been anticipated.

Structure of the Neocortex

While presenting a plausible theory for the dynamics of a brain, Hebb's classic book contains but one lapse into mathematical notation: he discusses in some detail the ratio

$$\frac{A}{S} \equiv \frac{\text{total association cortex}}{\text{total sensory cortex}}$$

for various mammalian species [34]. This ratio relates the area of the neocortex that is not directly tied to sensory inputs—the *associative* (A) regions—to the area of the *sensory* (S) regions, which are under direct environmental control from eyes, ears, and senses of touch and smell. If this ratio is zero, all of the cortex is under sensory control, and necessary conditions for behaviorist psychology are satisfied. On the other hand, larger values of the ratio imply increasing opportunities for the cortex to construct abstract cell assemblies with dynamics beyond direct control of the senses.

In general, Hebb pointed out, this A/S ratio increases as one moves through mammalian species from rat to dog to primate to human, in general agreement with two aspects of brains' behaviors. First, as most would agree, the character of a human's inner life is significantly more intricate than that of a chimp, which in turn is more than for a dog or a rat. Second, the time required for *primary learning* (until adulthood is reached) increases with the A/S ratio. Human infants are essentially helpless and remain so for several years as they slowly build the myriad assemblies upon which the complexities of their lives will eventually be based.

11.3 Elementary Assembly Dynamics

In this section, some simple models of cell-assembly dynamics are presented that describe the average behavior of a relatively large number of interacting model neurons. Because these descriptions are restricted to very simple representations of the neurons—little like the more realistic picture that was developed in Chapter 9—they should be viewed as indicating lower bounds on the possible behaviors of real neural systems. The generalization of such analyses to more realistic neural models is a challenge for current neuroscience research, and some such attempts are described in Section 11.5.

11.3.1 Ignition of an Assembly

To model the dynamics of an individual neural assembly as it turns on (ignites) or turns off (becomes extinguished), we can imagine a large mass of randomly connected McCulloch–Pitts (M–P) neurons as described by Equation (2.10), a problem that goes back to the 1950s [3, 26, 28, 79, 86, 87, 90]. In developing a simple formulation, it is convenient to make the following assumptions and definitions of additional variables.

- Time (t) is defined on a discrete lattice, with the duration of each interval equal to the *synaptic delay* τ.

- $F(t)$ represents the fraction of neurons that are firing at time t.

- I is the number of input connections to each neuron. These are received randomly from outputs of other neurons in the assembly.

- The refractory times of the neurons are shorter than the synaptic delay.

With these definitions, we can write the probability of a neuron receiving exactly j input signals at time t as

$$\left(\frac{I!}{j!(I-j)!} \right) F^j (1-F)^{I-j} ,$$

an expression that can be understood as follows.[1]

(1) $I!/j!(I-j)!$ is the number of different ways that j input signals can be selected from among I input channels.

(2) F^j is the probability of having signals appear on j of the input channels.

---

[1]The alert reader will recall that we met the same expression in Equation (2.5) of Chapter 2 describing the probability for k synaptic vesicles to release their transmitter substance through n presynaptic sites.

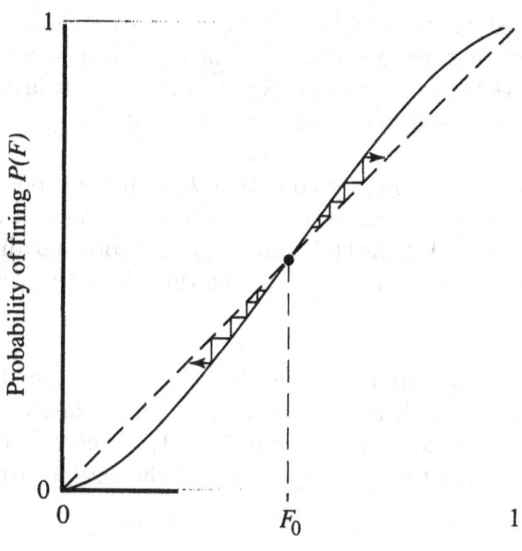

Figure 11.4. Qualitative behavior of the probability of a neuron firing in the next time increment $P(F)$ as a function of F, the current firing rate, assuming that $1 < \theta < I$.

(3) $(1 - F)^{I-j}$ is the probability of not having signals on the other $I - j$ input channels.

Because the M–P model neuron gives an output signal when its inputs are equal to or greater than the threshold θ, the probability of a neuron firing in the next increment of time is given by the summation

$$P(F) = \sum_{j=\theta}^{I} \left(\frac{I!}{j!(I-j)!} \right) F^j (1 - F)^{I-j} . \qquad (11.1)$$

Although this expression appears unwieldy, its qualitative behavior is straightforward; thus for

$$1 < \theta < I ,$$

$P(F)$ is the sigmoid function of F sketched in Figure 11.4.[2]

The condition

$$P(F) = F , \qquad (11.2)$$

---

[2]To see this, note that $P(F) \sim B(I, \theta)F^\theta$ near $F = 0$, where $B(I, \theta) \equiv I!/\theta!(I - \theta)!$ is a *binomial coefficient*. Similarly $P(F) \sim 1 - B(I, \theta - 1)(1 - F)^{I-\theta+1}$ near $F = 1$. Because direct calculation shows that $P(F)$ is a monotone increasing function, it must have the shape indicated in Figure 11.4.

which is satisfied for three values of F, indicates *stationary solutions* of the system because these are the values of F for which the probability of firing in the next time increment is equal to the present firing rate. Let us consider these three stationary solutions in detail.

1. The minimum stationary condition $F = 0$ corresponds to none of the neurons firing. This is a stable solution because if F is increased slightly from 0, Figure 11.4 shows that the corresponding increase in $P(F)$ is less than that of F, implying that the activity will relax back to zero.

2. The maximum stationary condition $F = 1$ corresponds to all of the neurons firing at their maximum rates. This is also a stable solution because if F is decreased slightly from 1, the corresponding value of $P(F)$ is greater than F, implying that the activity will rise back to one.

3. The stationary condition at $F = F_0$ corresponds to an intermediate firing rate, where F_0 increases from 0 to 1 as θ increases from 1 to I. In contrast to $F = 0$ and $F = 1$, this intermediate stationary level is *unstable*. To see this, note from Figure 11.4 that if F is increased slightly above F_0, the increase in $P(F)$ is greater than that of F, causing F to rise even more in the subsequent time increment. If F is decreased slightly below F_0, on the other hand, the decrease of $P(F)$ is more than that of F, causing F to fall even more in the subsequent time increment.

In the context of nonlinear system theory, therefore, a cell assembly shares properties of the Hodgkin–Huxley nerve impulse that were discussed in Section 4.6. Thus the stationary state at $F = 1$ can be viewed as an *attractor*, as can the null state at $F = 0$. In these terms, the intermediate stationary state at $F = F_0$ defines a *separatrix* lying on the boundary between the basins of these two attractors.

In other words, cell-assembly activity emerges from a net of interconnected neurons, much as a nerve impulse emerges from the Hodgkin–Huxley equations for a squid axon. Both exhibit the interrelated properties of all-or-nothing response and threshold, providing a basis for the hierarchical structures of assemblies shown in Figure 11.1 and to be considered in the following chapter.

From the perspectives of Chapter 1, the ignition of an assembly can be represented by the following positive feedback diagram:

Firing rate: F

↓   ↑

Probability of firing: $P(F)$

Above the level of ignition ($F = F_0$), positive feedback causes $P(F)$ to grow faster than F, so activity increases until the stable stationary state at $F = 1$ is reached. What is the time course of this growth?

Because the function $P(F)$ indicates the level of activity at time $t + \tau$, it was noted in the previous chapter that the discrete formulation of the dynamics is roughly equivalent to the ordinary differential equation

$$\frac{dF}{dt} = \frac{P(F) - F}{\tau},$$ (11.3)

where t is now considered to be a continuous variable.[3] For $F_0 < F < 1$, it is evident from Figure 11.4 that the right-hand side of this ODE has the same qualitative features as the right-hand side of Equation (1.3), which was used to derive the Verhulst curve for population growth shown in Figure 1.3.

Thus, $F(t)$—the dependence of the firing rate on time during assembly ignition—is given implicitly by the integral relation

$$\int_{F_{\text{init}}}^{F(t)} \frac{dF'}{P(F') - F'} = \frac{t}{\tau}.$$ (11.4)

Here, $F_{\text{init}} > F_0$ is the initial value of F at $t = 0$, which may have been established by inputs from other assemblies, external sensory inputs, or some combination of the two. (Although one actually calculates t as a function of F, it can be seen from Figure 11.4 that $F(t) \to 1$ as $t \to \infty$.)

To model its qualitative features, Equation (11.3) can be written as

$$\frac{dF}{dt} \approx -\frac{1}{\tau} F(F - F_0)(F - 1),$$ (11.5)

an ODE that is interesting to compare with the representation of a space-clamped patch of nerve membrane developed in Chapter 5. In that case, the reader will recall, transmembrane voltage obeys an ODE of the form

$$\frac{dV}{dt} = -\left(\frac{G}{C}\right)\left[\frac{V(V - V_1)(V - V_2)}{V_2(V_2 - V_1)}\right],$$ (11.6)

where C/G is an active time constant for the membrane, and a cubic approximation is used to model the transmembrane current that is plotted in Figure 5.1. Thus, we see a mathematical relationship between the switching of a patch of membrane and the switching of an assembly, although they are at quite different levels of description. This correspondence is of central importance for the perspectives being developed in this book and will be further discussed in the following chapter.

---

[3]Beware the analytic sleights of hand here. Time was assumed to be a discrete variable in order to derive an expression for $P(F)$ in Equation (11.1), and now it is redefined as a continuous variable in order to use that expression in an ODE.

Once an assembly has been ignited, Equation (11.5) indicates that it remains firing forever, but this overlooks habituation effects, inhibitory inputs from other assemblies, and external sensory inputs, all of which may reduce the firing rate and increase the ignition threshold F_0. (Similarly, Equation (11.6) neglects the recovery effects on a nerve fiber stemming from potassium ion current, which are treated in Chapter 6.) The time course of the extinction dynamics is again given more precisely by Equation (11.4), but now F_{init} is *less* than F_0 at $t = 0$, and it is seen from Figure 11.4 that $F(t) \to 0$ as $t \to \infty$.

This analytic formulation is tidy, but can we believe it? Should real nerve networks be expected to behave at all like the variables in these equations? Because the candid answer is that I do not know, it seems appropriate to underscore some areas of present concern with the hope that they will be selected for further study.

First, I repeat that we do not yet know how to accurately model a single nerve cell, thus the McCulloch–Pitts representation may miss essential neural properties. In particular, the preceding formulation reduces the communication among neurons to passing information about their average firing rates, an assumption that overlooks important aspects of neural dynamics. Perhaps real neurons talk to each other in languages that are based on time codes, space codes, or some subtle combinations thereof. Perhaps they use chemical or ephaptic interactions as a sort of body language. Over longer distances, cell assemblies might communicate via the information waves that were considered in the previous chapter. Finally, it could be that assemblies engage in activities beyond our present ken.

However assemblies interact, an important aspect of neural behavior that has been neglected in the preceding analysis is the fact that synaptic influences can be inhibitory as well as excitatory. We will see in the following section that inhibition plays a key role in determining the ways in which two or more cell assemblies behave.

11.3.2 Inhibition among Assemblies

At the time of Hebb's original formulation of the cell-assembly theory, there was no experimental evidence for inhibition among cortical neurons, so he conservatively assumed only excitatory interactions. By 1957, however, cortical inhibition had been observed, so Peter Milner, a colleague of Hebb's at McGill University, developed a "Mark II" version of the theory [63]. The most striking feature of this revised theory is that it allows independent assemblies to develop from an undifferentiated mass of model neurons.

To evaluate the effect that synaptic inhibition among cortical neurons might have on cell-assembly dynamics, it is convenient to represent the behavior of an individual assembly as simply as possible. To this end, let us set $\theta = 1$ in Equation (11.1), whereupon $P(F) = 1 - (1 - F)^I$. For $I = 2$

(two inputs for each neuron), this expression becomes

$$P(F) = 2F - F^2,$$

with the same qualitative behavior for larger values of I.

Under these simplifying assumptions ($\theta = 1$, $I = 2$), Equation (11.3) reduces to

$$\frac{dF}{dt} = F(1 - F),$$

where time is measured in units of the synaptic delay (τ). This is just the Verhulst equation with solution

$$F(t) = \frac{F(0)e^t}{1 + F(0)\,(e^t - 1)},$$

which follows from integration of Equation (11.4) and is displayed in Figure 1.3 for several initial values. The same growth equation describes both the firing rate of a cell assembly and the population of Belgium. Again, we find that identical mathematical formulations are useful at widely different levels of description.

Thus motivated, let us model the dynamics of two identical neural assemblies with inhibitory interactions by the coupled ODE system

$$\frac{dF_1}{dt} = F_1(1 - F_1) - \alpha F_2,$$

$$\frac{dF_2}{dt} = F_2(1 - F_2) - \alpha F_1,$$

(11.7)

where $0 \leq F_1 \leq 1$ and $0 \leq F_2 \leq 1$ because F_1 and F_2 represent the fraction of neurons in each assembly that are firing. When positive, the parameter α introduces an inhibitory interaction between the two assemblies because the $-\alpha F_2$ term in the first equation reduces dF_1/dt and similarly for the second equation.

To see how these equations model the role that inhibition plays in the formation of cell assemblies, let us recall a bit of history. As digital computers became available for scientific problems in the mid-1950s, Frankel reviewed several approaches to the numerical studies of brains, concluding that Hebb's cell-assembly theory was the most promising [17]. Rochester et al. [82] then began to study the growth of cell assemblies in a group of 99 McCulloch–Pitts style model neurons, allowing only excitatory interactions as had originally been proposed by Hebb [34]. Although they found a diffuse reverberation with a period on the order of the synaptic delay, assemblies did not develop.

This disappointing result follows directly from Equations (11.7). How? If we let α be negative, only excitatory interactions among the neurons are allowed. In this case, as is seen from Figure 11.5(a), all points on the (F_1, F_2)

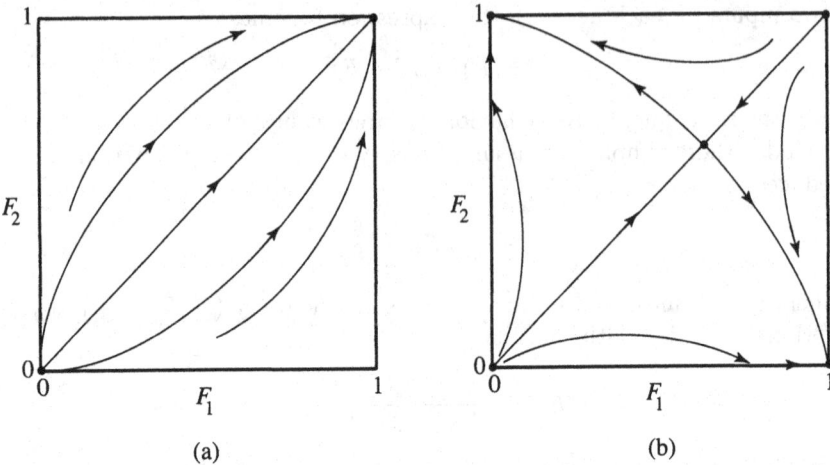

Figure 11.5. (a) A phase-plane plot from Equations (11.7) with $\alpha < 0$ (only excitatory interactions). (b) A similar plot for $\alpha > 1/3$ (excitatory and inhibitory interactions).

phase plane move to $(1,1)$, and no individual assemblies are permitted to ignite. In other words, all neurons end up firing at their maximum rates.

Rochester et al. then talked with Milner, who was revising Hebb's theory to include inhibition [63]. Thus inspired, they modified their computer model to include the growth of both excitatory and inhibitory interactions among 512 M–P neurons, with six neurons being externally driven [82]. Cell assemblies were then observed to form with excitatory interactions developing among cells in the same assembly and inhibitory interactions among different assemblies. How can this be seen in the context of our model?

Upon introducing inhibition in Equations (11.7) by making $\alpha > 0$, one finds a singular point at

$$F_1 = F_2 = 1 - \alpha,$$

where the time derivatives are zero. For $0 < \alpha < 1/3$, this singular point is stable, but for $\alpha > 1/3$, it becomes *unstable,* as shown in Figure 11.5(b). Stable states of the system are then at either

$$(F_1, F_2) = (1, 0) \ \text{ or } \ (0, 1).$$

Thus, with sufficiently large inhibition, Equations (11.7) suggest that assemblies can be individually ignited in accord with both the numerical observations of Rochester et al. [82] and the theoretical considerations of Milner's "Mark II" cell-assembly theory [28, 63].

At this point in the discussion, you should revisit Figure 11.2 and experience how your perception switches back and forth between the two orientations of the Necker cube. Although it is easy to see the cube in either orientation, note that you cannot perceive both orientations at the same time. (How rapidly can you switch between perceptions of the two orientations? Might the speed of these transitions be taken as a measure of how well your brain is working?)

Now, consider Equations (11.7) with $\alpha > 1/3$ and the corresponding phase-plane diagram shown in Figure 11.5(b). Evidently, these equations model the switching on and off of assemblies that correspond to the dynamics of those in your head as you regard the Necker cube.

From an engineering perspective, the interactive dynamics of two assemblies are like a "flip-flop" circuit widely used in the design of information storage and processing systems [27]. With a cell assembly, however, the bit of information being switched on or off is not the voltage level of a transistor but an intricate psychological perception embodied in the connections among thousands of neurons scattered about the brain that have developed in response to the lifelong experiences of the organism. Although this has been a "bottom-up" discussion of the brain's dynamics, it suggests the utility of "top-down" approaches. Regarding assembly firing rates as *order parameters* for higher level representations of the brain's dynamics, for example, Haken and his colleagues have been able to model a variety of psychological experiments [29, 30, 31].

To represent more than two assemblies, Equations (11.7) can be generalized to

$$\frac{dF_1}{dt} = +F_1(1 - F_1) - \alpha F_2 - \alpha F_3 - \cdots - \alpha F_n,$$
$$\frac{dF_2}{dt} = -\alpha F_1 + F_2(1 - F_2) - \alpha F_3 - \cdots - \alpha F_n, \qquad (11.8)$$
$$\cdots$$
$$\frac{dF_n}{dt} = -\alpha F_1 - \alpha F_2 - \alpha F_3 - \cdots + F_n(1 - F_n),$$

where $0 \leq F_j \leq 1$ for $j = 1, 2, \ldots, n$. In this n-assembly model, interestingly, all of the previous analysis (for $n = 2$) can be carried through. Thus, there is a singular point for positive α (the inhibitory case) at

$$F_1 = F_2 = \cdots = F_n = 1 - (n - 1)\alpha,$$

which is stable for

$$\alpha < \alpha_c = 1/(2n - 1)$$

and unstable for

$$\alpha > \alpha_c = 1/(2n - 1).$$

Below this critical value of inhibition (α_c), all of the assemblies can become simultaneously active. It turns out that the switching time (τ_{sw}) of this instability is

$$\tau_{sw} = \frac{1}{(2n - 1)\alpha - 1},$$

counterintuitively implying that the rate at which a neural system can change from one perception to another *increases* with inhibition (α). This result is in accord with Hebb's suggestion that we humans are more intelligent than our fellow mammals in part because we can switch our attention more quickly from one assembly to another [35, 36].

Another aspect of intelligence, however, is the total number of assemblies that can be remembered.

11.4   How Many Assemblies Can There Be?

Having considered some of the evidence for the existence of cell assemblies, it is interesting to ask how many of them can be stored in a human brain. This is a difficult question to answer because—as we have seen in Chapter 9—there is not yet a clear understanding of what the individual neurons are doing, but it is possible to make certain lower estimates. To this end, let us review three considerations.

First, it is presently necessary to use a McCulloch–Pitts style model in which each neuron is represented by a single switch. Evidently, conclusions based on this unrealistic assumption can provide only lower bounds on the possible number of assemblies.

Second, it is not correct to estimate the number of assemblies by dividing the number of neurons in the brain by the number of neurons in an assembly. Why not? Recall the social analog for cell assemblies, which was presented in Chapter 1. Just as a particular person in a city may be a member of more than one social assembly, so may a single neuron participate in several different cell assemblies.

Finally, any estimate of the maximum number of assemblies should account for the fact that the brain is hierarchically structured. Thus, complex assemblies comprise simpler assemblies, which in turn are composed of yet simpler ones, and so on.

In an important paper that appeared in the mid-1960s, Charles Legéndy assessed human brain capacity from a simple model [47]. Although the basic structure of his work is presented here, additional statistical details are in the original publications [48, 49].

To introduce hierarchical character, Legéndy assumed that the brain is already organized into subassemblies and modeled their organization into assemblies. In the spirit of Hebb's theory, an assembly and one of its subassemblies variously represent

> a setting and a person who is part of it, a word and one of its letters, an object and one of its details.

To avoid complications of spatial organization, interconnections among assemblies are taken to be evenly distributed over the brain. (Following a familiar caricature of a mathematician's approach to biology, this is the assumption of a "spherical brain".)

Like individual neurons, subassemblies and assemblies have excitation thresholds that must be exceeded for ignition. Whereas the threshold for a subassembly is assumed to be a certain number of active neurons, the threshold for an assembly is a certain number of active subassemblies. Legéndy considered the subassemblies to be already formed by *weak* contacts, whereas assemblies develop from subassemblies through the development of *latent* into *strong* contacts among neurons.

To proceed further, let us introduce the following notation.

- N is the number of neurons in the brain.

- A is the maximum number of assemblies that can form in the brain.

- n is the number of neurons in a subassembly.

- y is the number of subassemblies in an assembly.

- a is the number of strong (latent) contacts per neuron.

- m is the maximum number of strong contacts from an assembly to one of its subassemblies.

Assuming that half of the strong (latent) contacts make output (axonal) connections and the other half make input (dendritic) connections, the number of output contacts from an assembly is $nya/2$. Those outputs connecting to a particular subassembly reach a fraction n/N of the neurons in the brain; thus

$$m = \frac{n^2 ya}{2N}.$$

The maximum number of assemblies are stored in the model when about half of the latent connections have been converted into strong contacts. Why half? Think of a black and white photograph. If all of the pixels are all white or all black, the image conveys very little information. It is when about half of the pixels are black and the others are white that the most information is being stored, and so it is with the conversion from latent to strong contacts. Thus

$$A \sim \frac{Na}{2my}.$$

In ordinary English, this equation says that the maximum number of assemblies in the brain is given by half of the total number of strong (latent) connections in the brain ($Na/2$) divided by the number of strong (latent) contacts in a single assembly (my).

Combining the previous two equations yields an estimate for the maximum number of assemblies that can be stored in the brain:

$$A \sim \left(\frac{N}{ny}\right)^2. \tag{11.9}$$

Table 11.1. The number of cell assemblies (A) in a brain versus the number of neurons in the brain (N) and the number of neurons in an assembly (ny). These values are estimated from Equation (11.9).

	$N = 10^{10}$	$N = 10^{11}$
$ny = 10^3$:	10^{14}	10^{16}
$ny = 10^4$:	10^{12}	10^{14}
$ny = 10^5$:	10^{10}	10^{12}

Some values of the maximum number of assemblies (A) implied by this estimate for different values of the number of neurons in a brain (N) and in a subassembly (n) are given in Table 11.1. Because the number of neurons in the brain is variously estimated as from ten to a hundred billion [9, 12, 39], these two values are selected in the upper row of the table. The values for ny are not empirically established and are expected to vary widely according to the intricacy of the concept perceived. (Palm has suggested that "a total assembly should have somewhere around 10^4 neurons with a working range from a few thousand to several tens of thousands" [73].) Lower values for ny would increase estimates of the number of assemblies that can be stored in a brain.

From these approximate values, it appears that

$$A > 10^9$$

is a comfortable lower bound on the maximum number of assemblies stored in the human brain. Equal to the number of seconds in 30 years, 10^9 is also in accord with estimates by Griffith based on the rate at which the brain is able to absorb information [28].

Finally, it is interesting to compare Equation (11.9) with the maximum number of patterns (p_m) that can be stored in an attractor network from Equation (10.9) of Section 10.2.2. Although 10^9 is again a rough lower bound on the number of attractors that emerge for a brain comprising 10^{10} to 10^{11} neurons, the bases for these two estimates differ; in particular, $A \propto N^2$, whereas $p_m \propto N$.

An explanation for this difference is that under the analysis of Section 10.2.2, every neuron is assumed to be firing 50% of the time. Thus, p codes of length N were found to introduce noise of amplitude $\sqrt{p/N}$ into the retrieval task, which limits the number of stored codes to $O(N)$. Under Legéndy's analysis, on the other hand, a particular neuron fires only when assemblies in which it participates are ignited, which leads to smaller average firing rates in closer accord with empirical observations or cortical activity.

11.5 Cell Assemblies and Associative Networks

As most have now seen, holograms use a well-defined reference beam (usually a laser source) to translate information from a distributed memory (the hologram) into a family of three-dimensional images. A small piece of the hologram is able to reproduce the entire image, albeit with reduced resolution. Inspired by the realization of holographic memories in laser laboratories of the 1960s, it was suggested that similar nonlocal storage principles might apply to memory in the neocortex [6, 22, 23, 52, 75].

Because several requirements of a holographic memory are not satisfied in the neocortex (e.g., well defined reference beam, stable wave medium), a memory principle was sought that would capture the distributed features of holographic storage in a realistic neural context [53, 95]. Thus, it emerged that the neocortex might operate as an *associative memory* [40, 53, 95].

The basic element of an associative memory is a connection matrix relating two sets of patterns. Feeding a portion of one pattern into the matrix and introducing threshold discrimination often allows aspects of the corresponding pattern to be recovered. Such a system can be useful for a variety of information-processing tasks, including feature extraction, pattern reconstruction, pattern identification, and sequential association [42].

To make contact with the previous section, think of the connection matrix as $N \times N$, with elements indicating interconnection ("synaptic") strengths between N neocortical neurons, and take the fraction of nonzero elements in the patterns to be $O(ny/N)$. Then, the fraction of matrix elements (or synapses) used up in the learning of a pattern pair is $O[(ny/N)^2]$. Because the number of unactivated synapses after the learning of r random code pairs will be of the order

$$\left[1 - \left(\frac{ny}{N}\right)^2\right]^r \approx 1 - r\left(\frac{ny}{N}\right)^2,$$

the maximum number of pattern pairs that can be learned is $O[(N/ny)^2]$ [43, 69].

It was recognized in the 1970s that Hebb's brain model can be regarded as an *autoassociative memory*, where the paired patterns can be the same [69, 93]. To see this, turn back to Figure 11.4—which shows the dynamics of a single assembly—and consider what the network is doing as the firing rate (F) increases from its threshold value of F_0. Outputs from a fraction F of the neurons are fed back as inputs to all neurons of the assembly, further increasing the firing fraction until the entire assembly is firing ($F = 1$). In this manner, it may be said, an ignited assembly has recognized itself.

Noting that the maximum number of assemblies (A) is equal to the maximum number of pattern pairs that can be related by the synaptic matrix $(N/ny)^2$, Legéndy's Equation (11.9) is confirmed.

Since the 1970s, the relationship between autoassociative memories (or *associative networks*, as they are coming to be called) and Hebb's cell assemblies has been an increasingly active area of neuroscience research, which comprises mathematical [69, 70, 72], neurological [7, 8, 9, 59, 73], and numerical components [13, 16, 41, 42, 69, 74, 92, 94].

Although much of this work supports the idea that Hebb's cell assemblies "provide an intermediate description of the brain between the psychological and the electrophysiological level" (as Günther Palm, a leader in associative nets research, has put it [71]), further tests of the theory depend on more realistic neural models.

11.6   More Realistic Assembly Models

As we have learned in Chapter 9, the dynamic behavior of a real neuron is far more intricate than that of an M–P model; thus, the "elementary assembly dynamics" formulated in Section 11.3 are suspect. To move in the direction of more realistic models of cell-assembly dynamics, descriptions of the basic neurons must be improved.

An early attempt in this direction modeled the basic units on *motor neurons* (MN), with disappointing results [55]. Interconnected populations of model neurons showed little tendency for activity to continue after their initiating inputs were turned off, at variance with Hebb's original concept of an assembly "acting briefly as a closed system." Since the 1980s, however, more biologically based models of assembly dynamics have been studied by investigators at the Royal Institute of Technology in Stockholm, leading to positive results.

The research group (called Studies of Artificial Neural Systems, or SANS, in the Department of Numerical Analysis and Computing Science, with a web site at www.nada.kth.se/sans) stems from the doctoral research of Anders Lansner, which was published in 1986 [42]. From the beginning, this work concentrated on the development of flexible models that could be incorporated into system studies with nuanced tradeoffs between neural realism and the numerical demands of large networks.

Currently, the best introduction to this effort is the doctoral thesis published in 1996 by Erik Fransén under the direction of Lansner [18]. A key feature of these investigations was the assumption of excitatory neurons based on cortical *pyramidal* (P) cells, with Hodgkin–Huxley style parameters differing from those of MN-cells as follows [44]:

- Less negative resting potential (-50 mV rather than -70 mV).

- Larger "depolarizing after potential" and smaller "after hyperpolarization." (In Section 4.7, these effects are referred to as "enhancement zones" and "refractory zones," respectively.)

- Smaller repolarizing voltage (V_K in Table 4.1).

- Spikes of smaller amplitude and duration.

In addition to either P-cells or MN-cells, the numerical representation included inhibitory fast-spiking (FS) cells, which are modeled after cortical interneurons. The earliest simulations consisted of 50 pairs of an excitatory cell and an FS-cell with 408 excitatory synapses to excitatory cells, 1538 excitatory synapses to inhibitory interneurons, and 50 inhibitory synapses to excitatory cells.

The numerical model was a neural simulator called SWIM, which is based on biologically plausible compartmental models for the neurons and synapses [16]. To reduce the overall computational task, excitatory neurons (MN-cells or P-cells) comprised four compartments each, whereas the inhibitory interneurons (via FS-cells) had only two.

The system of 50 cell pairs was taught eight different assemblies consisting of eight cells each. Thus some of the excitatory cells (MN or P) were necessarily members of more than one assembly. As is suggested by the analysis of Section 11.3.2, interconnections among cells of the same assembly were excitatory, whereas inhibitory interconnections (FS-cells) were established between different assemblies.

Differences between the behaviors of MN-cells and P-cells in such studies of Hebb's cell-assembly theory indicate that details of neural modeling are qualitatively important. The salient results are now discussed.

After Activity and Reaction Time

As Hebb assumed and the simple analysis of Figure 11.4 suggests, a cell assembly is expected to maintain its activity for a significant period of time after the stimulation is turned off. In studies with P-cells, such *after activity* was typically observed, with assemblies remaining active for periods of 350 to 400 ms after the termination of a 40-ms-long stimulation. Using MN-cells, on the other hand, after activity occurred only in exceptional cases, in accord with the previous work of MacGregor and McMullen [55]. During sustained firing of the P-cells, the frequency gradually decreased due to buildup of internal concentrations of Ca^{++} ions, which activate a Ca-dependent hyperpolarizing K^+ current. This is analogous to Hebb's cellular "partial exhaustion," and it eventually leads to extinction of the assembly activity.

Interestingly, these numerical experiments showed very short *reaction times* of about 50–70 ms, implying that each P-cell fires only about five times before an ignited assembly becomes fully active. This numerical observation blunts criticisms of the cell-assembly theory based on the suggestion that the turn-on process (indicated by the up-going arrows in Figure 11.4) might be significantly longer than typical perceptual response times.

Ignition Threshold and Pattern Completion

A critical firing level, denoted as F_0 in Figure 11.4, was readily observed numerically. Typically, the stimulation of three cells was sufficient to ignite an assembly, whereas two was not, implying an *ignition threshold* in the range

$$\frac{1}{4} < F_0 < \frac{3}{8}.$$

The existence of a threshold for ignition is closely related to the phenomenon of *pattern completion*, under which stimulation of only part of a pattern is required for a correct response. In connection with the discussions of Figure 11.1, for example, it was noted that one need not examine every aspect of a geometric figure before its global form is perceived.

Competition and Noise Suppression

The strong lateral inhibition among neurons of different assemblies—parameterized by α in Equations (11.8)—implies that two or more mutually active assemblies will compete for dominance. Such competition (subjectively perceived for the Necker cube shown in Figure 11.2) was readily observed in Lansner and Fransén's numerical studies, with the winning assembly both activating its missing members and suppressing spurious activity (noise) of other cells [44].

Influence of Time Delay

Introduction of variable *time delays* in the firing of excitatory cells mimics the propagation of signals over extended axonal pathways of varying lengths. With P-cells, it was found that such axonal delays could be increased up to an average value of about 10 ms without significant changes in assembly behavior [19]. How far apart does this delay allow neurons of an assembly to be located?

Assuming that long cortical axons are myelinated (see Chapter 7) and of the order of a micron or more in diameter [80, 88], Equation (7.19) suggests an outside fiber diameter of at least 1.5 μm. The data on myelinated nerves of the cat summarized in Figure 7.3 then imply impulse speeds of more than 0.84 cm/ms. During 10 ms of axonal delay, therefore, a spike can travel at least 8.4 cm, or 3.3 inches, which is about the average distance between two randomly selected neurons in the human cortex. Thus, the SANS model seems to permit the extension of Hebb's "three-dimensional fishnet" throughout most of the brain.

Slow Firing Rates and the Role of Inhibition

One discrepancy between the foregoing results and the behavior of real brains involves the maximum firing rate of an ignited assembly. As is suggested by Figure 11.4 and observed numerically, neurons in an active assembly are expected to fire at their maximum rates, which can be as high as 300 Hz (every 3.3 ms) for typical pyramidal cells. Under normal physiological conditions, however, cortical neurons are observed to fire at about 20–60 Hz but seldom higher.

In response to this objection to Hebb's theory, Fransén and Lansner show that reduced firing rates are observed for fully active assemblies when synapses are assumed to be realistically slow and also *saturating*, implying an upper limit on its peak conductance [20]. From the discussion in Section 2.3.1, saturation of postsynaptic membrane conductance is a reasonable constraint because the density of channels in this membrane is limited [2].

During these simulations, inhibitory neurons were not included because only maximum firing rates were under investigation. Thus, the authors concluded [20]:

> Cortical inhibition may not be as critically involved in regulating firing rates of individual cells and producing oscillatory activity as has often previously been assumed. From the perspective of the cell-assembly theory, the role for inhibition in preventing spread of activity among overlapping assemblies and in the shaping of cellular response properties could be emphasized. In fact, in the neocortex a reduction of the inhibition by only 30% (Lindström, personal communication, 1994) leads to epileptiform seizures. This may be an example of activity spreading uncontrollably when inhibition no longer separates the partly overlapping assemblies.

These remarks are in accord with the preceding analysis of the system described by Equation (11.8), where a reduction in the inhibiting parameter (α) below a critical value (α_c) allows all of the assemblies to become simultaneously active. They are also relevant to an evaluation of the field theory models of epilepsy discussed in Section 10.3.

Modeling of Cortical Columns

In a more recent study, Fransén and Lansner have extended their numerical simulations to include *columns* of cells, which corresponds more closely to the structure of the neocortex [21]. In this model, there are 50 functional units (columns) comprising 12 pyramidal neurons and 3 fast-spiking (FS) inhibitory interneurons each (rather than individual cells) for a total of 750 neurons. Pyramidal cells were modeled with six compartments each and FS-cells with three for a total of 4050 compartments.

All of these properties (significant afteractivity, short reaction times, ignition thresholds, pattern completion, competition, and noise suppression) were observed in this more realistic context while rendering the neural interconnections more realistic. Thus, the interconnection probability between pyramidal cells from different columns is both sparse and asymmetric, as is observed in cortical tissue, whereas the interconnection between columns is symmetric, in closer correspondence with the attractor neural networks discussed in Section 10.2.2.

In response to certain qualitative objections raised by Malsburg [58], the numerical studies of Fransén and Lansner have established that Hebb's cell-assembly hypothesis is in approximate accord with both the elementary analysis of Section 11.3 and with present knowledge of cortical structure.

How has the theory fared in current electrophysiology laboratories?

11.7   Recent Evidence for Cell Assemblies

In the half century since Hebb's theory of cell assemblies was first proposed, the experimental techniques of electrophysiology have greatly improved. Classical methods have been refined and new techniques introduced, leading Nicolelis, Fanselow, and Ghazanfar to comment in 1997 [66]:

> What we are witnessing in modern neurophysiology is increasing empirical support for Hebb's views on the neural basis of behavior. While there is much more to be learned about the nature of distributed processing in the nervous system, it is safe to say that the observations made in the last 5 years are likely to change the focus of systems neuroscience from the single neuron to neural ensembles. Fundamental to this shift will be the development of powerful analytical tools that allow the characterization of encoding algorithms employed by distinct neural populations. Currently, this is an area of research that is rapidly evolving.

In assessing this optimistic perspective, it is important to remember that observing the dynamic behavior of a "three-dimensional fishnet" comprising several thousand neurons (each receiving several thousand synaptic inputs) and spread over much of the brain is a daunting task, yet not hopeless. Although there is presently no possibility of taking microelectrode readings from most of the neurons in an assembly, records from as few as two may offer interesting opportunities for research because the experimenter can ask whether the recorded voltages are *correlated* and observe how the degree of correlation depends upon the global behavior of the organism.

Suppose that voltages $V_1(t)$ and $V_2(t)$ are measured from two different neurons of a brain. To learn whether these signals are related to each other, one can compute their correlation as

$$C(\tau) = \int V_1(t + \tau) \times V_2(t) dt\,, \tag{11.10}$$

where the integration is over the greatest practical temporal range. If, for example, $V_1(t)$ is defined between 0 and T_1 and $V_2(t)$ is defined between 0 and T_2, with $T_1 \gg T_2$, then appropriate limits of integration would be from $t = 0$ to T_2. Thus, varying τ over the range

$$0 \leq \tau \leq T_1 - T_2$$

effectively slides V_1 past V_2.

With data sets of moderate length and currently available computing equipment, this is a straightforward numerical task that indicates how much alike the two signals look. If they are totally unrelated, $C(\tau)$ will be small and random, and one would judge the signals to be uncorrelated. If $C(\tau)$ exhibits some reproducible structure, the signals are partially correlated. Finally, a large peak at some value of τ suggests that part of V_1 looks much like a temporal translation of V_2.

To avoid unrealistic computing times in applying correlation analysis to neural data, it is important to choose discrete approximations of Equation (11.10) that place in evidence the features of interest. Thus an impulse train might be represented by a series of times $(t_1, t_2, \cdots, t_n)$ at which spikes are observed to occur. Dividing the time axis into B "bins" that are larger than the minimum interpulse intervals but much less than the total recording time, an impulse train can then be approximated as the B-dimensional vector $(v_1, v_2, \cdots, v_B)$, where v_j is the number of spikes appearing in the jth bin.

Two such vectors take the form

$$\mathbf{V_1} = (v_{11}, v_{12}, \cdots, v_{1b}, \cdots, v_{1B_1})$$
$$\mathbf{V_2} = (v_{21}, v_{22}, \cdots, v_{2b}, \cdots, v_{2B_2}),$$

where B_1 is not necessarily equal to B_2 because there may be a longer run of reliable data from one measurement than from another.

Assuming that $B_1 \gg B_2$, Equation (11.10) can be approximated as

$$C(\beta) = \sum_{b=1}^{B_2} v_{1(b+\beta)} \times v_{2b}. \tag{11.11}$$

Informally, this equation says to slide the longer vector ($\mathbf{V_1}$) along the shorter vector ($\mathbf{V_2}$) by β bins, where β is an integer lying within the range

$$0 \leq \beta \leq B_1 - B_2,$$

multiply the number of spikes in the (B_2) overlapping bins, and add the (B_2) products. A large peak of $C(\beta)$ at some value of β suggests that part of $\mathbf{V_1}$ looks much like a temporal translation of $\mathbf{V_2}$.

As noted in Chapter 1, it is now feasible to measure voltages from several dozen microelectrodes while the subject is undergoing behavioral tests [14, 60, 61, 67, 68, 96]. Because each electrode may indicate the dynamics of several neurons, up to 100 or more individual signals can be simultaneously recorded, analyzed, and compared with concomitant behavior.

Groups of neurons producing correlated signals might be acting as members of a common cell assembly, a possibility that can be checked by comparing correlation functions with behavioral observations. In maze experiments on rats, for example, the experimenter might notice whether such correlated signals both turn on when a certain behavior begins and turn off when it ceases. If so, it would be reasonable to suspect that correlated neurons are acting as part of an assembly related to that behavior.

Using such techniques, multiple-electrode recordings from a variety of animal species tend to confirm the hypothesis that neurons act not "as single spies but in battalions." Although far from an exhaustive survey of this work, the examples that follow give the flavor of current activities. Many more such results are expected to appear in the next few years.[4]

Mollusk

Multineuronal optical studies of the abdominal ganglion of the mollusk (*Aplysia*) were carried out by Wu, Cohen, and Falk, who recorded from up to 30% of the 900 neurons involved and related this activity to global behavior (gill withdrawal reflex, respiratory pumping, and so on) [98]. Instead of finding neural circuitry developed for specific tasks, these researchers observed that

> different behaviors appear to be generated by altered activities of a single, large distributed network rather than by small dedicated circuits.

Locust

Laurent and colleagues have used several glass microelectrodes to record from projection neurons (PNs) in the antennal lobe (a structural and functional analog of the vertebrate olfactory bulb) of the locust (*Schistocerca americana*) [45, 91]. Focusing on 1 s bursts of stimulants to which the locust has been previously exposed, odor-specific oscillatory responses were observed at 20 Hz, which suggest that memories of different odors are encoded as stimulus-specific assemblies (or "ensembles") of coherently firing neurons [46]. However,

> each odor appears to be represented not simply by an ensemble of synchronized neurons but by a progressive and odor-specific transformation of that ensemble, so that each neuron synchronizes with several others only during one or more precise epochs of the ensemble response.

In a style of interdisciplinary research that one hopes to see more often in coming years, Bazhenov et al. have reported numerical modeling of Laurent's experiments on the locust olfactory system by a team of biologists, neuroscientists and applied mathematicians [4, 5]. Similar to the modeling of the SANS group (described previously in Section 11.5), the

---

[4]Although some question the ethics of the work, there have been several multiple-electrode experiments on monkeys that also draw conclusions that support the cell-assembly hypothesis [1, 10, 15, 24, 51, 83].

model comprised 90 PNs and 30 inhibitory local neurons (LNs) randomly interconnected with biologically realistic representations of the membrane conductances. Although only one compartment was included in each PN and LN, numerical simulations of about 0.5 s duration showed 20–30 Hz oscillations and pattern discrimination in accord with the foregoing biological measurements. Detailed analyses of the model dynamics led this group to speculate

> that a stimulus does not simply set the initial state of a fixed dynamical system, but instead that each stimulus creates a new and unique dynamical system. This dynamical system has a stimulus-specific global attractor that determines its spatiotemporal response patterns or trajectory.

How is this possible? The global nature of the oscillatory activity, it seems, is rather sensitive to the stimulus pattern delivered to the inhibitory LNs, with different LN patterns igniting different assemblies. Stimulations of the PNs by a certain odor are then translated into a specific PN oscillatory response pattern mediated by the particular assembly that has been ignited.

Moth

Evidence for neural assemblies in the antennal lobe of the moth (*Manduca sexta*) has also emerged from correlation studies of recordings on silicon microprobe arrays published by Christensen et al. [11]. To better represent the brevity of natural odors in the context of turbulent air flow, stimulating pulses were only 0.1 s in duration, yielding data implying that

> the patterns of synchrony among different members of an odor-encoding ensemble are not the same for different concentrations of the same odor. Furthermore, the responses to odor blends cannot necessarily be predicted from the responses to the individual odors in the blend. We therefore propose that ensembles of olfactory PNs must use multiple and overlapping coding strategies to process olfactory information, and that these strategies are matched to the particular circumstances surrounding odor presentation.

At variance with the results of Laurent et al. because of the shorter time scale (0.1 s vs. 1 s), these multiunit recordings again suggest the importance of cell-assembly codes in insect olfactory systems.

Rat

How does a rat get around in the dark? Much as you or I would use our fingers, this little fellow employs his whiskers to interrogate his surroundings. Thus, the cheek (or trigeminal) nerves are of particular interest in understanding a rat's perceptions. With this in mind, Nicolelis and his colleagues have been recording from up to 48 (and more recently up to 100) cortical, thalamic, and brainstem neurons of freely moving rats [25, 65, 67]. Widespread oscillations in the range from 7 to 12 Hz were observed, which began when the animals were still but alert and predicted the onset of whisker twitching. Starting as a traveling wave of activity in the cortex (see Figure 10.3), this action spread to the thalamus and the trigeminal brainstem complex. Correlation calculations between pairs of these signals indicate that

> the coding of sensory information in most cortical and subcortical relays of the trigeminal pathways occurs at the ensemble rather than at the single unit level and involves both spatial and temporal domains.

Deadwyler et al. have studied the relationships among recordings from ten different locations in the rat's hippocampus while the animal was undergoing a behavioral learning task [14]. Because the hippocampus is regarded as essential for storage and readout of cerebral information, these experiments were expected to shed light on the nature of neural dynamics during learning. From observations on seven animals, these authors show that

> ensemble encoding and retrieval of "functionally relevant" information are represented as distinct firing patterns in hippocampal networks.

If neuronal assemblies exist in the neocortex, one would hope to be able to switch them on and off—as suggested by Figure 11.5(b) and the dynamic properties of Equations (11.7)—and this is what Maldonado and Gerstein have managed by inserting ten tungsten microelectrodes into the rat's auditory cortex [57]. Both intracortical microstimulation (ICMS) through the electrodes and acoustic stimuli were used as probes during 15 different experiments. Based on correlation analyses of their data, these researchers conclude as follows.

> We have identified neuronal assemblies in two ways, defined through similarity of receptive field properties and defined through correlated firing. Close anatomical spacing between neurons was conducive to, but not sufficient for membership in, the same assembly with either definition. ICMS changed cortical organization by altering assembly membership. Our data showed that neuronal assemblies in the rat's auditory cortex can be established transiently in time and that their membership is dynamic.

Finally, it is interesting to note recent evidence in support of Hebb's phase sequence in which a series of assemblies are ignited one after another to comprise a train of thought [54].

To this end, Louie and Wilson used implanted multielectrodes to record from hippocampal CA1 pyramidal cells of rats (see Figure 9.1), which are known to be "place cells" that tend to fire when the animal is in a particular location [96]. The rats were trained to run around a circular track in search of food, and recordings were made during the actual awake activity (RUN) and also during shorter periods of "rapid eye movement sleep" (REM) [97].

Only those cells judged to be "active" (with firing rates greater than 0.2 Hz) were included in the analysis, leading to impulse train recordings from between 8 and 13 electrodes for a particular experiment. With bin sizes of 1 s and RUN recording times up to 4 minutes, the RUN-REM correlation was computed for each electrode as in Equation (11.11) and then averaged over the electrodes.

Such computations of RUN-REM correlation showed no similarity between the two measurements, but this fails to account for the possibility that the time scale of the REM signal could differ from that of awake activity (RUN). Stretching out (or slowing down) the REM data by a factor of about 2, on the other hand, gave sharply defined correlation peaks that could not be ascribed to happenstance. The authors claim that these results demonstrate that "long temporal sequences of patterned multineuronal activity suggestive of episodic memory traces are reactivated during REM sleep."

11.8 Recapitulation

This chapter opened with a survey of Donald Hebb's seminal formulation of the cell-assembly hypothesis for the robust storage and retrieval of information in the human brain and emphasized key aspects of the theory. Early evidence in support of Hebb's theory was reviewed, including the hierarchical nature of learning, perceptions of ambiguous figures, stabilized image experiments, sensory deprivation experiments, and anatomical data from the structure of the neocortex.

A simple mathematical model for interacting cell assemblies was then developed that describes ambiguous perceptions and suggests the importance of inhibitory interactions among cortical neurons for assembly formation and switching.

This model implies that cell assemblies emerge from intricate closed causal loops (subnetworks) of positive feedback threading sparsely through the neural system. Assemblies exhibit all-or-nothing response and threshold properties (just like the Hodgkin–Huxley impulse or an individual neuron); thus, an assembly is also an attractor. Interestingly, speed of switching from one assembly to another is found to increase with the level of interassembly inhibition. Under simple assumptions, a generous lower bound on the number of complex assemblies that can be stored in a human brain is estimated as about one thousand million—the number of seconds in 30 years.

Conclusions drawn from the simple analytic model are in accord with numerical studies on more realistic neural representations, which predict several hundred milliseconds of significant afteractivity (Hebb's "acting briefly as a closed system"), psychologically reasonable reaction times (less than 100 ms), and pattern recognition (or completion) from imperfect data.

Finally, the concept of correlation was defined and some experimental observations were cited that appear to confirm Hebb's cell-assembly theory in neuronal activities of a mollusk, locust, moth, and rat.

References

[1] M Abeles, H Bergman, I Gat, I Meilijson, E Seidemann, N Tishby, and E Vaadia, Cortical activity flips among quasi-stationary states, *Proc. Nat. Acad. Sci. USA* 92 (1995) 8616–8620.

[2] P Andersen, Factors influencing the efficiency of dendritic synapses on hippocampal pyramidal cells, *Neurosci. Res.* 3 (1986) 521–530.

[3] WR Ashby, H von Foerster, and CC Walker, Instability of pulse activity in a net with threshold, *Nature* 196 (1962) 561–562.

[4] M Bazhenov, M Stopfer, M Rabinovich, HDI Abarbanel, TJ Sejnowski, and G Laurent, Model of transient oscillatory synchronization in the locust antenna lobe, *Neuron* 30 (2001) 553–567.

[5] M Bazhenov, M Stopfer, M Rabinovich, HDI Abarbanel, TJ Sejnowski, and G Laurent, Model of cellular and network mechanisms for odor-evoked temporal patterning in the locust antennal lobe, *Neuron* 30 (2001) 569–581.

[6] A Borsellino and T Poggio, Holographic aspects of temporal memory and optomotor responses, *Kybernetik* 10 (1972) 58–60.

[7] V Braitenberg, Cell assemblies in the visual cortex. In *Theoretical Approaches to Complex Systems,* Springer-Verlag, Berlin, 1978.

[8] V Braitenberg and F Pulvermüller, Entwurf einer neurologischen Theorie der Sprache, *Naturwissenschaften* 79 (1992) 102–117.

[9] V Braitenberg and A Schüz, *Anatomy of the Cortex,* Springer-Verlag, Heidelberg, 1991.

[10] M Castelo-Branco, R Goebel, S Neuenschwander, and W Singer, Neural synchrony correlates with surface segregation rules, *Nature* 405 (2000) 685–689.

[11] TA Christensen, VM Pawlowski, H Lei, and JG Hildebrand, Multi-unit recordings reveal context-dependent modulation of synchrony in odor-specific neural ensembles, *Nat. Neurosci.* 3 (2000) 927–931.

[12] PS Churchland and TJ Sejnowski, *The Computational Brain,* MIT Press, Cambridge, MA, 1992.

[13] GJ Dalanoort, In search of the conditions for the genesis of cell assemblies: A study in self-organization, *J. Soc. Biol. Struct.* 5 (1982) 161–187.

[14] SA Deadwyler, T Bunn, and RE Hampson, Hippocampal ensemble activity during spatial delayed-nonmatch-to-sample performance in rats, *J. Neurosci.* 16 (1996) 354–372.

[15] G Deco, K Laskey, M Diamond, W Freiwald, and E Vaadia, Neural coding: Higher-order temporal patterns in the neurostatistics of cell assemblies, *Neural Comput.* 12 (2000) 2621–2653.

[16] Ö Ekeberg, P Wallén, A Lansner, H Travén, L Brodin, and S Grillner, A computer based model for realistic simulations of neural networks, I: The single neuron and synaptic interaction, *Biol. Cybern.* 65 (1991) 81–90.

[17] S Frankel, On the design of automata and the interpretation of cerebral behavior, *Psychometrika* 20 (1955) 149–162.

[18] E Fransén, Biophysical simulation of cortical associative memory, Doctoral thesis, Royal Institute of Technology, Stockholm, 1996.

[19] E Fransén, A Lansner, and H Liljenström, A model of cortical memory based on Hebbian cell assemblies. In *Computation and Neural Systems,* FH Eeckman and JM Bower (eds), Kluwer, Boston, 1993.

[20] E Fransén and A Lansner, Low spiking rates in a population of mutually exciting pyramidal cells, *Network* 6 (1995) 271–288.

[21] E Fransén and A Lansner, A model of cortical associative memory based on a horizontal network of connected columns, *Network* 9 (1998) 235–264.

[22] D Gabor, Holographic model of temporal recall, *Nature* 217 (1968) 584.

[23] D Gabor, Improved holographic model of temporal recall, *Nature* 217 (1968) 1288.

[24] AP Georgopoulos, AB Schwartz, and RE Kettner, Neuronal population coding of movement direction, *Science* 233 (1986) 1416–1419.

[25] AA Ghazanfar and MAL Nicolelis, Nonlinear processing of tactile information in the thalamocortical loop, *J. Neurophysiol.* 78 (1997) 506–510.

[26] JS Griffith, On the stability of brain-like structures, *Biophys. J.* 3 (1963) 299–308.

[27] JS Griffith, *A View of the Brain,* Oxford University Press, Oxford, 1967.

[28] JS Griffith, *Mathematical Neurobiology,* Academic Press, New York, 1971.

[29] H Haken, *Synergetics,* third edition, Springer-Verlag, Berlin, 1983.

[30] H Haken, *Advanced Synergetics,* Springer-Verlag, Berlin, 1983.

[31] H Haken, *Principles of Brain Functioning: A Synergetic Approach to Brain Activity,* Springer-Verlag, Berlin, 1996.

[32] DO Hebb, Intelligence in man after large removals of cerebral tissue: Report of four left frontal lobe cases, *J. Gen. Psychol.* 21 (1939) 73–87.

[33] DO Hebb, On the nature of fear, *Physiol. Rev.* 53 (1946) 259–276.

[34] DO Hebb, *Organization of Behavior: A Neuropsychological Theory,* John Wiley & Sons, New York, 1949.

[35] DO Hebb, The structure of thought. In *The Nature of Thought,* PW Jusczyk and RM Klein, (eds), Lawrence Erlbaum Associates, Hillsdale, NJ, 1980.

[36] DO Hebb, *Essay on Mind,* Lawrence Erlbaum Associates, Hillsdale, NJ, 1980.

[37] W Heron, The pathology of boredom, *Sci. Am.* January 1957.

[38] RM Klein, D.O. Hebb: An appreciation. In *The Nature of Thought*, PW Jusczyk and RM Klein, (eds), Lawrence Erlbaum Associates, Hillsdale, NJ, 1980.

[39] C Koch, *Biophysics of Computation*, Oxford University Press, New York, 1999.

[40] T Kohonen, *Associative Memory*, Springer-Verlag, Berlin, 1977.

[41] T Kohonen, P Lehtiö, J Rovamo, J Hyvärinen, K Bry, and L Vainio, A principle of neural associative memory, *Neuroscience* 2 (1977) 1065–1076.

[42] A Lansner, Investigations into the pattern processing capabilities of associative nets, Doctoral thesis, Royal Institute of Technology, Stockholm, 1986.

[43] A Lansner and Ö Ekeberg, Reliability and recall in an associative network, *Trans. IEEE Pattern Anal. Mach. Intell.* PAMI-7 (1985) 490–498.

[44] A Lansner and E Fransén, Modelling Hebbian cell assemblies comprised of cortical neurons, *Network* 3 (1992) 105–119.

[45] G Laurent, Dynamical representation of odors by oscillating and evolving neural assemblies, *Trends Neurosci.* 19 (1996) 489–496.

[46] G Laurent, M Wehr, and H Davidowitz, Temporal representations of odors in an olfactory network, *J. Neurosci.* 16 (1996) 3837–3847.

[47] CR Legéndy, On the scheme by which the human brain stores information, *Math. Biosci.* 1 (1967) 555–597.

[48] CR Legéndy, The brain and its information trapping device. In *Progress in Cybernetics* 1, J Rose (ed), Gordon and Breach, New York, 1969.

[49] CR Legéndy, Three principles of brain structure and function, *Int. J. Neurosci.* 6 (1975) 237–254.

[50] CA Lindbergh, *The Saturday Evening Post*, June 6, 1953.

[51] C Lee, WH Rohrer, and DL Sparks, Population coding of saccadic eye movements by neurons in the superior colliculus, *Nature* 332 (1988) 357–360.

[52] HC Longuet-Higgins, Holographic model of temporal recall, *Nature* 217 (1968) 104.

[53] HC Longuet-Higgins, DJ Willshaw, and OP Buneman, Theories of associative recall, *Q. Rev. Biophys.* 3 (1970) 223–244.

[54] K Louie and MA Wilson, Temporally structured replay of awake hippocampal ensemble activity during rapid eye movement sleep, *Neuron* 29 (2001) 145–156.

[55] R MacGregor and T McMullen, Computer simulation of diffusely connected neuronal populations, *Biol. Cybern.* 28 (1978) 121–127.

[56] M MacMillan, *An Odd Kind of Fame: Stories of Phineas Gage*, MIT Press, Cambridge, MA, 2000.

[57] PE Maldonado and GL Gerstein, Neuronal assembly dynamics in the rat auditory cortex during reorganization induced by intracortical microstimulation, *Exp. Brain Res.* 112 (1996) 431–441.

[58] C von der Malsburg, Synaptic plasticity as basis of brain organization. In *The Neural and Molecular Basis of Learning,* JP Changeux and M Konishi (eds) John Wiley & Sons, New York, 1987.

[59] D Marr, Simple memory, *Philos. Trans. R. Soc. London* 262 (1971) 23–82.

[60] EM Maynard, CT Nordhausen, and RA Normann, The Utah intracortical electrode array: A recording structure for potential brain-computer interfaces, *Electroencephalogr. Clin. Neurophysiol.* 102 (1997) 228–239.

[61] TJ McHugh, KI Blum, JZ Tsien, S Tonegawa, and MA Wilson, Impaired hippocampal representation of space in CA1-specific NMDAR1 knockout mice, *Cell* 87 (1996) 1339–1349.

[62] R Melzak and TH Scott, The effects of early experience on the response to pain, *J. Comp. Phys. Psychol.* 50 (1957) 155–161.

[63] PM Milner, The cell assembly: Mark II, *Psychol. Rev.* 64 (1957) 242–252.

[64] PM Milner, The mind and Donald O. Hebb, *Sci. Am.* January 1993, 124–129.

[65] MA Nicolelis, LA Baccala, RCS Lin, and JK Chapin, Sensorimotor encoding by synchronous neural ensemble activity at multiple levels of the somatosensory system, *Science* 268 (1995) 1353–1358.

[66] MAL Nicolelis, EE Fanselow, and AA Ghazanfar, Hebb's dream: The resurgence of cell assemblies, *Neuron* 19 (1997) 219–221.

[67] MAL Nicolelis, AA Ghazanfar, BM Faggin, S Votaw, and LMO Oliveira, Reconstructing the engram: Simultaneous, multisite, many single neuron recordings, *Neuron* 18 (1997) 529–537.

[68] CT Nordhausen, EM Maynard, and RA Normann, Single unit recording capabilities of a 100 microelectrode array, *Brain Res.* 726 (1996) 129–140.

[69] G Palm, On associative memory, *Biol. Cybern.* 36 (1980) 19–31.

[70] G Palm, On the storage capacity of an associative memory with randomly distributed storage elements, *Biol. Cybern.* 39 (1981) 125–127.

[71] G Palm, Toward a theory of cell assemblies, *Biol. Cybern.* 39 (1981) 181–194.

[72] G Palm, *Neural Assemblies: An Alternative Approach to Artificial Intelligence,* Springer-Verlag, Berlin, 1982.

[73] G Palm, Cell assemblies, coherence, and corticohippocampal interplay, *Hippocampus* 3 (1993) 219–226.

[74] G Palm and T Bonhoeffer, Parallel processing for associative and neural networks, *Biol. Cybern.* 51 (1984) 201–204.

[75] KH Pribram, The neurophysiology of remembering, *Sci. Am.* January 1969, 73–85.

[76] RM Pritchard, W Heron, and DO Hebb, Visual perception approached by the method of stabilized images, *Can. J. Psychol.* 14 (1960) 67–77.

[77] RM Pritchard, Stabilized images on the retina, *Sci. Am.* June 1961, 72–79.

[78] MS Rabinovitch and HE Rosvold, A closed field intelligence test for rats, *Can. J. Psychol.* 4 (1951) 122–128.

[79] A Rapoport, "Ignition" phenomena in random nets, *Bull. Math. Biophys.* 14 (1952) 35–44.

[80] JM Ritchie, On the relation between fiber diameter and conduction velocity in myelinated nerve fibres, *Proc. R. Soc. London* B217 (1982) 29–35.

[81] AH Riesen, The development of visual perception in man and chimpanzee, *Science* 106 (1947) 107–108.

[82] N Rochester, JH Holland, LH Haibt, and WL Duda, Tests on a cell assembly theory of the action of a brain using a large digital computer, *Trans. IRE Inf. Theory* IT-2 (1956) 80–93.

[83] E Seidemann, I Meilijson, M Abeles, H Bergman, and E Vaadia, Simultaneously recorded single units in the frontal cortex go through sequences of discrete and stable states in monkeys performing a delayed localization task, *J. Neurosci.* 16 (1996) 752–768.

[84] M von Senden, *Space and Sight: The Perception of Space and Shape in the Congenitally Blind Before and After Operation*, Methuen & Co., London, 1960 (a republication of *Raum-und Gestaltauffassung bei Operierten vor und nach der Operation*, Barth, Leipzig, 1932).

[85] CS Sherrington, *The Integrative Action of the Nervous System*, Yale University Press, New Haven, 1906.

[86] A Shimbel and A Rapoport, A statistical approach to the theory of the central nervous system, *Bull. Math. Biophys.* 10 (1948) 41–45.

[87] DR Smith and CH Davidson, Maintained activity in neural nets, *J. Assoc. Comput. Mach.* 9 (1962) 268–279.

[88] A Surkis, B Taylor, CS Peskin, and CS Leonard, Quantitative morphology of physiologically identified and intracellularly labeled neurons from the guinea-pig laterodorsal segmental nucleus *in vitro*, *Neuroscience* 74 (1996) 375–392.

[89] WR Thompson and W Heron, The effects of restricting experience on the problem-solving capacity of dogs, *Can. J. Psychol.* 8 (1954) 17–31.

[90] E Trucco, The smallest value of the axon density for which "ignition" can occur in a random net, *Bull. Math. Biophys.* 14 (1952) 365–374.

[91] M Wehr and G Laurent, Odor encoding by temporal sequences of firing in oscillating neural assemblies, *Nature* 384 (1996) 162–166.

[92] H Wigström, Associative recall and formation of stable modes of activity in neural network models, *J. Neurosci. Res.* 1 (1975) 287–313.

[93] G Willwacher, Fähigkeiten eines assoziativen Speichersystems im Vergleich zu Gehirnfunktion, *Biol. Cybern.* 24 (1976) 181–198.

[94] G Willwacher, Storage of a temporal pattern sequence in a network, *Biol. Cybern.* 43 (1982) 115–126.

[95] DJ Willshaw, OP Buneman, and HC Longuet-Higgins, Non-holographic associative memory, *Nature* 222 (1969) 960.

[96] MA Wilson and BL McNaughton, Dynamics of the hippocampal ensemble code for space, *Science* 261 (1993) 1055–1058.

[97] MA Wilson and BL McNaughton, Reactivation of hippocampal ensemble memories during sleep, *Science* 265 (1994) 676–679.

[98] JY Wu, LB Cohen, and CX Falk, Neuronal activity during different behaviors in *Aplysia*: A distributed organization? *Science* 263 (1994) 820–823.

12

The Hierarchical Nature of Brain Dynamics

In previous pages of this book, we have considered mathematical formulations at several levels of neuroscience, from the Newtonian dynamics of individual membrane proteins, through the switching of isolated patches of membrane and the interactions among propagating nerve impulses, to the intricate dynamics of cell assemblies, extending over much of the brain. The picture of the brain that arises from this survey is a *cognitive hierarchy* of distinct dynamic levels in which each level of description is built upon—or emerges from—those below. Because the brain's hierarchical structure is a matter of observation, little is debatable about the preceding statement, but the implications of this perspective for the social sciences are not yet fully appreciated.

Since the demise of behaviorism as a credible theory of the human brain, a variety of alternative formulations have been advanced and are currently the subject of intense discussions among neuroscientists, psychologists, philosophers, and humanists [31]. This final chapter briefly surveys aspects of these debates, paying particular attention to the claims of reductive materialism and closing with a few modest suggestions for future research on the brain's dynamics.

12.1 The Biological Hierarchy

Before taking up the cognitive hierarchy, let us fix ideas by considering a related structure, the *biological hierarchy,*

Biosphere
Species
Organisms
Organs
Cells
Processes of replication
Genetic transcription
Biochemical cycles
Biomolecules
Molecules

with respect to which several comments are in order.

First, it is only the general nature of this hierarchy that is of interest here, not the details. One might include fewer or more levels in the diagram or account for branchings into (say) flora and fauna or various phyla. Although such refinements may be useful in particular discussions, the present aim is to become acquainted with the general nature of a nonlinear dynamic hierarchy, so a relatively simple diagram is appropriate.

Second, the nonlinear dynamics at each level of description generate *emergent structures*, and nonlinear interactions among these structures provide a basis for the dynamics at the next higher level [32].

Third, as we have seen throughout this book, the emergence of a new dynamic entity stems from the presence of a closed causal loop, which leads to positive feedback and exponential growth that is ultimately limited by nonlinear effects (as in the Verhulst curves of Figure 1.3).

Finally, it should be noted that philosophers disagree about the ontological nature of emergent levels. Are they mere designations convenient for academic organization, or do they mark qualitatively different realms of reality? In attempting to answer this question, it is important to know whether the upper levels can be derived from lower levels, which brings us to a consideration of *reductionism*.

12.1.1 Biological Reductionism

Since the days of Galileo and Newton, the reductive program has been surprisingly successful in providing explanations for the behavior of the natural world. Thus this perspective is now widely accepted by the scientific community as the fundamental way to pose and answer questions. Basically, the reductive approach to understanding proceeds in three steps.

- *Analysis.* Assuming some higher-level phenomenon is to be explained, separate the dynamics of that phenomenon into *components*, the behaviors of which are to be individually investigated.

- *Theoretical formulation.* Through empirical studies and an exercise of imagination, develop a *theory* of how the components interact.

- *Synthesis.* In the context of this theory, *derive* the higher-level phenomenon.

Among the many aspects of nature that have fallen to this approach, one can mention planetary motion (based on the concepts of mass and gravity and on Newton's laws of motion), electromagnetic radiation (based on the concepts of electric charge, electric fields, and magnetic fields related through Maxwell's equations), atomic and molecular structures (based on the concepts of mass, electric charge, and Schrödinger's equation), hydrodynamics (based on the concepts of mass density, viscosity, compressibility, and the Navier–Stokes equations), and nerve impulse propagation (based on the concepts of voltage, membrane permeability, ionic current, and the Hodgkin–Huxley equations). Generalizing from such specific examples, the philosophical perspective of reductionism asserts that *all* natural phenomena can be understood in this manner [36].

Some, on the other hand, believe there exist natural phenomena that cannot be completely described in terms of lower-level entities—life and the human mind being outstanding examples. In its more extreme form, this position is called *substance dualism*: the view that important aspects of the natural world do not have a physical basis. A less salient position is *property dualism*, which asserts aspects of the physical world that cannot be explained in terms of atomic or molecular dynamics.

To a statement of belief there is no scientific response, but if we can agree on the physical basis of natural phenomena, the scope of the discussion narrows. Let us assume, therefore, that all natural phenomena *supervene* on the physical in the following sense. If the constituent matter is removed, the phenomenon in question disappears, or as philosopher Jaegwon Kim puts it ([26], p. 12): "Any two things that are exact physical duplicates are exact psychological duplicates as well." This position is called *physicalism*.

Among biologists, it is now widely accepted that the physicalist position holds for the phenomenon of life. If the atoms comprising a living organism are removed one by one, it will surely die. Most also believe that a person's mind would not survive a detailed deletion of the molecules of his or her brain. Thus two interesting questions are

- Does reductionism follow from physicalism?

and

- Does physicalism allow property dualism?

Since the 1980s, such questions have been carefully considered by Kim, who reluctantly concludes that physicalism does indeed imply reductionism and sits uneasily with property dualism [26]. Let us briefly review his central argument with reference to Figure 12.1.

This figure represents a higher-level phenomenon (M_1) that supervenes on the lower-level physical properties (P_1), where supervenience is indicated by the vertical dashed line. In other words, if the properties P_1 are re-

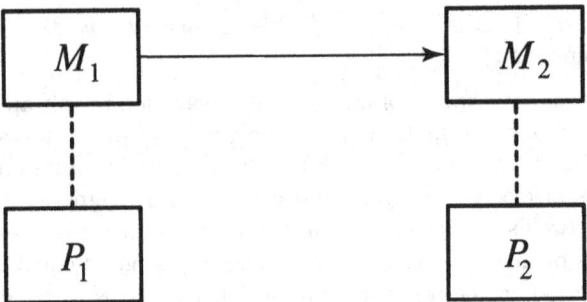

Figure 12.1. The causal interaction of higher-level phenomena (M_1 and M_2) that supervene on lower-level properties (P_1 and P_2).

moved, then the phenomenon M_1 will disappear, with a similar relationship between P_2 and M_2.

Now suppose that there is observed to be a *causal relationship* between M_1 and M_2 [10], indicated by the horizontal arrow in Figure 12.1. Thus the initial upper-level observation of M_1 always leads to a corresponding upper-level observation of M_2. Because under the assumption of physicalism P_1 (P_2) must be present to provide a basis for M_1 (M_2), we could as well say that P_1 causes P_2, which is a formulation of the upper-level causality in terms of the lower-level properties. Furthermore, one could interpret the phenomenon M_1 (M_2) in terms of P_1 (P_2), thereby undercutting a position of property dualism.

In the view of physicist Steven Weinberg [36], the dashed lines in Figure 12.1 can be replaced by upward-directed arrows at every level of description that show the direction of reductive implication. These arrows ultimately emanate from the most fundamental element of physical reality (nowadays known as the "Higgs boson"). Such a perspective does not suppose it to be practical or currently possible to describe the dynamics of (say) a bacterium in terms of the fundamental fields and particles of physics but that it can be done "in principle."

Finally, it can be argued that even if reductionism turns out not to hold in all aspects of biological organization, it is still a prudent strategy for the majority of biologists to take as a working hypothesis. Why? Often the riddles of one generation become standard knowledge of the next, including the nerve impulse, which so mystified Hermann Helmholtz in the nineteenth century. Thus the dualist (substance or property) is ever in danger of giving up too soon on the search for reductive explanations.

12.1.2 Objections to Reductionism

Although most social scientists reject the notion that physics is relevant to psychology, the previous section may have shown the reader that biological reductionism based on physicalism is a serious philosophical position meriting careful response. Those who object to the reductionist position must offer more than mere intuition that it does not make sense. Concrete objections and alternative suggestions must be provided. What are some of these objections?

Constructionism Versus Reductionism
Interestingly, the physics community is itself divided on the merits of reductionism. In general, theoretical physicists agree with Weinberg, whereas condensed-matter physicists—those who grapple with the details of understanding aggregates of matter—tend toward a somewhat different view. Thus Philip Anderson asserts that [3]:

> the reductionist hypothesis does not by any means imply a "constructionist" one: The ability to reduce everything to simple fundamental laws does not imply the ability to start from those laws and reconstruct the universe. In fact the more the elementary-particle physicists tell us about the nature of the fundamental laws, the less relevance they seem to have to the very real problems of the rest of science, much less to those of society. The constructionist hypothesis breaks down when confronted with the twin difficulties of scale and complexity.

What is it about "scale and complexity" that creates problems for the constructionist hypothesis?

Immense Numbers of Possibilities
As we have seen in Section 10.2.1, computational difficulties arise from the fact that the number of possible emergent structures at each level of the biological hierarchy is too large to be counted. To sharpen ideas in theoretical biology, physicist Walter Elsasser introduced the term *immense* to characterize a number that is both finite and greater than a *googol* (10^{100}) and thus inconveniently large for numerical studies [11, 13].

To see this in detail, consider the proteins. These biochemical workhorses are valence-bonded strings of amino acids, each of which is designated by an underlying DNA code. Because there are 20 different amino acids and a typical protein comprises some 200 of them, the number of possible proteins is greater than 20^{200}, which in turn is greater than a googol. Thus the number of possible protein molecules is immense.

What does this mean? All the matter in the myriad galaxies of the universe falls far short of that required to construct but one example of each possible protein molecule [13]. Throughout the eons of life on earth, in other words, most of the possible protein molecules have never been constructed and never will be. Those particular proteins we know were selected in the course of evolution through a succession of historical accidents that are consistent with but not governed by the laws of physics and chemistry.

So it goes at other levels of the biological hierarchy. The possible number of new entities that can emerge from each level—to form a basis for the dynamics of the next level—is immense, suggesting that happenstance, rather than basic laws of physics, guides important aspects of the evolutionary process [21].

It follows that biological science is fundamentally different from physics. As was noted in Section 10.2.1, the physical scientist deals with *homogeneous* sets, in which all of the elements are identical. Thus a physicist has the luxury of performing as many experiments as are needed to establish *laws* governing the interactions among electrons, protons, neutrons, and atoms. In the biological and social sciences, on the other hand, the number of possible members in any empirical category is immense, so experiments are necessarily performed on *heterogeneous subsets* of the classes of interest. Because the elements of heterogeneous sets are never exactly the same, causal laws cannot be determined with the same degree of certainty in the biological and social sciences as in the physical sciences.

In other words, psychologists establish *rules* rather than laws for interpersonal interactions, and your doctor can only give you the probability that a certain pill will make you feel better. At the levels of biology and social science, therefore, the horizontal arrow in Figure 12.1 should often be drawn fuzzy or labeled with a percentage of reliability.

The Nature of Causality

Whether one is concerned with establishing dynamical laws in the physical sciences or seeking corresponding rules in the biological and social sciences, the notion of *causality* requires careful consideration [10]. Some 24 centuries ago, Aristotle put it thus [4]:

> We have to consider in how many senses 'because' may answer the question 'why'.

As a "rough classification of the causal determinants (*aitiai*) of things," he suggested four types of cause.

- *Material cause.* Material cause stems from the presence of some physical substance that is needed for a particular outcome. Aristotle suggested that bronze is an essential factor in the making of a statue, but the concept is more general. As an example, note that the epidemic of gunshot wounds in the United States is materially caused by the large number of loaded handguns in private homes, just as alcoholism in Russia is materially caused by the availability of vodka.

- *Formal cause.* The material necessary for some outcome must be given the appropriate form. Thus, "the interval between two notes is not an octave unless they stand in the ratio of 2 to 1." Other examples of formal cause are easily imagined: the blueprints of a house are necessary for its construction, the DNA sequence of a particular gene is required for synthesis of the corresponding protein, and a violinist needs the score to play a concerto.

- *Efficient cause.* For something to happen, according to Aristotle, there must be an *"agent* that produces the effect and starts the material on its way."* Students of physical science deal primarily with efficient causes during their introductory courses in dynamics. Thus, a golf ball moves through the air in a certain trajectory because it was struck at a particular instant of time by the head of a club, and a radio wave is emitted in response to the current flowing through an antenna.

- *Final cause.* Often, things come about because they are desired by some intentional organism: a house is built—involving the assembly of materials, reading of plans, and pounding of nails—because someone wishes to have shelter from the elements. Although purposive answers to the question "why" are problematic in the biological sciences, they emerge as central issues at upper levels of the *cognitive hierarchy*, which we will soon consider.

Because we are presently interested in viewing causality from a mathematical perspective, the following paraphrasing of Aristotle's definitions may be helpful.

(1) At a particular level of the biological hierarchy, *material causes* might be time or space averages over dynamic variables at lower levels of description that enter as slowly varying *parameters* at the level of interest.

(2) Again, at a particular level of the biological hierarchy, *formal causes* might arise from values of dynamic variables at higher levels of description that enter as *boundary conditions* at the level of interest.

(3) An *efficient cause* is represented by the *stimulation-response* formulation analyzed in Appendix C. Following Galileo, this is the primary sense in which physical scientists currently use the term causality [10].

(4) In mathematical terms, it is not clear (to me, at least) how one might formulate a *final cause*.

Although this classification seems tidy, reality is more intricate. Thus Aristotle noted that causes may be difficult to sort out in particular cases, with several often "coalescing as joint factors in the production of a single effect" [4].

Distinctions among these "joint factors" are not always easy to make. A subtle difference between formal and efficient causes arises from consideration of the metaphor for Norbert Wiener's *cybernetics*: the steering mechanism of a ship [37]. If the wheel is connected directly to the rudder (via cables of some type), then the forces exerted by the helmsman's arms are the efficient cause of the ship executing a change of direction. For larger vessels, however, control is established through a servomechanism in which the position of the wheel merely sets a pointer that indicates the desired position of the rudder. The forces that move the rudder are generated by a feedback control system that minimizes the difference between the actual and desired positions. In this case, one might say that the position of the pointer is a formal cause of the ship's turning, with the servomotor of the control system acting as the efficient cause.

Another example—yet more relevant to the themes of this book—is provided by the conditions needed to cause the firing of a neuron, as represented by the McCulloch–Pitts (M–P) model of Equation (2.10). If the dendritic weights and threshold (α_{jk} and θ_j) are supposed to be constants, they can be viewed as formal causes of a firing event. On a longer time scale associated with learning, however, these parameters can be viewed collectively as a *weight vector* responding to the learning dynamics described in Section 10.1.2. From this perspective, components of the weight vector might be classified as efficient causes of neuron ignition. Of course, as we learned in Chapter 9, the switching of a real neuron is far more intricate than that of an MP model, but the point remains valid: neural switching is an intricate dynamic process involving the merging of many joint factors.

When a particular protein molecule is constructed within a living cell, for a final example of joint causality, sufficient densities and varieties of amino acids in the vicinity of the messenger RNA are material causes. The DNA code, controlling which amino acids are to be arranged in what order, is a formal cause. Lastly, the chemical (electrostatic and valence) forces acting among the constituent atoms are efficient causes. Because far more intricate situations are readily imagined, the reductionist should remain aware that the causal relations sketched in Figure 12.1 are not at all simple in the biological and social sciences.

For mathematicians, it is not surprising for several different types of causes to be involved in a single event. We expect that parameter values, boundary conditions, and forcing functions will all combine to influence the outcome of a given computation.

What other complications of causality are anticipated?

Nonlinear Causality

In mathematics, the term "nonlinear" is defined in the context of relationships between causes and effects. Suppose that a series of experiments on a certain system have shown that cause C_1 gives rise to effect E_1; thus

$$C_1 \to E_1 \,,$$

and similarly

$$C_2 \to E_2$$

expresses the relationship between cause C_2 and effect E_2. This relation is *linear* if

$$C_1 + C_2 \rightarrow E_{12} = E_1 + E_2 . \tag{12.1}$$

If, on the other hand, E_{12} is *not* equal to $E_1 + E_2$, the effect is said to be a *nonlinear* response to the cause.

Equation (12.1) indicates that for a linear system the cause can be arbitrarily divided into convenient components $(C_1, C_2, \ldots, C_n)$, whereupon the effect will be correspondingly divided into $(E_1, E_2, \ldots, E_n)$. Although convenient for analysis, this property is seldom found in the biological world.

Far more common is the nonlinear situation, where the effect from the sum of two causes is not equal to the sum of the individual effects. The whole is not equal to the sum of its parts. Nonlinearity is less convenient for the analyst because multiple causes interact among themselves, allowing possibilities for many more outcomes and confounding the constructionist. For this reason, however, nonlinearity plays a key role in the course of biological evolution.

The Nature of Time

Causality is intimately connected with the way we view time—thus, the statement "C causes E" implies that E does not precede C in time [10]—yet the properties of time may depend on the level of description [18, 19, 38]. As we have seen in Section 2.2, the dynamics underlying molecular vibrations are based on Newton's laws of motion, in which time is *bidirectional*. In other words, the direction of time in Newton's theoretical formulation can be changed without altering the qualitative behavior of the system. At the level of a nerve impulse, on the other hand, time is unidirectional, with a change in its direction making an unstable nerve impulse stable and vice versa. In appealing to Figure 12.1, therefore, the reductionist must recognize that the nature of the time used in formulating the causal relationship between P_1 and P_2 may differ from that relating M_1 and M_2.

Downward Causation

The doctrine of reductionism assumes that causality acts upward through the biological hierarchy, where the causality can be interpreted as both efficient and material. Formal causes, on the other hand, can act *downward* because variables at the upper levels of a hierarchy can place constraints on the dynamics at lower levels [2].

A dramatic example of downward causation occurred eons ago when certain bacteria began to harvest and store energy from the sun's light, creating atmospheric oxygen as a poisonous waste [28]. The presence of oxygen in the atmosphere led to the emergence of the animal kingdom, in which we humans participate. Other examples of downward causation include modifications of DNA codes caused by interactions among species, germination of an ovum following sexual activity, the disintegration of an organism upon death, and so on.

Although such examples seem to provide convincing evidence of downward causation, the means through which it acts are not well understood. To this end, Claus Emmeche and his colleagues have defined three sorts of downward causation, as follows [14].

- *Strong downward causation* (SDC). Under SDC, it is supposed that upper-level phenomena can act as efficient causal agents in the dynamics of lower levels. In other words, upper-level organisms can modify the physical and chemical laws governing their molecular constituents. Presently, there is no empirical evidence for the downward action of efficient causation, so SDC is almost universally rejected by biologists.

- *Weak downward causation* (WDC). WDC assumes that the molecules comprising an organism are governed by some nonlinear dynamics in a phase space, having attractors—which include the living organism—each with a corresponding basin of attraction. Death, in this formulation, is but another of the attractors shared by the interacting molecules, and a physician's job is to keep the molecules of a patient within the basin of the living state. (Unfortunately, the basin shrinks as we age, making the task ever more difficult.)

  Because many examples of such nonlinear systems have been carefully studied both experimentally and theoretically [32], there is little doubt about the scientific credibility of this means for downward causation. Building on the seminal suggestions of Alan Turing [34], biologists Stuart Kauffman [25] and Brian Goodwin [20] have presented detailed discussions of ways that WDC influences the development and behavior of living organisms.

- *Medium downward causation* (MDC). Although accepting WDC, supporters of MDC go further in supposing that higher-level dynamics (e.g., the emergence of a higher-level structure) can modify the local features of an organism's phase space through the downward actions of formal causes. (An example of MDC is provided by the *automatic frequency control* of an FM radio receiver. Here, a time average amplitude of the demodulated signal is used to adjust the input tuning capacitor, leading to the familiar experience of locking onto a particular signal.)

  In biology, MDC opens the possibility of closed causal loops spanning several layers of the hierarchy. In this picture, an organism emerges from the underlying phase space, which it in turn modifies. Using the positive feedback diagram introduced in Chapter 1, such closed causal loops can be represented as

Emergent organism

↓ ↑

Underlying phase space

Over two decades ago, biochemists Manfred Eigen and Peter Schuster suggested that closed causal loops around at least three layers of dynamic description were necessary for the emergence of living organisms from the oily scum of the Hadean oceans [12].

Open Systems

Biological organisms are *open systems*, as described in Appendix A, requiring a steady input of energy (sunlight or food) to maintain their metabolic activities. As a simple example of an open system, consider the flame of a candle. Computing the propagation velocity of the flame (v) as in Chapter 5, it is possible to establish a rule for where the flame will be located at a particular time. Corresponding to

$$M_1 \to M_2 \,,$$

in Figure 12.1, one such rule is the following. If the flame is at position x_1 at time t_1, then it will be at position

$$x_2 = x_1 + v(t_2 - t_1)$$

at time $t_2 > t_1$.

Because the flame is an open system, a corresponding relation

$$P_1 \to P_2$$

cannot be written—even "in principle"—for the physical substrate. This follows from the fact that the physical substrate is *continually changing* [7]. The molecules of air and wax vapor comprising the flame at time t_2 are entirely different from those at time t_1. Thus, the detailed positions and speeds of the molecules present in the flame at time t_2 are unrelated to those present at time t_1. What remains constant is the flame itself: a *process*.

Closed Causal Loops.

In his analysis of reductionism, Kim also fails to grasp the concept of a closed causal loop, asking: "How is it possible for the whole to causally affect its constituent parts on which its very existence and nature depend?" [27]. Causal circularity, he claims, is unacceptable because it violates the following "causal-power actuality principle."

> For an object, x, to exercise, at time t, the causal/determinative powers it has in virtue of having property P, x must already possess P at t. When x is being caused to acquire P at t, it does not already possess P at t and is not capable of exercising the causal/determinative powers inherent in P.

There are two replies to this argument, one theoretical and the other empirical.

(1) From a theoretical perspective, Kim is led astray by supposing that a coherent structure somehow pops into existence at time t, which would indeed be surprising. To see how a coherent structure actually organizes itself, return to Figure 1.3, which shows a biological population growing to a steady amplitude. Because both the population (N) and its rate of growth (dN/dt) are functions of time, related by Equation (1.3), solutions of this ODE yield the growth curves of Figure 1.3.

Similarly, in Kim's notation, both x and P are functions of time (t), which may be related as

$$\frac{dx}{dt} = F(x, P),$$
$$\frac{dP}{dt} = G(x, P),$$

where F and G may be general nonlinear functions of both x and P. (For example, the time scales of F and G might be very different, allowing P to remain approximately constant during the dynamics of x.) The emergent structure is not represented by $x(t)$ and $P(t)$ (which are functions of time), but by x_0 and P_0, satisfying

$$0 = F(x_0, P_0),$$
$$0 = G(x_0, P_0).$$

Assuming that x_0 and P_0 are an asymptotically stable solution of this system,

$$x(t) \rightarrow x_0,$$
$$P(t) \rightarrow P_0,$$

as $t \rightarrow \infty$, exemplifying the establishment of a dynamic balance between downward and upward causations.

Thus, Kim's causal-power actuality principle is recognized as an artifact of his static analysis of an essentially dynamic situation.

(2) Applied science offers many examples of positive feedback and subsequent emergence of coherent structures [32]. Engineers employ negative feedback to control the performance of amplifiers, routinely designing closed causal loops in which a signal from the output terminals is carried back to the input. Occasionally, this feedback signal becomes positive rather than negative and leads to unwanted oscillations (called "singing") that can be viewed as emergent structures.

In the physical sciences, corresponding emergent structures include tornadoes, tsunamis, and Jupiter's Great Red Spot, among many others [32]. A biological example is provided by cellular reproduction, wherein a DNA code is necessary to produce protein molecules and proteins are needed for transcription of the code. Finally, this book offers several examples of closed causal loops of positive feedback in neuroscience, including both the nerve impulse and the cell assembly.

These emergent structures are essential elements in the *cognitive hierarchy*, which shares many features of the biological hierarchy.

12.2   The Cognitive Hierarchy

The preceding discussion of the hierarchical nature of biological science was presented in some detail as an introduction to the main subject of this final chapter: the cognitive hierarchy of the human brain. With corresponding caveats about including fewer or more levels and allowing for branchings, a convenient version of the cognitive hierarchy takes the following form.

Human cultures
Phase sequences
Complex assemblies
· · ·

· · ·
Assemblies of assemblies of assemblies
Assemblies of assemblies
Assemblies of neurons
Neurons
Nerve impulses
Nerve membranes
Membrane proteins
Molecules

Although this diagram differs from the biological hierarchy in important ways, many of the previous comments carry over into the present discussion. In particular, each cognitive level has its own nonlinear dynamics, involving closed causal loops of positive feedback, out of which can emerge an immense number of possible entities. A necessarily small subset of these possibilities does in fact emerge and provides a basis for the nonlinear dynamics of the next higher level.

Perhaps the most significant difference between the biological and cognitive hierarchies is seen from consideration of the *internal levels,* which were introduced in the preceding chapter. Extracted from the cognitive hierarchy, these levels are

Complex assemblies
· · ·

· · ·
Assemblies of assemblies of assemblies
Assemblies of assemblies
Assemblies of neurons

the existence of which is based on theoretical speculation and circumstantial evidence rather than direct observation [30].

Because individual assemblies share the basic dynamic properties of a neuron (threshold behavior and all-or-nothing response), Hebb proposed that they can organize themselves into higher-level assemblies of assemblies (called "second-order assemblies"), which in turn become components of third-order assemblies and so on up to the complex assemblies that form the basis for normal thought [23, 24]. As we have seen, empirical support for these speculations is beginning to appear, but more evidence is needed before they can be regarded as firmly established.

Because it is not known how many internal cell-assembly levels there are or how they are organized, this region of the brain is presently a *terra incognita* of science—one of its more interesting unexplored frontiers.

Just as in the biological hierarchy, we expect to find formal causation acting downward also in the cognitive hierarchy. Indeed, the attractor neural networks described in Section 10.2.2 exemplify weak downward causation (WDC) [14]. The phenomenon of learning—whereupon strengths of interconnections among neurons are altered in response to experiences of the organism—provides an example of medium downward causation (MDC) because the local character of the underlying phase space is altered by global dynamics. As with the biological hierarchy, strong downward causation (SDC) is rejected by neuroscientists as both unproven and implausible.

Reductionism, in the context of neuroscience, is often interpreted as the view that all of the brain's behavior can be formulated in terms of local membrane dynamics. Referring back to Figure 12.1, P_1 (P_2) now represents the membrane states upon which a particular mental phenomenon M_1 (M_2) supervenes. Take away the P and the corresponding M disappears, according to the doctrine of physicalism to which most neuroscientists subscribe.

The doctrine of *cognitive reductionism*, therefore, holds that any causal relationship between M_1 and M_2 (which is indicated by the horizontal arrow in Figure 12.1) can "in principle" be formulated in terms of the underlying membrane states, P_1 and P_2. Proponents of this view should formulate responses to all of the objections raised in Section 12.1.2 against biological reductionism.

As in the biological hierarchy, downward causation (either WDC, MDC, or both) leads to additional opportunities for closed causal loops. These multilevel loops are far more intricate than most of the examples presented in this book, which can be represented by the diagram

$$\mathbf{A}$$
$$\downarrow\uparrow$$
$$\mathbf{B}$$

Implying that **A** causes **B**, which in turn causes **A**, this simple picture is appropriate for describing the emergence of coherent structures at a

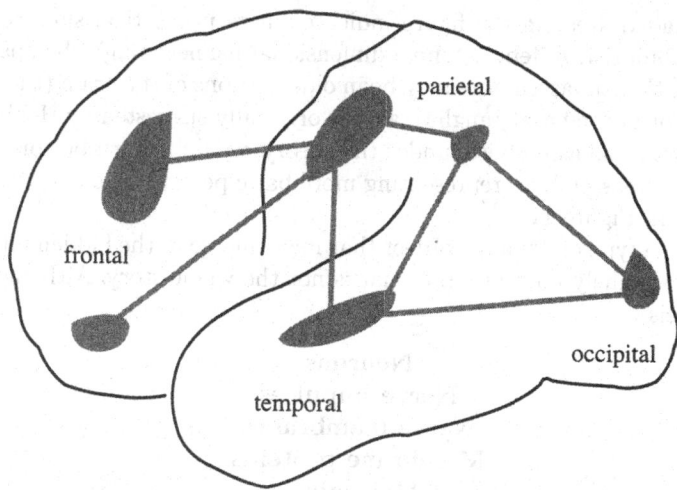

Figure 12.2. A sketch of the left-hand side of a human brain suggesting how the subassemblies (shaded areas) of a complex cell assembly might be distributed over various lobes of the neocortex.

particular dynamic level—the nerve impulse being a clear example. In the context of modern nonlinear science, each such diagram would correspond to the presence of an attractor in the phase space describing the system dynamics [32].

Complex cell assemblies, on the other hand, comprise subassemblies or attractors emerging at many different levels of the cognitive hierarchy, which can in turn become interconnected in an immense number of different ways.[1] Such interconnections might be effected via long axonal processes, the waves of information discussed in Section 10.3, or perhaps both.

Figure 12.2 suggests how various subassemblies of a complex cell assembly might be distributed over the principal lobes of the left hemisphere of the human neocortex. The *occipital* lobes, at the rearmost tip of each hemisphere, are related to vision, because they accept signals from the eyes. The *parietal* lobes, on the upper rear of each hemisphere, handle judgments of weight, size, shape, and feel.[2] The *temporal* lobes, near the temples, deal with language and the perception of sound, among other things, which is not surprising given their proximity to the ears. Finally, the *frontal* lobes, immediately behind the forehead, govern voluntary movements and some logical processes, which is why a frontal lobotomy tends to alter personality.

---

[1]Although the attractor neural network formulation of Section 10.2.2 seems to miss this feature, Amit shows how hierarchical structure can be introduced into the Hopfield theory. (See Chapter 8 of [1].)

[2]Hierarchical organization of tactile information processing in the human brain has been demonstrated by Bodegård et al. [8] using both PET and MRI scanning techniques.

The shaded areas in the figure indicate where neocortical subassembly neurons comprising Hebb's "three-dimensional fishnet" might be sparsely located. Other subassemblies may be in older regions of the brain (the diencephalon or in the basal ganglia), as was originally suggested by Hebb [22]. Each of these subassemblies, under the theory, would in turn be composed of further subassemblies representing more basic perceptual elements, as is indicated in Figure 11.3.[3]

Thus, the type of dynamic object that may emerge in the hidden internal levels is extremely intricate, but that is not the whole story. Although the lower levels

Neurons
Nerve impulses
Nerve membranes
Membrane proteins
Molecules

are open to direct empirical investigation, puzzles remain, as we have seen throughout this book. At the higher levels

Human cultures
Phase sequences

it is again possible to directly observe the dynamics, but new difficulties appear.

By the term "phase sequence," as the reader will recall from the first chapter, Hebb implied a "thought process" in which [22][4]:

> Each assembly action may be aroused by a preceding assembly, by a sensory event, or—normally—by both. The central facilitation from one of these activities on the next is the prototype of "attention."

We all experience ongoing trains of thought in every waking moment, but these processes do not occur in a psychic vacuum.

Throughout such trains, our individual thoughts are both guided by and constituent elements of the particular human culture in which we are immersed. We are, in other words, molded by levels of cultural reality of which we are often unaware. Thus corresponding cultural levels

---

[3]Some may be concerned that neurons separated by the distances indicated in Figure 12.2 could not effectively interact in the same Hebbian assembly. These readers should return to Section 11.5, which presented results of fairly realistic numerical simulations of cortical assembly dynamics. For random axonal delays up to about 10 ms, the assembly dynamics were unchanged [17], suggesting an allowed radius for neuronal interactions of several inches or more.

[4]If each "neuron" in Figure 10.2(b) is taken to represent a complex assembly, an elementary phase sequence corresponds to the transient $(001) \rightarrow (000) \rightarrow (010)$.

of reality—described by American anthropologist Ruth Benedict as "patterns of culture"—should be included in realistic models of the human brain [6, 31].

Last, but not least, the possibility of causal interactions among the various levels of the cognitive and biological hierarchies should be included in the overall theoretical perspective. At lower levels, this is obvious because the physiological state of a neuron must surely effect the manner in which it relates incoming and outgoing streams of information, but higher cognitive levels also have causal biological effects. Cultural imperatives to ingest a psychoactive substance, for example, can alter the dynamics of membrane proteins and lead to mental changes that influence bodily health with subsequent psychological effects in a wending path of branching causes and effects that boggles the analytic mind.

Presently, the sorts of phenomena that could emerge from such intricate causal networks—spanning several levels of both the biological and cognitive hierarchies—are only dimly imagined. How might mathematics help to sort out these speculations?

12.3 Some Outstanding Questions

Although the foregoing comments may seem pessimistic about the future of neuroscience research, that is not my intent. On the contrary, the awesome intricacy of the human brain presents us with unmatched challenges, making this a most exciting time in the history of neuroscience. Awaiting the assaults of vigorous young minds are many interesting problems, including the following.

Hierarchical Formulations
Throughout this book, the hierarchical nature of biological and cognitive systems has been noted, but in a merely descriptive manner leaving many unresolved issues. Building on the work of Eigen and Schuster [12], Voorhees [35], Fontana and Buss [16], Baas [5], and Nicolis [29], can one formulate the various causal relationships among levels of a nonlinear dynamic hierarchy (weak and medium downward causation, for example) in a manner that is suitable for mathematical analyses?

This is not at all a trivial matter because the time and space scales for models of living creatures differ by many orders of magnitude as one goes from the biochemical levels to the whole organism, creating a daunting challenge for the numerical modeler. Are there ways to evade such computational constraints? Might, for example, hierarchically organized functions be defined on nested sets of points, with different types of averaging at various stages of the computations? Is it possible to resolve issues without resorting to numerical computation?

Not only the scale but—as we have seen—the *nature* of time differs at biomolecular and cellular levels of description [18, 19]. Thus Newton's second law, governing the dynamics of the atoms within a molecule, is bidirectional in time, meaning that the direction of time's arrow can be reversed in the mathematical formulation without changing the qualitative behavior. In living systems, this is not the case.

Because the cognitive hierarchy is even more intricate than the biological hierarchy, what can be said about the nature of time governing the global behavior of the brain? If the nature of time is determined by the ongoing dynamic processes at a particular level of description, might some yet to be defined concepts of "psychological time" and "social time" also play useful roles?

Meaningful Information

Several possibilities for information processing in the dendritic and axonal trees of real neurons were sketched in Chapter 9, and some means for higher-level information processing were described in Chapters 10 and 11. How can these various models be incorporated into a mathematical formulation for the global activity of a real brain? Indeed, what should we intend by "information" in the context of living brain?

In engineering science, this term derives from the need to store and transmit a series of zeros and ones representing an arbitrary photograph, a computer code, a musical recording, or whatever [9, 33]. A system with p different configurations is said to store an amount of information equal to $\log_2(p)$ *bits*. (For example, each of N switches can be set in either the on or off position, leading to 2^N possible configurations. Conveniently, $\log_2(2^N) = N$, the total number of switches.)

Living organisms, on the other hand, are *intentional*, with self-determined programs of activity. Thus the importance of a particular fact is related to what the creature is concerned about [15], and an item of meaningful information can involve many details of the type that are captured by Hebb's cell assemblies.

After the learning experience sketched in Figure 11.1, for example, a person might recognize a triangle, and this recognition could be viewed as one item of information that might or might not be meaningful according to his or her current interests. In more intricate situations, a person might recognize that those red berries are poisonous, it is going to rain, the car he just hit is driven by a policeman, and so on. The vast collection of such assemblies that each of us carries about were crafted moment by moment, day by day, and year by year as we became adults, providing the contexts in which we interpret experience.

Considering the little we know of internal assemblies and the ways they interact, how might the concept of "meaningful information" be mathematically defined for an intentional organism? Could a sharpened concept of meaningful information lead to a formulation for Aristotle's final cause?

Subjective Experience

Beyond intentionality lie other mysteries of the mind. Many of the higher-level species, including humans, exhibit subjective emotional states and experience *feelings*—mental phenomena that are difficult to square with the reductive formulations of neuroscience.

Even if our equations or computer codes were to describe exactly the dynamics of a person's brain, subjective experience appears to be left out of the picture, as was eloquently expressed by Charles Scott Sherrington and D'Arcy Wentworth Thompson in the quotations at the front of this book.

Having discussed this matter elsewhere [31], there is no need to go into detail, but it seems appropriate to close with two questions that continue to puzzle me. Do human feelings emerge from intricate networks of positive feedback winding through many levels of the interacting biological, cognitive, and social hierarchies? Or are there aspects of neuroscience for which mathematical formulations will never be useful?

References

[1] DJ Amit, *Modeling Brain Function: The World of Attractor Neural Networks*, Cambridge University Press, Cambridge, 1989.

[2] PB Andersen, C Emmeche, NO Finnemann, and PV Christiansen, *Downward Causation: Minds, Bodies and Matter*, Aarhus University Press, Aarhus, Denmark, 2000.

[3] PW Anderson, More is different: Broken symmetry and the nature of the hierarchical structure of science, *Science* 177 (1972) 393–396.

[4] Aristotle, *The Physics* (translated by PH Wicksteed and FM Cornford), Harvard University Press, Cambridge, and William Heinemann Ltd, London, 1953.

[5] NA Baas, Emergence, hierarchies, and hyperstructures. In *Artificial Life III*, CG Langton (ed), Addison–Wesley, Reading, 1994.

[6] R Benedict, *Patterns of Culture*, Houghton Mifflin, Boston, 1989 (first published in 1934).

[7] MH Bickhard and DT Campbell, Emergence. In [2].

[8] A Bodegård, S Geyer, C Grefkes, K Zilles, and PE Roland, Hierarchical processing of tactile shape in the human brain, *Neuron* 31 (2001) 317–328.

[9] L Brillouin, *Science and Information Theory*, Academic Press, New York, 1956.

[10] M Bunge, *Causality and Modern Science*, third edition, Dover, New York, 1979.

[11] RE Crandall, The challenge of large numbers, *Sci. Am.* February 1997, 72–78.

[12] M Eigen and P Schuster, *The Hypercycle: A Principle of Natural Self-Organization*, Springer-Verlag, Berlin, 1979.

[13] WM Elsasser, *Reflections on a Theory of Organisms: Holism in Biology*, The Johns Hopkins University Press, Baltimore, 1998 (first published in 1987).

[14] C Emmeche, S Køppe, and F Stjernfelt, Levels, emergence, and three versions of downward causation. In [2].

[15] WJ Freeman, *How Brains Make Up Their Minds*, Columbia University Press, New York, 2000.

[16] W Fontana and LW Buss, "The arrival of the fittest": Toward a theory of biological organization, *Bull. Math. Biol.* 56 (1994) 1–64.

[17] E Fransén, A Lansner, and H Liljenström, A model of cortical memory based on Hebbian cell assemblies. In *Computation and Neural Systems*, FH Eeckman and JM Bower (eds), Kluwer, Boston, 1993.

[18] JT Fraser, *The Genesis and Evolution of Time*, Harvester Press, Brighton, England, 1982.

[19] JT Fraser, *Of Time, Passion, and Knowledge: Reflections on the Strategy of Existence*, second edition, Princeton University Press, Princeton, 1990.

[20] B Goodwin, *How the Leopard Changed its Spots: The Evolution of Complexity*, Scribner's, New York, 1994.

[21] SJ Gould, *Wonderful Life: The Burgess Shale and the Nature of History*, W.W. Norton & Co., New York, 1989.

[22] DO Hebb, *Organization of Behavior: A Neuropsychological Theory*, John Wiley & Sons, New York, 1949.

[23] DO Hebb, The structure of thought. In *The Nature of Thought*, PW Jusczyk and RM Klein (eds), Lawrence Erlbaum Associates, Hillsdale, NJ, 1980, pp 19–35.

[24] DO Hebb, *Essay on Mind*, Lawrence Erlbaum Associates, Hillsdale, NJ, 1980.

[25] S Kauffman, *The Origins of Order: Self-Organization and Selection in Evolution*, Oxford University Press, Oxford, 1993.

[26] J Kim, *Mind in a Physical World*, MIT Press, Cambridge, 2000.

[27] J Kim, Making sense of downward causation. In [2].

[28] L Margulis and D Sagan, *What Is Life?* Simon & Schuster, New York, 1995.

[29] JS Nicolis, *Dynamics of Hierarchical Systems: An Evolutionary Approach*, Springer-Verlag, Berlin, 1986.

[30] AC Scott, Brain theory from a hierarchical perspective, *Brain Theory Newsl.* 3 (1978) 66–612.

[31] AC Scott, *Stairway to the Mind*, Springer-Verlag, New York, 1995.

[32] AC Scott, *Nonlinear Science: Emergence and Dynamics of Coherent Structures*, Oxford University Press, Oxford, 1999.

[33] CE Shannon and W Weaver, *The Mathematical Theory of Communication*, University of Illinois Press, Urbana, 1963.

[34] AM Turing, The chemical basis of morphogenesis, *Philos. Trans. R. Soc. London* B237 (1952) 37–72.

[35] BH Voorhees, Axiomatic theory of hierarchical systems, *Behav. Sci.* 28 (1983) 24–34.

[36] S Weinberg, *Dreams of a Final Theory: The Search for the Fundamental Laws of Nature,* Pantheon Books, New York, 1992.

[37] N Wiener, *Cybernetics,* John Wiley & Sons, New York, 1961.

[38] AT Winfree, *When Time Breaks Down: The Three-Dimensional Dynamics of Electrochemical Waves and Cardiac Arrhythmias,* Princeton University Press, Princeton, 1987.

Appendix A
Conservation Laws and Conservative Systems

Assume a system that is uniform in the x-direction $(-\infty < x < +\infty)$ and conserves some quantity, Q. In other words, Q is neither created nor destroyed in the course of the dynamics under investigation. Then it is convenient to introduce the following definitions.

- $F(x, t)$ is the *flow* of the conserved quantity, or the amount of Q that passes the point x at time t.

- $D(x, t)$ is the *density* of the conserved quantity, or the amount of Q per unit of x.

In the context of these definitions,

$$\Delta x \frac{d}{dt} D(x + \Delta x/2, t) = F(x, t) - F(x + \Delta x, t) + O(\Delta x^2).$$

Taking the limit of this expression as $\Delta x \to 0$ yields the *conservation law*:

$$\frac{\partial D}{\partial t} + \frac{\partial F}{\partial x} = 0.$$

The total quantity

$$Q = \int_{-\infty}^{\infty} D \, dx$$

is evidently conserved because

$$\frac{dQ}{dt} = \int_{-\infty}^{\infty} \frac{\partial D}{\partial t} \, dx = -\int_{-\infty}^{\infty} \frac{\partial F}{\partial x} \, dx$$

$$= F(-\infty, t) - F(+\infty, t).$$

Thus

$$Q = \text{constant}$$

if the flow into the system at $x = -\infty$ equals the flow out at $x = +\infty$. This condition is satisfied if the dynamic variables approach zero as $x \to \pm\infty$.

There are many physical examples of conserved quantities, including the number of automobiles in a study of highway traffic flow, water in a river, minority carriers in a semiconductor, electromagnetic energy in a pulse of radio wave transmission, and mechanical energy in an elastic wave.

In nonlinear science, it is useful to have a means of distinguishing between systems or subsystems that include the effects of *energy* conservation and those that do not. The former are often referred to in the engineering literature as "conservative" or "lossless" and are considered to be constructed from elements such as inductors and capacitors or their mechanical analogs, masses and springs, excluding resistors or "dashpots" (mechanical resistors). Systems that do not conserve energy are called "dissipative" or "open" and require the presence of batteries or amplifying devices (e.g., transistors) to maintain dynamic activity.

For some arbitrary collection of terms in a PDE, however, it is not always clear whether energy is conserved. In such a situation, one can proceed by checking whether the system can be derived from a *Lagrangian density* $\mathcal{L}$, which is a function of the dependent variable $u(x, t)$ and certain of its derivatives [1].

Basic to this perspective is the assumption that the *action integral*[1]

$$I = \int_{x_1}^{x_2} \int_{t_1}^{t_2} \mathcal{L}(u, u_x, u_t, \cdots) \, dx \, dt$$

takes a maximum or minimum value along the true solution $u(x, t)$. In other words, the variation of I (written δI) is equal to zero when $\mathcal{L}$ is evaluated on $u(x, t)$.

Now let

$$\mathcal{L} = \mathcal{L}(u, u_x, u_t)$$

and choose

$$\delta u(x, t)$$

to be a small change of $u(x, t)$ that is zero at x_1, x_2, t_1, and t_2. Under these assumptions

$$\delta I = \int_{x_1}^{x_2} \int_{t_1}^{t_2} \left[\frac{\partial \mathcal{L}}{\partial u} \delta u + \frac{\partial \mathcal{L}}{\partial u_x} \delta u_x + \frac{\partial \mathcal{L}}{\partial u_t} \delta u_t \right] dx \, dt$$

---

[1]Where typographically convenient, subscripts are used to indicate partial derivatives. Thus $u_x \equiv \partial u/\partial x$, $u_t \equiv \partial u/\partial t$, and so on.

$$= \int_{x_1}^{x_2} \int_{t_1}^{t_2} \left[\frac{\partial \mathcal{L}}{\partial u} - \frac{\partial}{\partial x} \frac{\partial \mathcal{L}}{\partial u_x} - \frac{\partial}{\partial t} \frac{\partial \mathcal{L}}{\partial u_t} \right] \delta u(x,t) \, dx \, dt \,,$$

after integrating by parts using the boundary conditions assumed for $\delta u(x,t)$.

Evidently the condition

$$\delta I = 0$$

requires that $u(x,t)$ must satisfy the *Lagrange–Euler equation*[2]

$$\frac{\partial \mathcal{L}}{\partial u} - \frac{\partial}{\partial x} \frac{\partial \mathcal{L}}{\partial u_x} - \frac{\partial}{\partial t} \frac{\partial \mathcal{L}}{\partial u_t} = 0 \,.$$

As a simple example, note that a Lagrangian density for the wave equation

$$\frac{\partial^2 u}{\partial x^2} - \frac{\partial^2 u}{\partial t^2} = 0$$

is

$$\mathcal{L} = \frac{1}{2} (u_x^2 - u_t^2) \,.$$

Systems of partial differential equations associated with a Lagrangian density in the manner just described are said to be *conservative* for the following reasons.

1. One can define a *momentum density* as

$$\pi \equiv \frac{\partial \mathcal{L}}{\partial u_t} \,.$$

2. Then, an energy density (often called a *Hamiltonian density*) can be defined through the transformation

$$\mathcal{H}(u, u_x, \pi) \equiv \mathcal{L}(u, u_x, u_t) - \pi u_t \,.$$

3. Direct calculation shows that

$$\int_{-\infty}^{\infty} \mathcal{H} \, dx$$

---

[2]If $\mathcal{L}$ depends upon higher derivatives of u, the Lagrange–Euler equation includes additional terms that are obtained through integration by parts in a similar manner. Thus for

$$\mathcal{L} = \mathcal{L}(u, u_x, u_t, u_{xx}) \,,$$

the corresponding Lagrange–Euler equation is

$$\frac{\partial \mathcal{L}}{\partial u} - \frac{\partial}{\partial x} \frac{\partial \mathcal{L}}{\partial u_x} - \frac{\partial}{\partial t} \frac{\partial \mathcal{L}}{\partial u_t} + \frac{\partial^2}{\partial x^2} \frac{\partial \mathcal{L}}{\partial u_{xx}} = 0 \,,$$

and so on.

is a conserved quantity obeying the conservation law

$$\frac{\partial \mathcal{H}}{\partial t} + \frac{\partial \mathcal{P}}{\partial x} = 0,$$

where

$$\mathcal{P} \equiv -\frac{\partial \mathcal{L}}{\partial u_x} u_t$$

is a flow of energy, or a *power*.

For the wave equation,

$$\pi = -u_t,$$

$$\mathcal{H} = \frac{1}{2}(u_x^2 + \pi^2),$$

and

$$\mathcal{P} = -u_x u_t.$$

Not all systems of interest can be formulated in this manner. For example, the linear diffusion equation

$$\frac{\partial^2 u}{\partial x^2} - \frac{\partial u}{\partial t} = 0,$$

which conserves

$$\int_{-\infty}^{\infty} u \, dx,$$

does not have a Lagrangian density. Thus it is not conservative in the present sense.

Interestingly, the *nonlinear* diffusion equation

$$\frac{\partial^2 u}{\partial x^2} - \frac{\partial u}{\partial t} = f(u),$$

which plays a central role in nerve impulse dynamics, has no conserved quantities at all.

References

[1] H Goldstein, *Classical Mechanics*, Addison–Wesley, Reading, MA, 1951, Chapter 11.

Appendix B
Hodgkin–Huxley Dynamics

In the Hodgkin–Huxley formulation of nerve impulse dynamics, the membrane turn-on and turn-off variables (m, h, and n) are assumed to be solutions of first-order rate equations with voltage-dependent parameters. Thus [1]

$$
\begin{aligned}
\frac{dm}{dt} &= \alpha_m(1-m) - \beta_m m\,, \\
\frac{dh}{dt} &= \alpha_h(1-h) - \beta_h h\,, \\
\frac{dn}{dt} &= \alpha_n(1-n) - \beta_n n\,.
\end{aligned}
\tag{B.1}
$$

At a temperature of 6.3°C, the voltage dependencies of the coefficients are given by

$$
\begin{aligned}
\alpha_m &= \frac{0.1(25-V)}{\exp[(25-V)/10]-1}\,, \\
\beta_m &= 4\,\exp(-V/18)\,, \\
\alpha_h &= 0.07\,\exp(-V/20)\,, \\
\beta_h &= \frac{1}{\exp[(30-V)/10]+1}\,, \\
\alpha_n &= \frac{0.01(10-V)}{\exp[(10-V)/10]-1}\,, \\
\beta_n &= 0.125\,\exp(-V/80)\,,
\end{aligned}
\tag{B.2}
$$

in units of milliseconds^{-1}. At other temperatures, these rates change in a manner that can be accounted for by multiplying by the factor

$$\kappa = 3^{(\text{Temp}-6.3)/10}, \tag{B.3}$$

where "Temp" is the Celsius temperature. The membrane voltage V is measured in millivolts with respect to the resting potential, and an increase in the potential *inside* the axon is taken to be positive.[1]

In Equation (4.5), the first (G_{Na}) term accounts for sodium ion current, the second (G_K) accounts for potassium ion current, and the last term (G_L) accounts for all other ions. To understand how this system works, note that Equations (B.1) can also be written in the form

$$\frac{dm}{dt} = -\frac{m - m_0(V)}{\tau_m(V)},$$

$$\frac{dh}{dt} = -\frac{h - h_0(V)}{\tau_h(V)},$$

$$\frac{dn}{dt} = -\frac{n - n_0(V)}{\tau_n(V)},$$

where, from Equations (B.2),

$$
\begin{aligned}
m_0(V) &= \alpha_m/(\alpha_m + \beta_m), \\
\tau_m(V) &= 1/(\alpha_m + \beta_m), \\
h_0(V) &= \alpha_h/(\alpha_h + \beta_h), \\
\tau_h(V) &= 1\,(\alpha_h + \beta_h), \\
n_0(V) &= \alpha_n/(\alpha_n + \beta_n), \\
\tau_n(V) &= 1/(\alpha_n + \beta_n).
\end{aligned}
$$

Thus, $m(t)$ strives to reach $m_0(V)$ at a rate $\tau_m(V)$ and similarly for $h(t)$ and $n(t)$.

References

[1] AL Hodgkin and AF Huxley, A quantitative description of membrane current and its application to conduction and excitation in nerve, *J. Physiol. (London)* 117 (1952) 500–544.

---

[1] The present formulation differs from that of Hodgkin and Huxley, who define a positive membrane potential in the opposite sense.

Appendix C
Fredholm's Theorem

Consider linear stimulus–response problems of the form

$$Lu(x) = f(x), \tag{C.1}$$

where $-\infty < x < \infty$, L is a linear differential operator, $f(x)$ is a specified source (or cause), and $u(x)$ is a resulting response (or effect) that is to be determined by solving the equation. Emphasizing the cause-and-effect relation between $f(x)$ and $u(x)$,

$$f \xrightarrow{\;L\;} u$$

is a symbolic form of the system.

Although it may seem that all such problems would have solutions, this is not so. For particular operators L, there are causes $f(x)$ that contradict themselves, rendering Equation (C.1) unsolvable.

To see this, consider two real functions with the boundary conditions

$$v(x) \;\to\; 0 \;\text{ as } x \to \pm\infty,$$
$$w(x) \;\to\; 0 \;\text{ as } x \to \pm\infty,$$

and define an *inner product* as

$$(v(x), w(x)) \equiv \int_{-\infty}^{\infty} v(x)\, w(x) dx\,.$$

Two functions with zero inner product are said to be *orthogonal*.

Next, define the *adjoint* of the operator L as the operator $L^\dagger$ satisfying the condition

$$(Lv, w) \equiv (v, L^\dagger w).$$

All functions ψ for which

$$L^\dagger \psi(x) = 0 \tag{C.2}$$

with

$$\psi(x) \to 0 \text{ as } x \to \pm\infty$$

are said to "span the *null space* of the adjoint operator."

Now, make the following two assumptions.

1. Equation (C.1) has a solution.

2. The inner product (f, ψ) is not zero for some ψ in the null space of $L^\dagger$.

Then the calculation

$$(f, \psi) = (Lu, \psi) = (u, L^\dagger \psi) = (u, 0) = 0$$

leads to the contradiction: zero is not equal to zero. Clearly, one of the assumptions must go. Because we do not want to give up assumption #1, a *necessary* condition for Equation (C.1) to have a solution is that the inner products of the source with all functions satisfying Equation (C.2) be zero.

If these conditions are satisfied, it is possible to construct a solution of Equation (C.1), leading to the following theorem [1, 2].

> **Fredholm's theorem:** For Equation (C.1) to have a solution, it is necessary and sufficient that the inner products of $f(x)$ with all solutions $\psi(x)$ of Equation (C.2) be zero. In other words, the source function must be orthogonal to the null space of the adjoint operator.

Driven PDE systems that satisfy the Fredholm theorem for particular sources are variously said to be "solvable," "compatible," or "consistent."

In Appendices E and F, this theorem is used to establish solvability conditions on equations for corrections arising from various perturbations of nerve models.

References

[1] JP Keener, *Principles of Applied Mathematics*, Addison–Wesley, Reading, MA, 1988.

[2] I Stackgold, *Green's Functions and Boundary Value Problems*, John Wiley & Sons, New York, 1979.

Appendix D
Stability of Axonal Impulses

In a traveling-wave analysis, one assumes that solutions are functions only of the variable $\xi = x - vt$, where v is a wave speed that enters into the resulting ODE system as an adjustable parameter. Here traveling-wave analysis is viewed as a special case of the *independent variable transformation*

$$V(x,t) \longrightarrow \tilde{V}(\xi, \tau),$$

where ξ and τ are related to the original independent variables (x and t) by

$$\begin{aligned} \xi &= x - vt, \\ \tau &= t. \end{aligned}$$

In the new (ξ, τ) system, ξ is measured on a distance scale (or meter stick) moving with velocity v in the x-direction, and time τ is measured on the same time scale (the same clock) as in the laboratory frame of reference. Thus the transformation is from the stationary or *laboratory system* (x, t) to a *moving system* (ξ, τ).

It is convenient to assign a different symbol for time in the moving system because partial derivatives transform as

$$\frac{\partial V(x,t)}{\partial x} \to \frac{\partial \tilde{V}(\xi, \tau)}{\partial \xi} \frac{\partial \xi}{\partial x} + \frac{\partial \tilde{V}(\xi, \tau)}{\partial \tau} \frac{\partial \tau}{\partial x}$$

and

$$\frac{\partial V(x,t)}{\partial t} \to \frac{\partial \tilde{V}(\xi, \tau)}{\partial \xi} \frac{\partial \xi}{\partial t} + \frac{\partial \tilde{V}(\xi, \tau)}{\partial \tau} \frac{\partial \tau}{\partial t}.$$

Because $\partial\xi/\partial x = 1$, $\partial\tau/\partial x = 0$, $\partial\xi/\partial t = -v$, and $\partial\tau/\partial t = 1$, the partial derivatives transform as

$$\frac{\partial}{\partial x} \longrightarrow \frac{\partial}{\partial \xi},$$

$$\frac{\partial}{\partial t} \longrightarrow \frac{\partial}{\partial \tau} - v\frac{\partial}{\partial \xi}.$$

As a specific example, consider the nonlinear diffusion equation

$$\frac{\partial^2 V}{\partial x^2} - \frac{\partial V}{\partial t} = f(V),$$

which transforms to

$$\frac{\partial^2 \tilde{V}}{\partial \xi^2} - \frac{\partial \tilde{V}}{\partial \tau} + v\frac{\partial \tilde{V}}{\partial \xi} = f(\tilde{V}). \tag{D.1}$$

Assuming that $\tilde{V}$ is independent of τ in the moving system, this PDE reduces to the ODE

$$\frac{d^2 V_0}{d\xi^2} + v\frac{dV_0}{d\xi} = f(V_0) \tag{D.2}$$

of traveling-wave analysis. In general, however, $\tilde{V}$ depends upon both ξ and τ, and we must study Equation (D.1) to learn about the *stability* of a traveling wave.

To investigate the stability of $V_0(\xi)$, write

$$\tilde{V}(\xi,\tau) = V_0(\xi) + \phi(\xi,\tau),$$

where $\phi(\xi,\tau)$ is an alteration of the traveling wave. Then from Equations (D.1) and (D.2), $\phi(\xi,\tau)$ satisfies the nonlinear PDE

$$\frac{\partial^2 \phi}{\partial \xi^2} + v\frac{\partial \phi}{\partial \xi} - \frac{\partial \phi}{\partial \tau} = f(V_0 + \phi) - f(V_0).$$

To this point, no approximations have been made—we have merely transformed the independent and dependent variables: a matter of bookkeeping.

A linear-stability analysis for this system was first carried out by Zeldovich and Barenblatt in 1959 [15].[1] Following these authors, we assume

$$|\phi(\xi,0)| \ll |V_0(\xi)|$$

and approximate

$$f(V_0 + \phi) - f(V_0) \doteq G[V_0(\xi)]\,\phi(\xi,\tau),$$

---

[1] *Nonlinear* stability analyses of this equation have been published by Lindgren and Buratti [8] and by Maginu [9, 12].

where

$$G[V_0(\xi)] \equiv G(\xi) \equiv \left.\frac{df(V)}{dV}\right|_{V=V_0(\xi)},$$

and ϕ satisfies the linear equation

$$\frac{\partial^2 \phi}{\partial \xi^2} + v\frac{\partial \phi}{\partial \xi} - \frac{\partial \phi}{\partial \tau} \doteq G(\xi)\,\phi. \tag{D.3}$$

This linear PDE has been obtained by linearizing the transformed nonlinear PDE—Equation (D.1)—about the traveling-wave solution $V_0(\xi)$.

At the price of assuming that $\phi(\xi, 0)$ is a sufficiently small initial alteration of the traveling wave, in other words, we have obtained a *linear* PDE for the evolution of $\phi(\xi, \tau)$ in time. From the perspective of *linear stability analysis*, we can say that our system is:

- *Asymptotically stable* if *all* solutions of Equation (D.3) approach zero as $\tau \to \infty$,

- *Unstable* if *any* solution of Equation (D.3) grows as $\tau \to \infty$, and

- *Stable* otherwise.

Equation (D.3) is conveniently analyzed by separating variables. Thus $\phi(\xi, \tau)$ is expressed as a generalized sum of elementary products of the form $\Phi(\xi)\,T(\tau)$. Here

$$T(\tau) = e^{\lambda \tau},$$

λ is can be complex, and $\Phi(\xi)$ is a solution of the stability equation

$$\frac{d^2 \Phi}{d\xi^2} + v\frac{d\Phi}{d\xi} - [\lambda + G(\xi)]\Phi = 0. \tag{D.4}$$

Each bounded solution of this equation is called an *eigenfunction* and the corresponding value of λ is an *eigenvalue*. All of the values of λ for which Equation (D.4) has bounded solutions are referred to as its *spectrum*. Depending upon the boundary conditions imposed as $\xi \to \pm\infty$, there are two types of eigenvalues.

Continuous Eigenvalues

If it is required that solutions of Equation (D.4) be bounded by some finite value as $\xi \to \pm\infty$, then $\Phi(\xi)$ has the asymptotic form

$$\Phi(\xi) \sim e^{\pm ik\xi}.$$

Such eigenfunctions represent *radiation* from the underlying traveling wave $V_0(\xi)$, and if the radiation grows with time, the underlying impulse is unstable.

As $\xi \to +\infty$, $G(\xi)$ approaches a positive constant, $G_1 > 0$. Similarly, as $\xi \to -\infty$, $G(\xi) \to G_2 > 0$. Substitution of $\Phi(\xi) = \exp(\pm ik\xi)$ into Equation

(D.4), therefore, leads to the eigenvalues

$$\lambda = -(G_1 + k^2) \pm ikv$$

as $\xi \to +\infty$. This set of eigenvalues is said to be *continuous* because it contains elements for all real values of k.

To see that these continuous eigenvalues correspond to eigenfunctions representing radiation from the traveling wave, note that in the $+\xi$-direction, $\phi(\xi, \tau)$ has the form

$$\phi_{\text{rad}}(\xi, \tau) \sim e^{-(G_1 + k^2)\tau} \cos(k\xi + kv\tau),$$

with a corresponding expression for radiation in the $-\xi$-direction.

For all real values of k, this radiative eigenfunction is damped with time, falling exponentially to zero as $\tau \to +\infty$. Thus the continuous spectrum does not contribute to instability of the underlying traveling wave.

Discrete (or "Point") Eigenvalues
With the boundary conditions

$$\Phi(\xi) \to 0 \quad \text{as} \quad \xi \to \pm\infty,$$

the corresponding eigenvalues occur at discrete (isolated) locations on the real axis of the complex λ-plane. These "point" eigenvalues are the particular values of λ for which the asymptotic behavior of Equation (D.4) as $\xi \to -\infty$ goes smoothly over into a solution of this same equation as $\xi \to +\infty$.

If Equation (D.4) has any eigenfunction with a negative eigenvalue ($\lambda > 0$), then the corresponding traveling-wave solution is unstable. If all eigenvalues are positive, then the traveling wave is asymptotically stable.

It is immediately evident that Equation (D.4) always has an eigenfunction for the eigenvalue $\lambda = 0$. To see this, differentiate Equation (D.2) with respect to ξ; thus

$$\frac{d^2}{d\xi^2}\left(\frac{dV_0}{d\xi}\right) + v\frac{d}{d\xi}\left(\frac{dV_0}{d\xi}\right) = G(\xi)\left(\frac{dV_0}{d\xi}\right),$$

which is identical to Equation (D.4) with

$$\lambda = 0 \quad \text{and} \quad \Phi(\xi) = dV_0/d\xi.$$

Such a striking property has physical significance, which can be understood as follows. Suppose that we start with a traveling-wave solution $V_0(\xi)$ and add a small amount of its derivative $dV_0/d\xi$. The result is

$$V_0(\xi) + \epsilon\left(\frac{dV_0}{d\xi}\right) = V_0(\xi + \epsilon) + O(\epsilon^2).$$

Thus adding the derivative of a traveling wave with respect to its traveling-wave variable merely translates the original solution in the ξ-direction.

Because a translated traveling wave is still an exact solution of the system, such a disturbance has no dependence on time.

Although we have obtained this result for a special case, it is generally true for all linear-stability analyses of traveling waves that[2]:

> The derivative of a traveling wave with respect to its traveling wave variable is an eigenfunction with zero eigenvalue of the corresponding stability equation.

Armed with this knowledge, the task is to determine whether Equation (D.4) has a negative eigenvalue. To this end, it is convenient to introduce the dependent-variable transformation

$$\Phi(\xi) = \psi(\xi)\, e^{-v\xi/2},$$

changing Equation (D.4) to

$$\frac{d^2\psi}{d\xi^2} - \left[\lambda + \frac{v^2}{4} + G(\xi)\right]\psi = 0. \tag{D.5}$$

Equation (D.5) is a second-order self-adjoint operator equation with [3]

$$U(\xi) = \frac{v^2}{4} + G(\xi).$$

Such systems share the following properties [13].

- Because $U(\xi)$ has the shape of a potential well, the eigenfunction (ψ_0) with the most positive eigenvalue (λ_0) has no finite zero crossing.

- The eigenfunction (ψ_1) with the next largest eigenvalue (λ_1) has one zero crossing.

- The eigenfunction (ψ_2) with the next largest eigenvalue (λ_2) has two zero crossings.

- And so on.

---

[2] Physicists have a special name for this translational eigenfunction: the "Goldstone boson."

[3] To check that the transformation from $\Phi(\xi)$ to $\psi(\xi)$ does not violate the null boundary condition on $\Phi(\xi)$ as $\xi \to -\infty$, assume that $G(\xi) \to G_2 > 0$ as $\xi \to -\infty$. This implies

$$\psi(\xi) \sim \exp\left(\xi\sqrt{G_2 + v^2/4 + \lambda}\right)$$

and therefore

$$\Phi(\xi) \sim \exp\left[\xi\left(-v/2 + \sqrt{G_2 + v^2/4 + \lambda}\right)\right] \quad \text{as } \xi \to -\infty.$$

Thus for $\lambda \geq 0$, $\psi(\xi) \to 0$ as $\xi \to -\infty$, implying $\Phi(\xi) \to 0$ as $\xi \to -\infty$.

Thus the discrete eigenvalues of Equation (D.5) can be ordered as

$$\lambda_0 > \lambda_1 > \lambda_2 > \cdots > \lambda_n > \text{etc.},$$

where eigenfunction ψ_n has n finite zero crossings.[4]

From these results, the following conclusions can be drawn. First consider the stability of a monotone decreasing leading-edge solution $V_0(\xi)$ as shown in Figure 5.2(c). The function $\psi(\xi) = e^{v\xi/2} dV_0/d\xi$ is the eigenfunction of Equation (D.5) corresponding to the eigenvalue $\lambda = 0$. Because this eigenfunction has no zero crossings, $\lambda_0 = 0$ is the most positive eigenvalue. In other words, all eigenvalues are less than or equal to zero, and the traveling-wave solution $V_0(\xi)$ is stable. It is not asymptotically stable because the perturbation can include components with the most positive eigenvalue, which do not decay with time. Because this argument uses only the qualitative shape of $V_0(\xi)$, the same conclusion holds for any monotone increasing or decreasing (level change) traveling-wave solution of the nonlinear diffusion equation.

As we have seen in Chapter 5, the leading-edge PDE also has a solution that is impulse-shaped, with a maximum amplitude (V_m) at some finite value of ξ, corresponding to homoclinic trajectories in the $(\tilde{V}, W)$ phase plane of Figure 5.4(b). Because the ξ-derivative of such a function has a zero crossing at a finite value of ξ, $\lambda = 0$ is not the most positive eigenvalue of Equation (D.5). In other words, $\lambda_0 > 0$, so the solution of Figure 5.4(c) is unstable.

Augmentations of the preceeding method to general systems that include both the Hodgkin–Huxley and FitzHugh–Nagumo formulations of nerve impulse dynamics have been described by Evans [1, 2, 3, 4] and by Sattinger [11], and there have been several more detailed analyses of the FitzHugh–Nagumo system [5, 6, 7, 10, 14]. Among the conclusions of these studies are the following.

- A sufficient condition for instability is that one mode of the linearized PDE has an eigenvalue with positive real part.

- Because there is always an eigenfunction of the linearized PDE for $\lambda = 0$, impulses are at most stable.

- Necessary conditions for stability are that no eigenvalues have positive real parts and that the zero eigenvalue ($\lambda = 0$) is nondegenerate.

To appreciate the requirement that the $\lambda = 0$ eigenvalue must be nondegenerate for stability, consider a Green function $\mathcal{G}(\xi, \tau)$ for the linearized PDE from which the total disturbance of the traveling-wave solution can

---

[4]An intuitive way to establish these results is to study Equation (D.5) in the $(\psi, d\psi/d\xi)$ phase plane.

be computed as [13]

$$\phi(\xi,\tau) = \int_0^\tau d\tau' \int_{-\infty}^{+\infty} \mathcal{G}(\xi - \xi', \tau - \tau') F(\xi', \tau') d\xi'.$$

In this formulation, $F(\xi', \tau')$ represents an arbitrary disturbance of the traveling wave, and $\mathcal{G}(\xi, \tau)$ is the response of the linearized PDE to a disturbance that is delta-function localized in both space and time. Because F is arbitrary, the properties of $\mathcal{G}(\xi, \tau)$ indicate whether any disturbance will grow with time.

To see how this comes about, consider the Laplace transform of $\mathcal{G}(\xi, \tau)$, which is defined as

$$\tilde{\mathcal{G}}(\xi,\lambda) \equiv \int_0^\infty \mathcal{G}(\xi,\tau)e^{-\lambda\tau}d\tau.$$

From this function of ξ and λ, $\mathcal{G}(\xi, \tau)$ can be recovered through the inverse transform

$$\mathcal{G}(\xi,\tau) = \int_C \tilde{\mathcal{G}}(\xi,\lambda)e^{\lambda\tau}d\lambda,$$

where for $\tau > 0$ the integration is over a closed curve (C) in the complex λ-plane that encloses all of the singularities of $\tilde{\mathcal{G}}(\xi, \lambda)$.

Notice that the inverse transform is a generalized sum of terms of the form $e^{\lambda\tau}$, which appears in the separation of variables for the linearized PDE. Thus, the singularities of $\tilde{\mathcal{G}}(\xi, \lambda)$ comprise the spectrum of the linearized PDE. The implications of a degenerate eigenvalue at $\lambda = 0$ can now be appreciated by recalling the Laplace transform pair

$$\frac{\tau^{n-1}}{(n-1)!} \longleftrightarrow \frac{1}{\lambda^n}.$$

Thus if the $\lambda = 0$ eigenvalue is doubly degenerate, $\tilde{\mathcal{G}}(\xi, \lambda)$ contains the factor $1/\lambda^2$, implying a corresponding temporal response that is proportional to τ. Linear growth with τ implies instability.

With the exception of the translation-mode eigenvalue at $\lambda = 0$, all eigenvalues of Equation (D.4) have negative real parts of finite magnitude. Thus if the eigenvalues of a F–N system are assumed to depend continuously on ε, there is a parameter range of some $\varepsilon > 0$ over which the upper curve in Figure 6.3 is stable and the lower curve is unstable [6, 7, 14]. Similarly for an H–H system with some

$$\tau_m > 0,$$
$$1/\tau_h > 0,$$
$$1/\tau_n > 0,$$

the faster impulse should be stable and the slower one unstable.

To extend an analytic proof of stability out to the critical values of these time constants (beyond which traveling waves are not found), it is necessary to show that no pair of eigenvalues at $\lambda = \sigma \pm i\omega$ has a real part (σ) that becomes positive. If so, the impulse would have an unstable internal mode of oscillation with the temporal behavior $e^{\sigma\tau} \cos \omega\tau$. It is not known presently whether such unstable internal modes can occur for some values of F–N or H–H parameters.

References

[1] JW Evans, Nerve axon equations: I. Linear approximations, *Indiana Univ. Math. J.* 21 (1972) 877–885.

[2] JW Evans, Nerve axon equations: II. Stability at rest, *Indiana Univ. Math. J.* 22 (1972) 75–90.

[3] JW Evans, Nerve axon equations: III. Stability of the nerve impulse, *Indiana Univ. Math. J.* 22 (1972) 577–593.

[4] JW Evans, Nerve axon equations: IV. The stable and unstable impulse, *Indiana Univ. Math. J.* 24 (1975) 1169–1190.

[5] JA Feroe, Temporal stability of solitary impulse solutions of a nerve equation, *Biophys. J.* 21 (1978) 103–110.

[6] CKRT Jones, Some ideas in the proof that the FitzHugh–Nagumo pulse is stable. In *Nonlinear Partial Differential Equations,* J Smoller (ed), Contemporary Mathematics 17, American Mathematical Society, Providence, 1984, pp 287–292.

[7] CKRT Jones, Stability of the travelling wave solution of the FitzHugh–Nagumo system, *Trans. Am. Math. Soc.* 286 (1984) 431–469.

[8] AG Lingren and RJ Buratti, Stability of waveforms on active nonlinear transmission lines, *Trans. IEEE Circuit Theory,* CT–16 (1969) 274–279.

[9] K Maginu, On asymptotic stability of waveforms on a bistable transmission line, *IECE Professional Group on Nonlinear Problems,* NLP, Institute of Electrical and Computer Engineering, Tokyo, Japan, 70–24 (in Japanese), 1971.

[10] K Maginu, Stability of periodic travelling wave solutions of a nerve conduction equation, *J. Math. Biol.* 6 (1978) 49–57.

[11] DH Sattinger, On the stability of waves of nonlinear parabolic systems, *Adv. Math.* 22 (1976) 312–355.

[12] AC Scott, The electrophysics of a nerve fiber, *Rev. Mod. Phys.* 11 (1975) 487–533.

[13] I Stackgold, *Green's Functions and Boundary Value Problem,* John Wiley & Sons, New York, 1979.

[14] E Yanagida, Stability of fast travelling pulse solutions of the FitzHugh–Nagumo equations, *J. Math. Biol.* 22 (1985) 81–104.

[15] YB Zeldovich and GI Barenblatt, Theory of flame propagation, *Combust. and Flame,* 3 (1959) 61–74.

Appendix E
Perturbation Theory for the F–N Impulse

How does the traveling-wave velocity of a FitzHugh–Nagumo impulse depend on ε? To answer this question when $0 < \varepsilon \ll 1$, consider a paper by Casten, Cohen, and Lagerstrom that appeared in 1975 [1].

The first step is to express v, $V(\xi)$, and $R(\xi)$ as power series in ε; thus

$$
\begin{aligned}
v &= v_0 + \varepsilon v_1 + \varepsilon^2 v_2 + \cdots, \\
V &= V_0 + \varepsilon V_1 + \varepsilon^2 V_2 + \cdots, \\
R &= R_0 + \varepsilon R_1 + \varepsilon^2 R_2 + \cdots.
\end{aligned}
$$

Substituting these expressions into Equations (6.13) (with $b = 0$ and $c = 0$) and equating terms that are independent of ε leads to

$$
\frac{d^2 V_0}{d\xi^2} + v_0 \frac{dV_0}{d\xi} - [f(V_0) + R_0] = 0, \tag{E.1}
$$

$$
\frac{dR_0}{d\xi} = 0.
$$

Equating terms that are first order in ε yields

$$
\frac{d^2 V_1}{d\xi^2} + v_0 \frac{dV_1}{d\xi} - V_1 f'(V_0) = R_1 - v_1 \frac{dV_0}{d\xi}, \tag{E.2}
$$

$$
\frac{dR_1}{d\xi} = -\frac{V_0}{v_0}. \tag{E.3}
$$

Assuming that the pulse is propagating into a region of zero recovery variable—as is indicated in Figure 6.5—then $R_0 = 0$, and from integration of Equation (E.3),

$$R_1(\xi) = \frac{1}{v_0} \int_\xi^\infty V_0(\xi')d\xi'.$$

Thus everything on the right-hand side of Equation (E.2) is known except the value of v_1, which represents the dependence of the traveling-wave speed to first order in ε. We can find v_1 by using the *Fredholm theorem* (see Appendix C), which provides conditions for Equation (E.2) to have a solution.

To see how this goes, write Equation (E.2) in the form

$$LV_1 = R_1 - v_1 \frac{dV_0}{d\xi},$$

where L is a linear differential operator defined as

$$L \equiv \frac{d^2}{d\xi^2} + v_0 \frac{d}{d\xi} - f'(V_0).$$

From differentiation of Equation (E.1) with respect to ξ, it is seen that

$$L\frac{dV_0}{d\xi} = 0.$$

In other words, $dV_0/d\xi$ is a *null function* of L. Because the *adjoint* of L is

$$L^\dagger = \frac{d^2}{d\xi^2} - v_0 \frac{d}{d\xi} - f'(V_0),$$

it has a null function ψ, where

$$L^\dagger \psi = 0,$$

and

$$\psi = e^{v_0 \xi} \frac{dV_0}{d\xi},$$

as may be verified by direct substitution.

From the Fredholm theorem, a necessary condition for Equation (E.2) to have a solution is that its right-hand side must be orthogonal to ψ. Thus

$$v_1 = \frac{\int_{-\infty}^\infty \left[\int_\xi^\infty V_0(\xi')d\xi'\right](dV_0/d\xi)e^{v_0\xi}d\xi}{v_0 \int_{-\infty}^\infty (dV_0/d\xi)^2 e^{v_0\xi}d\xi},$$

so the traveling-wave velocity depends on ε as

$$v = v_0 + \left(\frac{\varepsilon}{v_0}\right) \frac{\int_{-\infty}^\infty \left[\int_\xi^\infty V_0(\xi')d\xi'\right](dV_0/d\xi)e^{v_0\xi}d\xi}{\int_{-\infty}^\infty (dV_0/d\xi)^2 e^{v_0\xi}d\xi} + O(\varepsilon^2). \qquad \text{(E.4)}$$

The first term on the right-hand side of this expression gives the traveling-wave velocity when ε and the recovery variable (R) are both equal to zero. The second term, which is negative because $dV_0/d\xi$ is negative on the leading edge of an impulse, gives an $O(\varepsilon)$ correction to the traveling-wave velocity, assuming that the pulse retains its zero-order shape. The final (unevaluated) terms account for variations in the traveling-wave velocity that stem from changes in the pulse shape.

Evidently, this is the behavior observed on the upper curve in Figure 6.3 as $\varepsilon \to 0$. Thus Equation (E.4) provides an analytic benchmark for the numerical calculations that depends only on knowledge of v_0 and $V_0(\xi)$, which is the information provided by the leading-edge studies presented in Chapter 5.

Along the lower curve in Figure 6.3, $v(\varepsilon) \to 0$ as $\varepsilon \to 0$, implying

$$v_0 = 0.$$

In this case, the preceding perturbation expansion does not work because the $O(\varepsilon)$ term in Equation (E.4) diverges. To deal with this difficulty, introduce a perturbation expansion of the form

$$
\begin{aligned}
V &= V_0 + \sqrt{\varepsilon}V_1 + \varepsilon V_2 + \cdots, \\
R &= \sqrt{\varepsilon}R_1 + \varepsilon R_2 + \cdots, \\
v &= \sqrt{\varepsilon}v_1 + \varepsilon v_2 + \cdots.
\end{aligned}
$$

Then using similar arguments, it follows that

$$v = \sqrt{\varepsilon}\left(\frac{\int_{-\infty}^{\infty} V_0^2\, d\xi}{\int_{-\infty}^{\infty}(dV_0/d\xi)^2\, d\xi}\right)^{1/2} + O(\varepsilon). \tag{E.5}$$

Along the lower branch of Figure 6.3, therefore, Equation (E.5) provides a benchmark for the numerical calculations that depends only on $V_0(\xi)$ for the threshold impulse, which was derived in Section 5.3 and is displayed in Figure 5.4(c).

To restate the results of this appendix, the perturbation theory developed by Casten et al. shows how the traveling-wave speeds for an F–N impulse depend on ε as $\varepsilon \to 0$. To lowest order in ε, these dependencies can be explicitly computed from the corresponding solutions for $\varepsilon = 0$. Thus the leading-edge approximations presented in Chapter 5 gain additional status.

References

[1] RG Casten, H Cohen, and PA Lagerstrom, Perturbation analysis of an approximation to Hodgkin–Huxley theory, *Q. Appl. Math.* 32 (1975) 365–402.

Appendix F
Perturbation Analyses of Ephaptic Interactions

Here it is shown how perturbation theory can be used to investigate the synchronization (or "locking") of nerve impulses on parallel fibers. The impulses are described both as the leading-edge waveforms of Chapter 5 and in the FitzHugh–Nagumo approximation of Chapter 6.

F.1 Leading-Edge Interactions

From Equations (8.3), it is assumed that leading-edge waveforms are described by the coupled PDEs

$$(1-\alpha)\frac{\partial^2 V_1}{\partial x^2} - \alpha\frac{\partial^2 V_2}{\partial x^2} - \frac{\partial V_1}{\partial t} = f(V_1),$$

$$(1-\alpha)\frac{\partial^2 V_2}{\partial x^2} - \alpha\frac{\partial^2 V_1}{\partial x^2} - \frac{\partial V_2}{\partial t} = f(V_2),$$

where α is a small *coupling parameter* defined as the ratio of outside to total series resistance per unit length of each axon.

To begin the analysis, assume two traveling waves of the form

$$V_k(x,t) = V_k(\xi) = V_k(x - vt), \quad k = 1,2,$$

where v is the speed of two leading edges moving synchronously. Then the preceding PDEs become the ODE system

$$(1-\alpha)\frac{d^2 V_1}{d\xi^2} - \alpha\frac{d^2 V_2}{d\xi^2} + v\frac{dV_1}{d\xi} = f(V_1),$$

$$(1 - \alpha)\frac{d^2 V_2}{d\xi^2} - \alpha\frac{d^2 V_1}{d\xi^2} + v\frac{dV_2}{d\xi} = f(V_2).$$

A solution of these equations represents two leading edges ($V_1(\xi)$ and $V_2(\xi)$), one on each fiber and moving with the same speed. For α sufficiently small, these solutions can be written as power series

$$V_k = V_{k0} + \alpha V_{k1} + \alpha^2 V_{k2} + \cdots .$$

The corresponding traveling-wave velocities can also be expressed as

$$v = v^{(k)} = v_0 + \alpha v_1^{(k)} + \alpha^2 v_2^{(k)} + \cdots,$$

allowing solutions on the two fibers to have speeds that differ to first order in α.

Substituting into the ODE system and equating terms of zero order in α, one finds

$$\frac{d^2 V_{k0}}{d\xi^2} + v_0\frac{dV_{k0}}{d\xi} = f(V_{k0}),$$

for which exact solutions are known from Chapter 5.

Equating terms of first order in alpha yields two equations,

$$\frac{d^2 V_{k1}}{d\xi^2} + v_0\frac{dV_{k1}}{d\xi} - f'(V_{k0})V_{k1} = \frac{d^2 V_{10}}{d\xi^2} + \frac{d^2 V_{20}}{d\xi^2} - v_1^{(k)}\frac{dV_{k0}}{d\xi},$$

for $k = 1, 2$. These are linear operator equations of the form

$$L_1 V_{11} = F_1\left(V_{10}, V_{20}, v_1^{(1)}\right),$$

$$L_2 V_{21} = F_2\left(V_{20}, V_{10}, v_1^{(2)}\right),$$

with inhomogeneous (or *forcing*) terms on their right-hand sides. In Appendix C, it is shown that for such equations to have solutions, Fredholm conditions must be satisfied, which allow determination of $v_1^{(1)}$ and $v_1^{(2)}$ in terms of V_{10} and V_{20}.

In particular, the Fredholm theorem requires that the forcing terms must be orthogonal to solutions of the homogeneous adjoint equations

$$L_k^\dagger y_k(\xi) = 0, \tag{F.1}$$

where

$$L_k^\dagger = \frac{d^2}{d\xi^2} - v_0\frac{d}{d\xi} - f'(V_{k0}).$$

Because these homogeneous equations have the solutions

$$\psi_k(\xi) = e^{v_0\xi}\frac{dV_{k0}}{d\xi},$$

the Fredholm conditions can be written as

$$\int_{-\infty}^{\infty} \psi_k(\xi) F_k(\xi) d\xi = 0. \tag{F.2}$$

To evaluate these integrals, note that if the solution on fiber #2 leads the solution on fiber #1 by a distance δ, then

$$
\begin{aligned}
V_{20}(\xi) &= V_{10}(\xi - \delta), \\
\psi_2(\xi) &= \psi_1(\xi - \delta).
\end{aligned}
$$

Thus Equations (F.2) imply that

$$
\left[v_1^{(1)}(\delta) - v_1^{(2)}(\delta)\right] = \frac{1}{N} \int_{-\infty}^{\infty} e^{v_0 \xi} \frac{dV_{10}}{d\xi}(\xi) \left[\frac{d^2 V_{10}}{d\xi^2}(\xi - \delta) - \frac{d^2 V_{10}}{d\xi^2}(\xi + \delta)\right] d\xi,
$$

where

$$
N \equiv \int_{-\infty}^{\infty} e^{v_0 \xi} \left[\frac{dV_{10}}{d\xi}(\xi)\right]^2 d\xi.
$$

For the cubic representation of the sodium ion current

$$
f(V) = V(V - a)(V - 1),
$$

formulas are available for $V_0(\xi)$ and v_0. Thus these integrals can be evaluated, leading to the expression for the velocity difference

$$
\left[v^{(1)}(\delta) - v^{(2)}(\delta)\right] = \alpha \left[v_1^{(1)}(\delta) - v_1^{(2)}(\delta)\right]
$$

given in Equation (8.5) and plotted in Figure 8.3 [1].

F.2   The FitzHugh–Nagumo System

Consider impulses described by the coupled F–N system

$$
\begin{aligned}
(1 - \alpha)\frac{\partial^2 V_1}{\partial x^2} - \alpha \frac{\partial^2 V_2}{\partial x^2} - \frac{\partial V_1}{\partial t} &= f(V_1) + R_1, \\
\frac{\partial R_1}{\partial t} &= \varepsilon V_1, \\
(1 - \alpha)\frac{\partial^2 V_2}{\partial x^2} - \alpha \frac{\partial^2 V_1}{\partial x^2} - \frac{\partial V_2}{\partial t} &= f(V_2) + R_2, \\
\frac{\partial R_2}{\partial t} &= \varepsilon V_2,
\end{aligned}
$$

where ε takes a fixed value (of about 0.1) corresponding to a typical nerve impulse.

Again seeking synchronized traveling-wave solutions of the form

$$
\begin{aligned}
V_k(x, t) &= V_k(\xi) = V_k(x - vt), \\
R_k(x, t) &= R_k(\xi) = R_k(x - vt),
\end{aligned}
$$

where $k = 1, 2$ and v is the propagation speed of two impulses, leads to a set of ODEs that can be written

$$(1 - \alpha)\frac{d^2V_1}{d\xi^2} - \alpha\frac{dV_2}{d\xi^2} + v\frac{dV_1}{d\xi} = f(v_1) + R_1,$$

$$v\frac{dR_1}{d\xi} = -\varepsilon V_1,$$

$$(1 - \alpha)\frac{d^2V_2}{d\xi^2} - \alpha\frac{dV_1}{d\xi^2} + v\frac{dV_2}{d\xi} = f(v_2) + R_2,$$

$$v\frac{dR_2}{d\xi} = -\varepsilon V_2.$$

Just as in the previous section, V_k and v are expressed as power series in the small coupling parameter α, provisionally allowing v to differ by order α on the two fibers. Eliminating R_1 and R_2 and equating terms that are independent of α leads to the third-order nonlinear ODE

$$\frac{d^3V_{k0}}{d\xi^3} + v_0\frac{d^2V_{k0}}{d\xi} - f'(V_{k0})\frac{dV_{k0}}{d\xi} + \frac{\varepsilon}{v_0}V_{k0} = 0, \tag{F.3}$$

solutions to which are discussed in Chapter 6. Similarly equating terms that are first order in α yields the pair of coupled linear equations

$$\frac{d^3V_{k1}}{d\xi^3} + v_0\frac{d^2V_{k1}}{d\xi^2} - f'(V_{k0})\frac{dV_{k1}}{d\xi} - \left(f''(V_{k0})\frac{dV_{k0}}{d\xi} - \frac{\varepsilon}{v_0}\right)V_{k1}$$

$$= v_1^{(k)}\left(\frac{\varepsilon}{v_0^2}V_{k0} - \frac{d^2V_{k0}}{d\xi^2}\right) + \frac{d^3V_{10}}{d\xi^3} + \frac{d^3V_{20}}{d\xi^3}. \tag{F.4}$$

Here

$$f'(V_{k0}) \equiv \left[\frac{df(V_k)}{dV_k}\right]_{V_k=V_{k0}}$$

and

$$f''(V_{k0}) \equiv \left[\frac{d^2f(V_k)}{dV_k^2}\right]_{V_k=V_{k0}}.$$

Because each of Equations (F.4) can be written as a forced linear operator equation of the form

$$L_k V_{1k} = F_k,$$

Fredholm's theorem requires for solutions to exist that the integrals

$$\int_{-\infty}^{\infty} \psi_k(\xi)F_k(\xi)d\xi = 0,$$

where $\psi_k(\xi)$ is a solution of

$$L_k^\dagger \psi_k = 0, \tag{F.5}$$

and $L_k^\dagger$ is the adjoint of L_k.

Through integration by parts, it is seen that the adjoint of

$$-f'(V_{k0})\frac{d}{d\xi} - f''(V_{k0})\frac{dV_{k0}}{d\xi}$$

is

$$+f'(V_{k0})\frac{d}{d\xi}\,,$$

so Equations (F.5) become

$$-\frac{d^3\psi_k}{d\xi^3} + v_0\frac{d^2\psi_k}{d\xi^2} + f'(V_{k0})\frac{d\psi_k}{d\xi} + \frac{\varepsilon}{v_0}\psi_k = 0,\quad k=1,2.\qquad \text{(F.6)}$$

With these results in hand, a perturbation calculation proceeds as follows.

1. Solve Equation (F.3) for $V_{k0}(\xi)$. This calculation must be done numerically, as described in Chapter 6.

2. Solve Equation (F.6) for $\psi_k(\xi)$, the solutions of the homogeneous adjoint problem. This calculation must also be done numerically.

3. Assume that $V_{20}(\xi)$ differs from $V_{10}(\xi)$ by a translation of δ in the traveling-wave variable ξ. Thus,

$$V_{20}(\xi) = V_{10}(\xi - \delta),$$
$$\psi_2(\xi) = \psi_1(\xi - \delta),$$

implying that the impulse on fiber #2 is leading the impulse on fiber #1 by a distance δ.

4. The Fredholm solvability conditions for Equations (F.4) then require that the inner products of the right-hand side forcing functions with the ψ_k be zero. Thus,

$$v_1^{(1)} = \frac{1}{N}\int_{-\infty}^{\infty}\psi_1(\xi)\left(\frac{d^3V_{10}(\xi)}{d\xi^3} + \frac{d^3V_{10}(\xi-\delta)}{d\xi^3}\right)d\xi$$

and

$$v_1^{(2)} = \frac{1}{N}\int_{-\infty}^{\infty}\psi_2(\xi)\left(\frac{d^3V_{20}(\xi+\delta)}{d\xi^3} + \frac{d^3V_{20}(\xi)}{d\xi^3}\right)d\xi\,,$$

where

$$N \equiv \int_{-\infty}^{\infty}\psi_1(\xi)\left(\frac{d^2V_{10}(\xi)}{d\xi^2} - \frac{\varepsilon}{v_0^2}V_{10}(\xi)\right)d\xi$$
$$= \int_{-\infty}^{\infty}\psi_2(\xi)\left(\frac{d^2V_{20}(\xi)}{d\xi^2} - \frac{\varepsilon}{v_0^2}V_{20}(\xi)\right)d\xi.$$

For particular values of the parameters, numerical evaluations of the preceding integral expressions as functions of δ allow plots of $v_1^{(1)}(\delta)$ and $v_1^{(2)}(\delta)$ as in Figure 8.5(b) [3, 4, 5]. The dynamic implications of these perturbation results are in accord with integration of the original PDE [2].

References

[1] S Binczak, JC Eilbeck, and AC Scott, Ephaptic coupling of myelinated nerve fibers, *Physica D* 148 (2001) 159–174.

[2] JC Eilbeck, SD Luzader, and AC Scott, Pulse evolution on coupled nerve fibers, *Bull. Math. Biol.* 43 (1981) 389–400.

[3] SD Luzader, Neurophysics of parallel nerve fibers, Doctoral thesis at the University of Wisconsin, Madison, 1979.

[4] AC Scott, *Nonlinear Science: Emergence and Dynamics of Coherent Structures*, Oxford University Press, Oxford, 1999.

[5] AC Scott and SD Luzader, Coupled solitary waves in neurophysics, *Phys. Scr.* 20 (1979) 395–401.

Index